METHODS IN MOLECULAR BIOLOGY

Series Editor
John M. Walker
School of Life and Medical Sciences
University of Hertfordshire
Hatfield, Hertfordshire, UK

For further volumes:
http://www.springer.com/series/7651

For over 35 years, biological scientists have come to rely on the research protocols and methodologies in the critically acclaimed *Methods in Molecular Biology* series. The series was the first to introduce the step-by-step protocols approach that has become the standard in all biomedical protocol publishing. Each protocol is provided in readily-reproducible step-by-step fashion, opening with an introductory overview, a list of the materials and reagents needed to complete the experiment, and followed by a detailed procedure that is supported with a helpful notes section offering tips and tricks of the trade as well as troubleshooting advice. These hallmark features were introduced by series editor Dr. John Walker and constitute the key ingredient in each and every volume of the *Methods in Molecular Biology* series. Tested and trusted, comprehensive and reliable, all protocols from the series are indexed in PubMed.

Retinoid and Rexinoid Signaling

Methods and Protocols

Edited by

Swapan K. Ray

Department of Pathology, Microbiology, and Immunology, University of South Carolina School of Medicine, Columbia, SC, USA

Humana Press

Editor
Swapan K. Ray
Department of Pathology, Microbiology, and Immunology
University of South Carolina School of Medicine
Columbia, SC, USA

ISSN 1064-3745 ISSN 1940-6029 (electronic)
Methods in Molecular Biology
ISBN 978-1-4939-9587-5 ISBN 978-1-4939-9585-1 (eBook)
https://doi.org/10.1007/978-1-4939-9585-1

This Humana imprint is published by the registered company Springer Science+Business Media, LLC, part of Springer Nature.
The registered company address is: 233 Spring Street, New York, NY 10013, U.S.A.

Preface

It is very exciting that many recent investigators are exploring the retinoids and rexinoids in the preclinical models for demonstrating their diverse biological effects, especially neuroprotective and anticancer effects. This volume of the *Methods in Molecular Biology* series on *Retinoid and Rexinoid Signaling* provides the cutting-edge synthetic procedures for producing receptor-specific retinoids and rexinoids, molecular biology methods, and new technologies to demonstrate the therapeutic activities of these molecules and also indicates new research avenues that can be used by graduate students, postdoctoral fellows, and principal investigators, who are interested in further exploring the signaling mechanisms of these fascinating molecules in their specific preclinical models. I hope that the users of the protocols described in *Retinoid and Rexinoid Signaling* find clarity, confidence, and creativity in performing their own experiments and translate their discoveries into new tests, new treatments, and new prevention strategies for various diseases and injuries.

The natural and synthetic derivatives of vitamin A play many important roles in physiological processes such as vision, immune system, cell differentiation and apoptosis, and brain development in humans and other vertebrates. Two of the best-known natural metabolites of vitamin A are all-*trans* retinoic acid (ATRA) and 9-*cis* retinoic acid (9CRA). Biogenesis of 9CRA occurs from the intermediate ATRA. Retinoids act as ligands for activation of the retinoic acid receptors (RARs) that have three isoforms such as RARα, RARβ, and RARγ. Rexinoids are ligands for activation of the retinoid X receptors (RXRs) such as RXRα, RXRβ, and RXRγ. Both RARs and RXRs are nuclear receptors. To participate as a functional nuclear transcription factor, an isoform of RARs heterodimerizes with an isoform of RXRs, but an isoform of RXRs can homodimerize with itself and also heterodimerize with an isoform of RARs. These nuclear receptors act as the ligand-activated transcription factors that bind to specific DNA elements in the promoter region for transcriptional activation or repression of the target genes in specific tissues. The functional responses of these ligands and their receptors are modulated by many coactivators and corepressors. Coactivators and corepressors are known to modify the chromatin and/or interact with the typical transcriptional machinery for modulating transcription of the target genes. It should be noted that some retinoids and rexinoids may work via receptor-independent mechanisms as well for some physiological functions and therapeutic effects. In addition to their physiological functions, retinoids and rexinoids have shown high potentials as the therapeutic agents for prevention, amelioration, and successful treatment of different diseases, disorders, and injuries in preclinical models.

An example of the natural retinoids is ATRA that, for its biological functions as an endogenous RAR-selective agonist, binds to the ligand-binding domain of one of the three isoforms of the RARs such as RARα, RARβ, and RARγ. Thus, ATRA is a naturally occurring pan-RAR agonist. Fenretinide, also known as N-(4-hydroxyphenyl) retinamide (4-HPR), is a synthetic pan-RAR agonist that can also exert biological effects via RAR-independent manner. Aromatic retinoids, such as Am80 and Am580, containing an amide in the linker region are synthetic RARα agonists. Synthetic retinoids, such as BMS185411 and AGN193639, with a larger hydrophobic region have been shown to be RARβ-specific agonists. Synthetic compounds, such as SR11254, CD666, and BMS270394, which contain a hydrogen-bond-donating group near the hydrophobic-region aromatic ring, are

RARγ agonists. Adapalene and Tazarotene are synthetic RARβ and RARγ agonists. An example of the natural rexinoids is 9CRA, which is a major endogenous RXR-selective agonist, as it potently activates all isoforms of RXRs. It should be noted that 9CRA is also capable of activating all isoforms of RARs. So, 9CRA is a naturally occurring RXR agonist and a pan-RAR agonist. Bexarotene (LGD1069) and SR11237 are synthetic pan-RXR agonists. However, more potent or more RXR-selective rexinoids need to be developed for avoiding adverse events and significant therapeutic efficacy in treatment of cancers and other diseases in the clinics.

The versatility of the retinoid and rexinoid signaling mechanisms is derived from the capabilities of these molecules to activate mainly RARs and RXRs, respectively, and due to the diverse pairing of each isoforms of RARs and RXRs. Over the years, molecular studies have firmly established that RXR can combine with at least 20 different receptor partners, in addition to RAR, and thereby modulate the transcription of several dozens of genes for controlling cell growth, cell differentiation, cell death, energy metabolism, inflammation, and immune responses. Synthetic methods are currently used for generation of agonists and antagonists of specific isoforms of RAR and RXR so as to avoid toxic side effects, such as skin rash and liver damage, that hinder the use of retinoids and rexinoids in clinical settings. Multifunctional effects of RAR-specific and RXR-specific agonists are expected to be highly useful in the treatment of different cancers that occur due to dysfunction of multiple genes. Multifunctionality of rexinoids seems to be extremely useful as a single drug therapy to deal with a vast majority of human cancers. Also, emerging evidence suggests that activities of a combination of RAR-specific or RXR-specific agonist and another chemotherapeutic drug at low doses avoid adverse side effects and work synergistically to induce apoptosis, while neither drug alone induces apoptosis in cancers in preclinical models. The combination of two drugs makes the tumors disappear due to inhibition of the cell survival pathways and induction of differentiation and apoptotic pathways.

Each chapter in this volume contains an Abstract followed by four major sections: Introduction, Materials, Methods, and Notes. The "Abstract" is a brief summary of the research technique(s) related to the retinoid or rexinoid for revealing its synthesis, signaling mechanisms, or behavioral effects. The first section, "Introduction," describes the background information, principle of the procedure with references to the work of the author (s) and other investigators in the field, sketches the plan for major procedures of the protocol, and presents data (Figures and Tables) to show feasibility and validity of the procedures. The second section, "Materials," is an important part of the chapter describing the chemical and biochemical reagents, buffers, solutions, supplies, and major equipment needed to successfully perform the procedures. This section also prudently mentions the special requirements for storage, stability, purity, temperature and light sensitivity, harmful effects of any reagents and solutions, treatment, protection, and disposal. The third section, "Methods," very carefully describes how to execute the individual steps in the protocol, mentions details of the practical tips, and explains logical aspects of these steps in a decent layout. The forth section, "Notes," may mean a lot to a new user of the protocol as it clearly points out the root causes of the problems and failures and how to sense and overcome the pitfalls to achieve success with the protocol.

This volume on *Retinoid and Rexinoid Signaling* contains 18 chapters. Here is a brief content of each chapter: proteomics to specify loss of RXRγ during progression of epithelial ovarian cancer (Chapter 1); methodology for identification of acacetin as a ligand and regulator of non-genomic signaling of RARγ (Chapter 2); lentiviral delivery of shRNA constructs into acute promyelocytic leukemia cells for ATRA-induced differentiation and autophagy (Chapter 3); generation of new ligands for modulation of RXR functions (Chapter 4); retinoic acid signaling in regulation of angiogenesis in vitro (Chapter 5); methods to analyze RAR signaling in colorectal cancer cells (Chapter 6); rapid luciferase-based assays to assess activity and potency of rexinoids (Chapter 7); structure-activity relationship (SAR) study to generate novel rexinoids on a potent RXR agonist (Chapter 8); differentiation of primary myoblasts by using RXR agonist (Chapter 9); methylase-assisted bisulfite sequencing (MAB-seq) assay to evaluate generation of 5-formylcytosine (5-fC) and 5-carboxylcytosine (5-caC) in response to ATRA in mouse embryonic fibroblasts (Chapter 10); development of RARα-/RXRα-specific agonists as hemodynamics-based therapeutic components for induction of miR-10a and inhibition of atherosclerotic lesion (Chapter 11); the use of the Seahorse XF Analyzer to evaluate the role of RARβ signaling in regulation of cellular metabolism in melanoma (Chapter 12); reporter cell-based method to determine retinoic acid levels, synthesis, and catabolism in embryonic tissue (Chapter 13); methodology for analyzing effects of retinoid treatment on nervous system development and larval swimming behavior (Chapter 14); protocol for culturing embryonic chick lung explants and testing impact of retinoic acid in branching and patterning using morphometric and molecular analyses (Chapter 15); quality assessment of RNA for chemopreventive effects of retinoids and rexinoids in formalin-fixed tissues (Chapter 16); protocols for assessment of autophagic flux in ATRA-treated 2D and 3D breast cancer cultures (Chapter 17); and non radioactive and radioactive telómerase assays with retinoid-treated cancer cells (Chapter 18).

I would like to thank all the authors for contributing their excellent chapters with the modern molecular biology protocols and innovative technologies to this volume for helping other investigators who are interested and involved in research in retinoid and rexinoid signaling. I would also like to thank the series editor, Dr. John Walker, for encouraging me for successful completion of this volume for its production. My special thanks also go to the editorial staff at Springer Nature for their all efforts in publishing this volume.

Columbia, SC, USA ***Swapan K. Ray***

Contents

Contributors

CECILE ABILDGAARD • *Danish Cancer Society Research Center, Copenhagen, Denmark*
AYMAN AL HAJ ZEN • *British Heart Foundation Centre of Research Excellence, Division of Cardiovascular Medicine, Radcliffe Department of Medicine, Wellcome Trust Centre for Human Genetics, University of Oxford, Oxford, UK; Nuffield Department of Medicine, Target Discovery Institute, University of Oxford, Oxford, UK; Qatar Foundation, College of Health and Life Sciences, Hamad Bin Khalifa University, Doha, Qatar*
D. ARENDT • *Developmental Biology Unit, European Molecular Biology Laboratory, Heidelberg, Germany; Centre for Organismal Studies, University of Heidelberg, Heidelberg, Germany*
SHARMILA A. BAPAT • *National Centre for Cell Science, Savitribai Phule Pune University Campus, Pune, Maharashtra, India*
DALYIA BENJAMIN • *Department of Pharmacology, Weill Cornell Medicine, New York, NY, USA; Cork Cancer Research Centre, University College Cork, Cork, Ireland*
MARY CAHILL • *Cork Cancer Research Centre, University College Cork, Cork, Ireland; Department of Haematology, Cork University Hospital, Cork, Ireland*
JIHONG CHEN • *Faculty of Medicine, Department of Pathology and Laboratory Medicine, University of Ottawa, Ottawa, ON, Canada*
JENG-JIANN CHIU • *Institute of Cellular and System Medicine, National Health Research Institutes, Miaoli, Taiwan; Institute of Biomedical Engineering, National Tsing Hua University, Hsinchu, Taiwan; Institute of Biomedical Engineering, National Cheng-Kung University, Tainan, Taiwan; College of Pharmacy, Taipei Medical University, Taipei, Taiwan; Institute of Polymer Science and Engineering, National Taiwan University, Taipei, Taiwan*
CHRISTINA DAHL • *Danish Cancer Society Research Center, Copenhagen, Denmark*
ANGEL R. DE LERA • *Facultade de Química, Departamento de Química Orgánica, CINBIO and IBIV, Universidade de Vigo, Vigo, Spain*
ENRICO GARATTINI • *Laboratory of Molecular Biology, Istituto di Ricerche Farmacologiche Mario Negri IRCCS, Milan, Italy*
LORRAINE J. GUDAS • *Department of Pharmacology, Weill Cornell Medicine, New York, NY, USA*
PER GULDBERG • *Danish Cancer Society Research Center, Copenhagen, Denmark*
M. HANDBERG-THORSAGER • *Max Planck Institute of Molecular Cell Biology and Genetics, Dresden, Germany*
HAIDER M. HASSAN • *Department of Biochemistry, Western University, London, ON, Canada; Department of Oncology, The London Regional Cancer Program and the Lawson Health Research Institute, London, ON, Canada*
JIAN-GANG HUANG • *School of Pharmaceutical Sciences, Xiamen University, Xiamen, China*
MASAMICHI IMAJO • *Laboratory of Bioimaging and Cell Signaling, Graduate School of Biostudies, Kyoto University, Kyoto, Japan*
JENNIE N. JEYAPALAN • *Faculty of Medicine and Health Sciences, School of Veterinary Medicine and Science, University of Nottingham, Nottingham, UK*
PETER W. JURUTKA • *School of Mathematical and Natural Sciences, New College of Interdisciplinary Arts and Sciences, Arizona State University, Glendale, AZ, USA*

Rajkumar S. Kalra • *Drug Discovery and Assets Innovation Lab, DBT-AIST International Laboratory for Advanced Biomedicine (DAILAB), National Institute of Advanced Industrial Science and Technology, Tsukuba, Ibaraki, Japan*
Loretta László • *Faculty of Medicine, Department of Clinical Oncology, University of Debrecen, Debrecen, Hungary*
Ding-Yu Lee • *Departments of Food Science and Biological Science and Technology, China University of Science and Technology, Taipei, Taiwan*
Leo M. Y. Lee • *Faculty of Medicine, School of Biomedical Sciences, The Chinese University of Hong Kong, Hong Kong, Hong Kong*
Qiao Li • *Faculty of Medicine, Department of Pathology and Laboratory Medicine, University of Ottawa, Ottawa, ON, Canada; Faculty of Medicine, Department of Cellular and Molecular Medicine, University of Ottawa, Ottawa, ON, Canada*
Jie Liu • *School of Pharmaceutical Sciences, Xiamen University, Xiamen, China*
Claudio Martínez • *Facultade de Química, Departamento de Química Orgánica, CINBIO and IBIV, Universidade de Vigo, Vigo, Spain*
Peter J. McCaffery • *Institute of Medical Sciences, University of Aberdeen, Aberdeen, UK*
Sharon McKenna • *Cork Cancer Research Centre, University College Cork, Cork, Ireland*
Nigel P. Mongan • *Department of Pharmacology, Weill Cornell Medicine, New York, NY, USA; Faculty of Medicine and Health Sciences, School of Veterinary Medicine and Science, University of Nottingham, Nottingham, UK*
Rute S. Moura • *Life Health Sciences Research Institute (ICVS), School of Medicine, University of Minho, Braga, Portugal; ICVS/3B's—PT Government Associate Laboratory, Braga/Guimarães, Portugal*
Manuele G. Muraro • *Department of Biomedicine, University of Basel and University Hospital Basel, Basel, Switzerland*
Tracey R. O'Donovan • *Corck Cancer Research Centre, University College Cork, Cork, UK*
Nina Orfali • *Department of Pharmacology, Weill Cornell Medicine, New York, NY, USA; Cork Cancer Research Centre, University College Cork, Cork, Ireland; Department of Haematology, Cork University Hospital, Cork, Ireland*
Sarah Parejo • *Division of Experimental Pathology, Institute of Pathology, University of Bern, Bern, Switzerland*
Swapan K. Ray • *Department of Pathology, Microbiology, and Immunology, University of South Carolina School of Medicine, Columbia, SC, USA*
Anna M. Schläfli • *Division of Experimental Pathology, Institute of Pathology, University of Bern, Bern, Switzerland*
M. Schubert • *Sorbonne Université, CNRS, Laboratoire de Biologie du Développement de Villefranche-sur-Mer, Institut de la Mer de Villefranche-sur-Mer, Villefranche-sur-Mer, France*
Alisa S. W. Shum • *Faculty of Medicine, School of Biomedical Sciences, The Chinese University of Hong Kong, Hong Kong, Hong Kong*
José A. Souto • *Facultade de Química, Departamento de Química Orgánica, CINBIO and IBIV, Universidade de Vigo, Vigo, Spain*
Giulio C. Spagnoli • *Department of Biomedicine, University of Basel and University Hospital Basel, Basel, Switzerland; National Research Council, Institute of Translational Pharmacology, Rome, Italy*
Selina T. K. Tam • *Faculty of Medicine, School of Biomedical Sciences, The Chinese University of Hong Kong, Hong Kong, Hong Kong*

MARIO P. TSCHAN • *Division of Experimental Pathology, Institute of Pathology, University of Bern, Bern, Switzerland; Graduate School for Cellular and Biomedical Sciences, University of Bern, Bern, Switzerland*
P. TOMANÇAK • *Max Planck Institute of Molecular Cell Biology and Genetics, Dresden, Germany*
JOSEPH TORCHIA • *Department of Biochemistry, Western University, London, ON, Canada; Department of Oncology, The London Regional Cancer Program and the Lawson Health Research Institute, London, ON, Canada*
V. ULMAN • *Max Planck Institute of Molecular Cell Biology and Genetics, Dresden, Germany*
T. MICHAEL UNDERHILL • *Department of Cellular and Physiological Sciences and the Biomedical Research Center, University of British Columbia, Vancouver, BC, Canada*
IVÁN P. URAY • *Faculty of Medicine, Department of Clinical Oncology, University of Debrecen, Debrecen, Hungary*
CARL E. WAGNER • *School of Mathematical and Natural Sciences, New College of Interdisciplinary Arts and Sciences, Arizona State University, Glendale, AZ, USA*
CORRINE L. WOODCOCK • *Faculty of Medicine and Health Sciences, School of Veterinary Medicine and Science, University of Nottingham, Nottingham, UK*
JIN-ZHANG ZENG • *School of Pharmaceutical Sciences, Xiamen University, Xiamen, China*

Chapter 1

Proteomics to Predict Loss of RXR-γ During Progression of Epithelial Ovarian Cancer

Rajkumar S. Kalra and Sharmila A. Bapat

Abstract

Retinoid and rexinoid receptors are known to regulate key processes during development, differentiation, and cell death in vertebrates. However, their contributions to progression of malignant disease remain largely elusive although it is realized that transformed cancer cells, which essentially evade apoptosis, may display altered molecular expressions or functions associated with retinoid signaling. Here, using a progression model of ovarian cancer, we describe a proteomics-based approach including experimental procedures toward identification and validation of altered protein profiles during transformation. Effectively, this specifies loss of RXR-γ during progression of epithelial ovarian cancer.

Key words RXR-γ, 2-Dimensional Gel Electrophoresis (2DE), Epithelial ovarian cancer, Transformation, Cancer progression model, MALDI TOF/TOF, Retinoid signaling

1 Introduction

Nuclear receptors are recognized as key drivers of diverse cellular processes during development, homeostasis, and metabolism [1]. Among these, retinoic acid and retinoid X receptors (RAR and RXR, respectively) are known to be conserved across species [2, 3]. Although retinoid signaling is identified to be vital in several cellular functions [4], it is often compromised in carcinogenesis [5], suggesting inevitability of its attenuated function for tumor development. Previous reports identify loss of RARβ levels in aggressive breast cancer [6]; such a loss is an early event in esophageal malignancies [7], while a decline in RARα levels in premalignant vs. malignant gastric tissues was demonstrated using in situ hybridization [8]. Importantly, RARβ/RARβ2 receptors could instigate tumor suppressive activities, while long-term exposure to retinoic acid was shown to lead to loss of receptor expression in several cancers [9–13]. Repressive chromatin modifications through methylation also contribute to loss of RARβ/RARRES1 levels in lung cancer and choriocarcinoma

Swapan K. Ray (ed.), *Retinoid and Rexinoid Signaling: Methods and Protocols*, Methods in Molecular Biology, vol. 2019,
https://doi.org/10.1007/978-1-4939-9585-1_1,

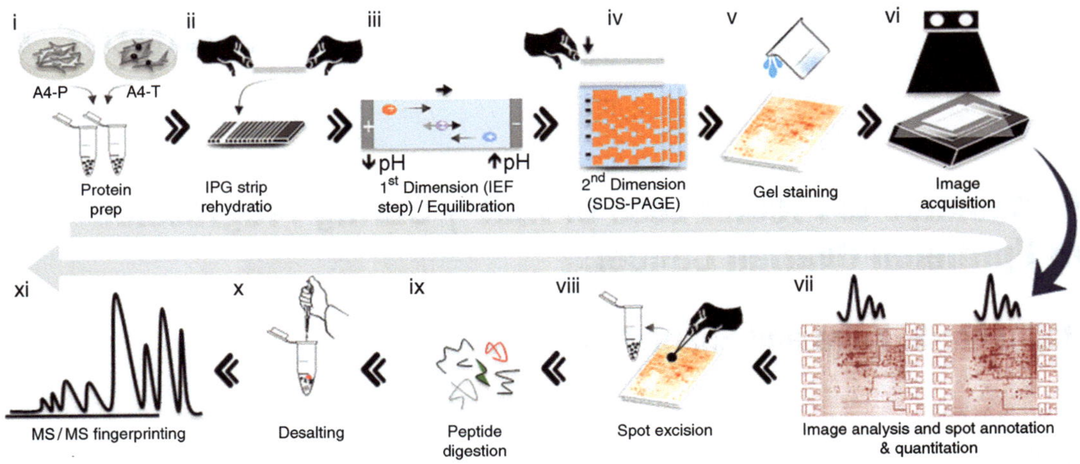

Fig. 1 Schematic steps of sequential workflow in proteomic profiling of ovarian cancer progression model. (**i**) Protein extraction following lysis of A4-P and A4-T cells; (**ii**) IPG strip rehydration; (**iii**) IEF run; (**iv**) 2nd dimension PAGE, (**v**) Gel staining, (**vi**) Image acquisition; (**vii**) Spot annotation and analysis; (**viii**) Spot excision; (**ix**) Protein digestion; (**x**) Desalting; (**xi**) MS/SM analysis

[14, 15]. Concordantly, decreased RARα or RARβ levels were also reported in non-small cell lung carcinoma [16]. Together, such reports based on epigenetic, genomic, and transcriptomic profilings highlight decline in RAR/RXR receptors in different cancers. However, understanding of such alterations at protein levels remained largely uncharacterized. Hence, proteomic profiling during malignant progression was deemed necessary to shed light on the etiology of such events.

Here, using a cancer progression model of epithelial ovarian cancer established earlier in our lab [17, 18], we outline our expression proteomics approach (Fig. 1), that validated loss of RXR-γ protein as a significant event during transformation (Figs. 2 and 3). Similar approaches could be enrolled to analyze/validate changes in protein levels of a marker of interest, in any designated cellular pathway.

2 Materials

Prepare all stock/working solutions stock using Milli-Q water unless otherwise indicated. Use proteomics-grade water to reconstitute trypsin in 50 mM Ammonium Bicarbonate (ABC) and to wet paper wick in Iso-Electric Focusing (IEF). Use analytical grade reagents for MS/MS sample prep and processing. Prepare and store all reagents at room temperature (unless otherwise specified) and dispose all categorized waste as per institutional policies and regulations.

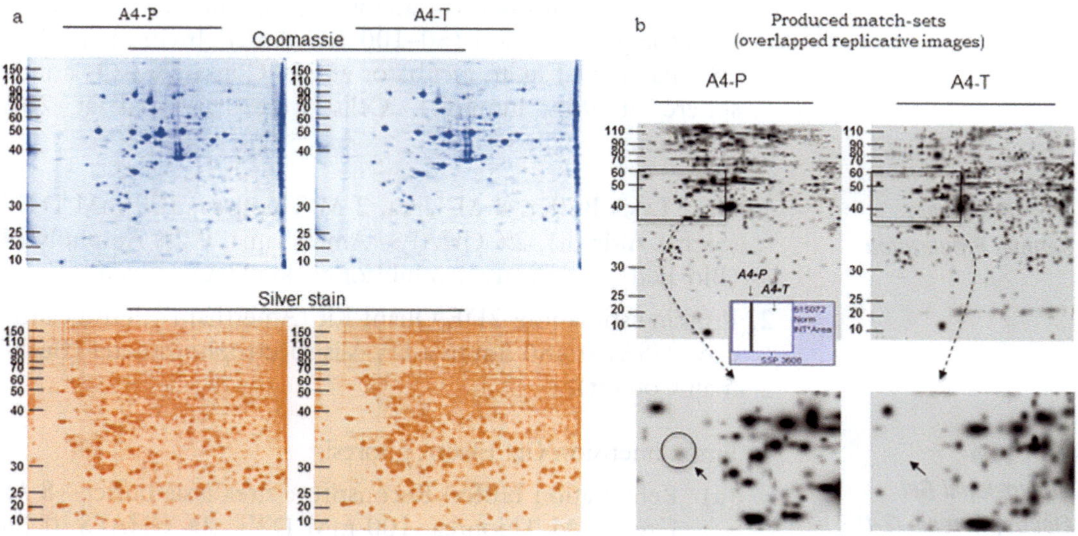

Fig. 2 Comparative 2DE match-set analysis-based identification of loss of RXR-γ in A4-T cells. (**a**) Resolved 2DE protein spots of A4-P (left) and A4-T (right) subclones visualized by Coomassie (Upper level) and silver stain (Lower level); (**b**) A4-P and A4-T match-sets produced form 3 overlapped 2DE replicates, inset (boxed; dotted arrow indicating section below) analyzed by PDQuest™ identified exclusive spot in A4-P which is absent in A4-T derived proteins (marked in circle and identified later as RXR-γ)

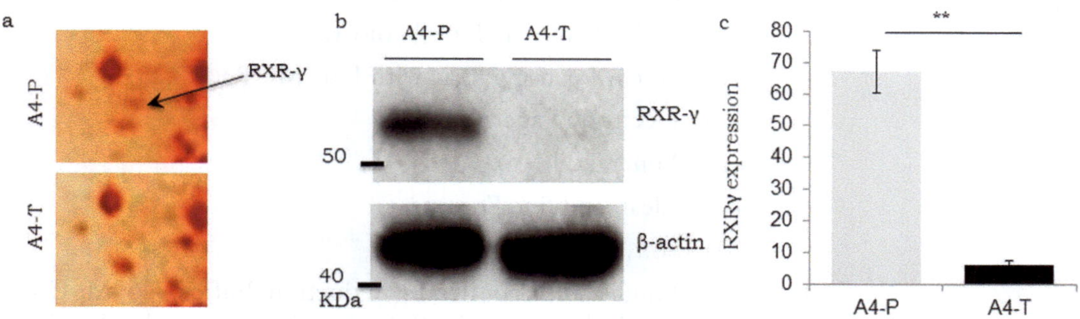

Fig. 3 Validation of differential expression of RXR-γ. (**a**) 2DE section indicating RXR-γ spot in a replicate of A4-P and A4-T silver stained gels; (**b**) Immunoblot validating differential expression of RXR-γ in above subclones; (**c**) showing quantitation of RXR-γ levels (%) in these subclones, as compared to β-actin that was taken as internal loading control

2.1 Cell System and Culture Conditions

1. Isolation and establishment of single cell clones from a patient diagnosed with Grade IV serous ovarian adenocarcinoma has been described earlier (*see* refs. [17, 18]). One of these, viz. the A4 clone, was initially nonaggressive immortal/untransformed but underwent spontaneous transformation into an aggressive, tumorigenic derivative. These two subclones, representative of Pre-transformed and Transformed epithelial ovarian malignant states, were termed as A4-P and A4-T, respectively.
2. Unless otherwise mentioned, A4-P and A4-T subclones were cultured in MEM(E) supplemented with 5% fetal bovine serum

(FBS), 1% non-essential amino acids, and 1% Penicillin-Streptomycin solution (50–100 I.U./mL) (Invitrogen, CA), and maintained in an incubator at 37 °C in a 5% CO_2 atmosphere at 95% humidity. Cells were harvested at 70% confluency.

2.2 Proteomic Sample Preparation

1. Urea Lysis Buffer: 8 M Urea, 2 M Thiourea, 100 mM DTT (Sigma Aldrich), 2% CHAPS (Amersham), 0.2% Ampholytes (Bio-Rad), and protease-inhibitor cocktail (Amersham).
2. Protein estimation: 2DE Quant Kit (Amersham™, GE healthcare, USA) at 480 nm in a UV-Vis spectrophotometer (Beckman Coulter).

2.3 Two-Dimensional Gel Electrophoresis (2-DE)

1. First dimension Gel Electrophoresis:
   (a) Rehydration buffer: Reconstitute urea lysis buffer—8 M Urea, 2 M Thiourea, 100 mM DTT, 2% CHAPS, 0.2% Ampholytes (pH 4–7), and finally add 0.005% bromophenol.
   (b) 18 cm immobilized pH gradient (IPG) strip (Bio-Rad, pH 4–7).
   (c) Dehydration Cassette 18 cm, clean forceps, and Proteomics Grade Water (Bio-Rad).
   (d) PROTEAN™ IEF Cell (Bio-Rad).
   (e) 18 cm Focusing Tray with Lid (Bio-Rad).
   (f) Electrode Paper Wicks.
   (g) Mineral oil.
   (h) Clean and fine Forceps.
   (i) Soft Tissue Paper.
   (j) Equilibration buffer-I (Reduction buffer)—6 M Urea, 0.375 M Tris (pH 8.8), 2% Sodium Dodecyl Sulfate (SDS), 20% glycerol, 2% (w/v) Dithiothreitol (DTT); all reagents were procured from Sigma Aldrich.
   (k) Equilibration buffer-II (Alkylation buffer)—same composition as of reduction buffer, except DTT is replaced with 2.5% (w/v) Iodoacetamide (Sigma Aldrich).
2. Second dimension Gel Electrophoresis:
   (a) SDS-PAGE Gel Composition:

      Resolving Gel (10%)—1.5 M Tris–HCl, pH 8.8, 30% Acrylamide, 10% SDS, 10% Ammonium per sulfate (APS) with 4 μL of N, N, N, N-Tetramethyl-ethylenediamine (TEMED; 14.4 M) to make up a 10 mL solution in Milli-Q.

      Stacking Gel (4%)—SDS-PAGE Buffer: 1.0 M Tris–HCl, pH 6.8, 30% Acrylamide, 10% SDS, 10% Ammonium per sulfate (APS) with adding 4 μL of N,

N, N, N-Tetramethyl-ethylenediamine (TEMED; 14.4 M) to make up a 4 mL solution in Milli-Q.

(b) PROTEAN™ II xi 2-D Cell (Bio-Rad) apparatus and Accessories.

(c) Protein ladder (Precision Plus Standards, Bio-Rad).

(d) Overlay/Low-melting Agarose for SDS-PAGE.

(e) 10× Tris Glycine SDS (TGS)/SDS-PAGE Running buffer—Weigh and dissolve 30 g of Tris base (250 mM), 144 g of Glycine (1.92 M), and 10 g of SDS (1%) in water. Adjust pH to 8.3. Make volume to 1000 mL; dilute tenfold to make 1×.

(f) Electrophoresis settings—10 mA for 15 min as an initial step, proceeded by electrophoresis at 150 V, constant.

2.4 Staining, Imaging, and Analysis

1. Coomassie blue—Mass spectrometry-compatible modified Coomassie blue (Pierce, Thermo-Fisher) staining solution—Dissolve 0.1% Coomassie R-250 in 40% ethanol with 10% acetic acid for staining. Add 10% ethanol and 7.5% acetic acid in Milli-Q water to prepare destaining solution.
2. Silver staining—Prepare a 250 mL volume of all required solutions, as enlisted below.
3. Fixing Solution—Mix 33.75 mL formalin (13.5%) and 100 mL methanol (40%) to make up 250 mL volume of fixing solution in Milli-Q water.
4. Sodium thiosulfate ($Na_2S_2O_3$)—Make a solution of 0.5 g (0.02%) of Sodium thiosulfate (Sigma Aldrich) in 250 mL Milli-Q separately to be used as sensitizing and in developing solution.
5. Silver Nitrate ($AgNO_3$)—Prepare it by dissolving 0.25 g $AgNO_3$ (Sigma Aldrich) in 250 mL Milli-Q in an amber colored or aluminum foiled-covered bottle.
6. Developing Solution—Weigh 7.5 g Sodium Carbonate (3%) to prepare developing solution and add 125 μL formalin (0.05%) and 200 μL of $Na_2S_2O_3$ (0.000016%) freshly from prepared stock.
7. Stop Solution—Weigh 10 g of Citric acid (2.3 M; Sigma Aldrich) to make up 20 mL stop solution for each gel.
8. Imaging and analyses: Quantity One® software of VersaDoc™ system (Bio-Rad Laboratories, USA) was used for acquisition of 2-DE stained gel images, while PDQuest™ (advanced version 8.0, Bio-Rad) was used for image analyses.

2.5 Gel Excision and Peptide Digestion

1. Spot excision and processing: Pick protein spots using a clean sharp spot cutter/blade. Collect each spot individually in a clean/sterile 1.5 mL eppendorf tube.
2. Destaining solutions: 25 mM Ammonium Bicarbonate (ABC; Sigma Aldrich) Solution for Coomassie Blue staining or Add 200 μL of Farmers reagent [49.4 mg $K_3Fe(CN)_6$ + 72 mg $Na_2S_2O_3$] in 5 mL Milli-Q water in case of silver staining.
3. Dehydration solution: 2:1 mixture of Acetonitrile (ACN; Sigma Aldrich) + 50 mM ABC.
4. Reduction/Alkylation solution: 20 mM DTT and 100 mM Iodoacetamide.
5. Trypsin Solution for protein digestion: Reconstitute 20 μG proteomics-grade trypsin (Sigma Aldrich) in 100 μL 1 mM HCl solution (pH 3). Dilute to a final volume of 1000 μL with diluent (9% Acetonitrile + 40 mM ABC + 51% Milli-Q) to make a 20 μG/mL stock.
6. Peptide extraction solution: Add 1% (v/v) formic acid to 20% Acetonitrile. Use ZipTip™ C_{18} (Millipore, USA) filter tips for desalting of peptide mixture in extraction solution (*see* **Note 1**).

2.6 MALDI TOF/TOF Analysis

1. A 4800 MALDI-TOF/TOF mass spectrometer (AB Sciex, Framingham, MA) linked to 4000 series explorer software (version 3.5.3) was used to acquire extracted peptides.
2. This instrument produces mass range from 800 to 4000 Da, with a Nd:YAG 355 nm laser at acceleration and extraction voltages of 20 kV and 18 kV, respectively.
3. GPSTM Explorer software version 3.6 (AB Sciex) was used to explore obtained MS and MS/MS peaks, while MASCOT (version 2.1) (http://www.martixscience.com; SwissProt database) enabled protein identification by MS/MS ion search.

2.7 Immunoblotting

1. Mini-PROTEAN® Electrophoresis System and gel casting accessories (Bio-Rad).
2. PVDF Membranes (Immobilon-P, Millipore).
3. Thick blot transfer pad and Mini Trans-Blot® Cell System.
4. Wet transfer buffer: (25 mM Tris, 192 mM glycine, pH 8.3, 20% methanol).
5. Tris-buffered saline (TBS; 10×): 1.5 M NaCl, 0.1 M Tris–HCl, pH 7.4. In 1× TBS buffer, add 0.05% Tween-20 to make up 1× TBS-T.
6. Blocking solution: Weigh 5 g Bovine Serum Albumin (BSA; 5%) and dissolve in 100 mL of 1× TBS-T.
7. ECL Plus Chemiluminescent reagent (GE Amersham™) and Gel Documentation unit (Bio-Rad).

8. Electrophoresis settings—Stacking gel, 60 V; resolving gel, 80 V constant.
9. Plastic staining containers.
10. HRP Conjugated Anti-Mouse/Rabbit-antibodies (GE Amersham™) and RXR-γ (Santa Cruz) antibodies were procured.

3 Methods

3.1 Protein Sample Preparation

1. Trypsinize and collect A4-P and A4-T cells at ~70% confluency followed by 1× PBS wash. Add 500 μL of Urea lysis buffer to (10^7) cells of A4-P and A4-T each in separate 1.5 mL eppendorf tubes (Fig. 1). Mix gently with the help of a micropipette (*see* **Note 2**). Incubate the mixture at room temperature for 30 min (*see* **Note 3**).
2. Centrifuge tube(s) containing lysed samples at 110,000 × *g* for 1 h at 4 °C and collect supernatant(s) in fresh tubes (*see* **Note 4**).
3. Estimate protein concentration with 2DE Quant Kit at 480 nm in UV/Vis spectrophotometer; see the procedure at https://www.gelifesciences.com).
4. Aliquot samples into multiple working vials and store at −70 °C till further use (*see* **Note 5**).

3.2 Two-Dimensional Gel Electrophoresis

1. Rehydration: Prepare fresh rehydration buffer. Take 350 μG of extracted protein and dilute to a volume of 340 μL in rehydration buffer. Clean the rehydration cassette with distilled water and dry. Remove the 18 cm IPG strip (pH 4–7) from −20 °C, thaw and peel off the protective plastic film from the end marked with + or arrow sign, carefully.
2. Load the protein sample in rehydration buffer into the cleaned cassette and place the IPG strip carefully into it (gel side down) with the help of a clean and dry forceps (Fig. 1) (*see* **Note 6**). Wait for 30 min to ensure absorption of the entire sample into the IPG strip. Add 1–1.5 mL of mineral oil to cover the strip to avoid dehydration and crystallization of urea. Incubate the IPG strip to rehydrate overnight at RT (*see* **Notes** 7 and **8**).
3. Iso-Electric Focusing (IEF)—Place the filter paper wick (trimmed) on electrode wire in well of IEF tray and wet by adding 10–20 μL proteomics-grade water. Remove the rehydrated IPG strip from the tray and wipe off excess oil with help of a soft tissue paper (*see* **Note 9**). Place the IPG strip on wet paper wick, aligning the marked + and − mark signs in the tray accordingly (Fig. 1). Add 2–3 mL fresh mineral oil to cover the strip (*see* **Note 10**). Place the IEF tray cover and close the lid.

4. Set up a 3-step IEF procedure using Protean IEF cell, as follows:
   (a) Step I—50 V for 20 min.
   (b) Step II—10,000 V for 2 h and.
   (c) Step III—10,000 V for 45,000 V-h.
5. In the meantime, prepare SDS-polyacrylamide gel (10–12%) without the upper stack, leaving ~50 mm space at the top wherein the IPG strip can be placed.
6. On completion of IEF, remove the IPG strip with clean forceps and place carefully in a clean well of rehydration tray (gel side up; *see* **Note 11**).
7. Equilibration of IPG strip: Reduction/alkylating equilibration buffers should be freshly prepared.
   (a) Reduction step—Add about 6 mL of freshly prepared Reduction buffer to the well in order to cover the 18 cm IPG strip and incubate for 15 min with gentle shaking.
   (b) Alkylation step—Transfer the IPG strip to another well containing 6 mL freshly prepared alkylation buffer. Place on orbital shaker for 15 min (*see* **Note 12**).
8. Remove the IPG strip from well and dip it in and out 2–3 times (5 s each) in 1× SDS-PAGE Running buffer in a measuring cylinder or long glass vessel appropriately sized to accommodate the IPG strip. Remove the IPG strip and wipe off excess buffer with tissue paper from the plastic backing side end (*see* **Note 13**).
9. Place the IPG strip in the space at the top of the resolving gel aligning the plastic base of the strip with the larger glass plate (*see* **Note 14**). Load molecular weight ladder on the right end and subsequently layer 1–2 mL of low-melting agarose on the top of the gel to seal the IPG strip (*see* **Note 15**).
10. Fix the gels in cassette and electrophoresis apparatus; add chilled 1× running buffer and carry out electrophoresis at 10 mA for 15 min (initial step), proceeded by a constant 150 volts continuation step (Fig. 1).

3.3 Gel Staining, Image Acquisition, and Analysis

1. Coomassie: Remove SDS-PAGE Gel(s) from glass plates and stain in mass spectrometry compatible modified Coomassie blue staining solution for 30 min. Subsequently place the 2DE gel in destaining solution with gentle shaking until visible protein spots with clear background are observed (Fig. 2a). Change destaining solution intermittently during the process.
2. Silver stain: Soak 2DE gel in fixing solution for 10 min. Wash twice in Milli-Q water, each wash for 5 min. Transfer gel to 0.02% $Na_2S_2O_3$ for 1 min followed by 2 quick washes with

Milli-Q water. Soak gel in 0.1% $AgNO_3$ for 10 min, then immerse in developing solution followed by Milli-Q water wash. Add freshly diluted formalin (13.5%) to developing solution and soak the gel in it until protein bands are clearly visible (Fig. 1 and 2a, at lower) (*see* **Note 16**). Add 20 mL stop solution to each gel and incubate with gentle shaking for 10 min. Wash gel with Milli-Q and proceed for imaging.

3. Image acquisition—Place stained gel under gel documentation unit (Fig. 1) and acquire image using Quantity One® software with equal parametric values.
4. Image analysis—Load 2DE image files to PDQuest™ software for demarcation and annotation of differentially expressed proteins (Fig. 2b) (*see* **Note 17**). For each cell type, i.e., A4-P and A4-T, a match-set/master set of ≥3 2DE gel replicates were prepared and analyzed. Each individual spot was marked and given a unique ID in replicative sets. Analysis with reference to the match-set by overlapping replicates was performed to identify differentially expressed proteins in A4-P and A4-T with a twofold quantity threshold. Thus, unique proteins as well as common proteins between different gels were identified (Fig. 2b). Comparison of A4-P and A4-T 2DE match sets identified protein spots, showing exclusive loss or gain in each cell types.

3.4 Spot Picking and In-Gel Peptide Digestion

1. Excise the spots in 2DE gels annotated with PDQuest™ using a sterile sharp spot cutter; a micro tip with its tip widened by cutting can also be used (Fig. 1). Collect each excised spot in individual sterile 1.5 mL tube, mince finely, perform labeling, and proceed for peptide digestion (*see* **Note 18**).
2. Flush the excised gel pieces with 50 μL of 25 mM ammonium bicarbonate (ABC) several times to destain Coomassie Blue in the gel. In case of silver stain, 100 μL of Farmer's reagent (49.4 mG $K_3Fe(CN)_6$ + 72 mG $Na_2S_2O_3$ in 5 mL Milli-Q water) is used for destaining.
3. Dehydrate gel slices with 50 μL of dehydration solution for 5 min. Remove supernatant and add 50 μL of 25 mM ABC for 2 min.
4. Repeat the above cycle three times for ~5 + 2 min each.
5. Add 50 μL of 10 mM DTT for 1 h at 60 °C (reduction), following which, rinse the gel slices with 50 μL of 25 mM ABC.
6. Alkylate the protein with 50 μL of 100 mM Iodoacetic acid (IAA) for 15 min at RT in dark.
7. Repeat the dehydration and rehydration steps one more time.
8. Decant the solutions and place tube carrying gel slices in a vacuum drier and dry the gel completely (*see* **Note 19**).

9. Add 20 μL of reconstituted trypsin (in 50 mM ABC; conc. –20 μG/mL; Proteomic Grade Trypsin) in each tube and place on ice for 30 min (Fig. 1).
10. As gel slices absorb the trypsin, add additional 50 μL volume of 25 mM ABC on the soaked gel to cover surface and incubate for at least 5 h, but preferably overnight at 37 °C (*see* **Note 20**).
11. Following digestion, collect the solution containing digested peptide into a fresh tube. Dry the peptide in solution in vacuum concentrator and resuspend in 10 μL of 20% ABC and 1% formic acid solution. Process extracted peptide through C_{18} resin Zip-Tip® tips for desalting the sample (Fig. 1).
12. Suspend extracted peptide mixtures in matrix solution at ~0.1–10 pmol/μL final concentration. For MALDI analysis, a-cyano-4-hydroxycinnamic acid-based matrix was prepared in 50% ACN, 0.1% trifluoroacetic acid at a 20 mG/mL concentration.
13. Spot about 1 μL of peptide in matrix solution on a stainless steel MALDI sample plate in equal volume.

3.5 MALDI-TOF/TOF Analyses and Protein Identification

1. Digested peptides were acquired on 4800 MALDI-TOF/TOF mass spectrometer linked to 4000 series explorer software. Reflector mode recorded the produced mass spectra within a mass range from 800 to 4000 Da, using a Nd:YAG 355 nm laser (Fig. 1). The acceleration and extraction voltages were set at 20 kV and 18 kV, respectively. Entire MS spectra were obtained from accumulation of 900 shots, while MS/MS spectra were acquired with a total of 1500 laser shots and collision energy of 1 kV. Instrument was calibrated with a 6-point method using a peptide standard kit.
2. Following MS survey scan, a list of precursor ions was generated on processing the data for interrogation by MS/MS. GPSTM Explorer software version 3.6 was used to explore the combined MS and MS/MS peaks. Subsequently MASCOT (version 2.1; http://www.martixscience.com) search engine with SwissProt database enabled protein identification (Fig. 1). Input search parameters were set as follows: all entries and human taxonomy, trypsin digestion and one missed cleavage, fixed modifications: carbamidomethylation of cysteine residues, mass tolerance: 150 ppm for MS and 0.4 Da for MS/MS.
3. Analysis led to selection of proteins having at least two unique matched peptides with a ≥95% interval threshold of confidence.

3.6 Immunoblotting

1. Immunoblotting was performed using standard procedures (*see* ref. [19, 20]). Briefly, 20 μG protein was probed with RXR-γ antibody for 3 h at RT.
2. Following 1× TBS-T washes, HRP-tagged anti-mouse antibody enabled detection of RXR-γ levels in A4-P and validated its loss in A4-T (Fig. 3b, c), as predicted in 2DE-based proteomic profiling (Fig. 2a).

4 Notes

1. Analytical grade reagents were procured for use in peptide extraction and processing for MS/MS analysis to avoid a nonspecific interference in mass spectroscopy procedure.
2. Proliferating cells (~70% confluency) possess optimal expression of nuclear receptors, i.e., RAR/RXR and other transcription factors (*see* ref. 21). A brief cell pellet wash with Milli-Q water is routinely advised to remove salts from PBS before adding the Urea lysis buffer. Foaming must be avoided while adding the lysis buffer to cell pellet and mixing.
3. Mixing lysis buffer at room temperature increase protein solubility. Nuclear receptors, i.e., RAR and RXR usually do not pose a solubility issue; however CHAPS content can be increased slightly (≥2%) to achieve complete solubilization.
4. Collecting clear suspension of cell-derived protein is crucial. Membrane raft and cytoskeleton proteins constitute a major proportion of insolubilized debris that settles down in the tube. Retinoic receptor proteins are usually readily accessible and therefore easily extracted in solution. Collect clear lysate and avoid contamination with debris, be ready to sacrifice some sample to avoid the same at this step.
5. Aliquot small volumes of lysate in multiple vials and store at −70 °C for later use; this avoids loss of sample integrity and quality by repeated cycles of thawing and freezing-down of stock.
6. Handle IPG strip carefully using dust-free gloves and make sure not to touch the gel anywhere or with naked hands.
7. While rehydrating IPG strip with protein sample, make sure no air bubble is trapped in the solution. For this, place one end of IPG strip first and then slowly lower the strip over the sample (Fig. 1).
8. Add sufficient mineral oil to cover IPG strip, starting from at one end. In case of using multiple strips, make sure to note-down the sample name in respective well to avoid ambiguity later.

9. While taking out the IPG strip with clean and dry forceps, avoid touching and damaging the gel matrix. Forceps should hold the plastic backing while transferring or wiping the excess mineral oil in IEF or during any of the later steps.
10. At this step, placing of IPG strip on wetted electrode wick is crucial. Toward a smooth IEF step, make sure no air bubble is trapped under the gel matrix before proceeding to IEF step.
11. Decant remaining oil on IPG strip by holding it vertically with the help of forceps for 20–30 s over a clear tissue paper. Place the plastic backing side down on tissue paper to remove oil from the upper side (If you plan to perform IEF later, IPG strip can be stored at −70 °C).
12. Decant the reducing buffer by tilting the tray vertically, making sure that the buffer does not spill over in other empty wells.
13. This step is incorporated to get rid of DTT and IAA from IPG Strip.
14. Loading the IPG strip correctly at the top of the gel is a crucial step. Pour 1× TGS buffer in the space over the gel and place IPG strip carefully by placing its plastic backing with large glass plate (Fig. 1).
15. Pour melted overlay/low-melting agarose to fix the IPG Strip in space and make sure that no air bubbles are trapped during overlay. Allow agarose to solidify for 5 min before beginning electrophoresis run.
16. Nuclear RAR and RXR receptors are not abundantly expressed proteins. Staining of gel is crucial to make faint spot just visible with no dark background. Make sure not to over-/under-stain the gel at this step with coomassie blue or silver stains (Fig. 2a). A good stained protein spot with contrasting/clear background is optimum for identifying and annotating exclusive/differentially expressed proteins (Fig. 1, steps v-vii). Staining variations from gel to gel in a batch, could result in unsynchronized spot analysis, erroneous annotation and faulty identification of exclusive/differentially expressed proteins.
17. The gel size should critically be kept uniform during generating master gels (Figs. 1-step vii and 2b). Uneven size, contrast or image variations could produce faulty results in the comparative analysis.
18. Spot-picking and peptide digestion are sensitive steps (Fig. 1, steps viii–ix). Wear clean dust-free gloves, mask, and specifically head cap to avoid keratin cross-contamination from skin and hair in samples. Avoid dust, cover skin, and perform above steps on a neat and clean bench during the process. Also, clean the reagents bottles and container surfaces if earlier handled with naked hands.

19. Once the excised gel spot dries, it shrinks remarkably in size and becomes light-weight. Make sure to check the micropipette tip for gel remnants, while removing from vial after adding reagents.
20. Addition of 25 mM ABC during incubation at 37 °C prevents possible dehydration of trypsin-rehydrated gel slices.

Acknowledgments

We thank Dr. R. Srikanth and Ms. Snigdha (Proteomics facility, NCCS) for providing support for MS analyses. Technical assistance by group members Avinash Mali, Mihir Metkar and Manish Kumar is also gratefully acknowledged.

References

1. Mangelsdorf DJ, Thummel C, Beato M et al (1995) The nuclear receptor superfamily: the second decade. Cell 83:835–839. https://doi.org/10.1016/0092-8674(95)90199-x
2. Lera ARD, Bourguet W, Altucci L, Gronemeyer H (2007) Design of selective nuclear receptor modulators: RAR and RXR as a case study. Nat Rev Drug Discov 6:811–820. https://doi.org/10.1038/nrd2398
3. Maire AL, Alvarez S, Shankaranarayanan P et al (2012) Retinoid receptors and therapeutic applications of RAR/RXR modulators. Curr Top Med Chem 12:505–527. https://doi.org/10.2174/156802612799436687
4. Huang P, Chandra V, Rastinejad F (2013) Retinoic acid actions through mammalian nuclear receptors. Chem Rev 114:233–254. https://doi.org/10.1021/cr400161b
5. Altucci L, Leibowitz MD, Ogilvie KM et al (2007) RAR and RXR modulation in cancer and metabolic disease. Nat Rev Drug Discov 6:793–810. https://doi.org/10.1038/nrd2397
6. Widschwendter M, Berger J, Daxenbichler G et al (1998) P11 loss of retinoic acid receptor β expression in breast cancer and morphologically normal adjacent tissue but not in the normal breast tissue distant from the cancer. Eur J Cancer. https://doi.org/10.1016/s0959-8049(97)89229-5
7. Qiu H, Zhang W, El-Naggar AK et al (1999) Loss of retinoic acid receptor-β expression is an early event during Esophageal carcinogenesis. Am J Pathol 155:1519–1523. https://doi.org/10.1016/s0002-9440(10)65467-3
8. Jiang S-Y, Shen S-R, Shyu R-Y et al (1999) Expression of nuclear retinoid receptors in normal, premalignant and malignant gastric tissues determined by in situ hybridization. Br J Cancer 80:206–214. https://doi.org/10.1038/sj.bjc.6690340
9. Xu X-C (2007) Tumor-suppressive activity of retinoic acid receptor-β in cancer. Cancer Lett 253:14–24. https://doi.org/10.1016/j.canlet.2006.11.019
10. Yang Q, Sakurai T, Kakudo K (2002) Retinoid, retinoic acid receptor β and breast Cancer. Breast Cancer Res Treat 76:167–173. https://doi.org/10.1023/a:1020576606004
11. Sabichi AL, Hendricks DT, Bober MA, Birrer MJ (1998) Retinoic acid receptor β expression and growth inhibition of gynecologic cancer cells by the synthetic retinoid N-(4-hydroxyphenyl) retinamide. JNCI 90:597–605. https://doi.org/10.1093/jnci/90.8.597
12. Lefebvre B, Brand C, Flajollet S, Lefebvre P (2006) Down-regulation of the tumor suppressor gene retinoic acid receptor β2 through the phosphoinositide 3-kinase/Akt signaling pathway. Mol Endocrinol 20:2109–2121. https://doi.org/10.1210/me.2005-0321
13. Stephen R, Darbre PD (2000) Loss of growth inhibitory effects of retinoic acid in human breast cancer cells following long-term exposure to retinoic acid. Br J Cancer 83:1183–1191. https://doi.org/10.1054/bjoc.2000.1388
14. Tahara E (2002) Histone acetylation and retinoic acid receptor beta DNA methylation as novel targets for gastric Cancer therapy. Drug

News Perspect 15:581. https://doi.org/10.1358/dnp.2002.15.9.840063

15. Huebner H, Strick R, Wachter DL et al (2017) Hypermethylation and loss of retinoic acid receptor responder 1 expression in human choriocarcinoma. J Exp Clin Cancer Res. https://doi.org/10.1186/s13046-017-0634-x
16. Inui N, Sasaki S, Suda T et al (2003) The loss of retinoic acid receptor? And alcohol dehydrogenase3 expression in non-small cell lung cancer. Respirology 8:302–309. https://doi.org/10.1046/j.1440-1843.2003.00481.x
17. Bapat SA, Mali AM, Koppikar CB, Kurrey NK (2005) Stem and progenitor-like cells contribute to the aggressive behavior of human epithelial ovarian Cancer. Cancer Res 65:3025–3029. https://doi.org/10.1158/0008-5472.can-04-3931
18. Kalra RS, Bapat SA (2013) Expression proteomics predicts loss of RXR-γ during progression of epithelial ovarian Cancer. PLoS One. https://doi.org/10.1371/journal.pone.0070398
19. Yang P-C, Mahmood T (2012) Western blot: technique, theory, and trouble shooting. N Am J Med Sci 4:429. https://doi.org/10.4103/1947-2714.100998
20. Kurien BT, Scofield RH (2009) Introduction to protein blotting. Meth Mol Biol:9–22. https://doi.org/10.1007/978-1-59745-542-8_3
21. Bunaciu RP, Yen A (2011) Activation of the aryl hydrocarbon receptor AhR promotes retinoic acid-induced differentiation of Myeloblastic Leukemia cells by restricting expression of the stem cell transcription factor Oct4. Cancer Res 71:2371–2380. https://doi.org/10.1158/0008-5472.can-10-2299

Chapter 2

Targeting Non-Genomic Activity of Retinoic Acid Receptor-Gamma by Acacetin

Jie Liu, Jian-gang Huang, and Jin-Zhang Zeng

Abstract

Retinoic acid receptors (RARs) are ligand-dependent transcription factors of nuclear hormone receptor superfamily (NR). They are important pharmacological targets and current drug development paradigms are largely based on their nuclear transcription mechanism (genomic action). However, the side effects and limited therapeutic efficacy of retinoid-like drugs with such strategy remain a problem in clinical practice. Increasing evidences have demonstrated that many NRs including RARs can act outside the nucleus in a transcription-independent manner (non-genomic action), which are often implicated in human pathological conditions, suggesting that targeting to the non-genomic signaling of NRs is an alternative method for drug discovery. We recently reported that acacetin could antagonize the non-genomic action of RARγ via tipping the balance of AKT-p53 driven by RARγ from tumor promoting to tumor suppressive effect. This chapter provides methodology for identification of acacetin as a ligand and regulator of non-genomic signaling of RARγ. These laboratory protocols should be helpful for those researchers and beginners who are passionate about identifying chemical leads to probe the non-genomic roles of RARs and other NRs for developing new therapeutic technologies.

Key words Retinoic acid receptor, Non-genomic activity, Protein–protein interaction, Fluorescence resonance energy transfer, Competitive ligand-binding, Confocal laser scanning microscopy, Methodology

1 Introduction

Retinoic acid receptors (RARs) include three subtypes RARα, RARβ, and RARγ, with structures conserved in DNA-binding and -ligand-binding domains but variant in their N-terminus [1, 2]. This class of receptors share functional redundancy but have distinct traits of their own in cell physiology and pathology [3]. They are essential for retinoid signaling depending on their transcriptionally genomic action by directly binding to RA-response elements in the promoters or enhancers of genes that are critical for cellular process and homeostasis [4]. Beyond, RARs can also be stimulated and translocated into cytoplasm where they function non-genomically through signal transduction cascade [5–7]. We recently

Swapan K. Ray (ed.), *Retinoid and Rexinoid Signaling: Methods and Protocols*, Methods in Molecular Biology, vol. 2019, https://doi.org/10.1007/978-1-4939-9585-1_2,

demonstrated that amplification of the non-genomic signaling of RARγ can be of pathological significance. In certain tumors, like hepatocellular carcinoma, RARγ is frequently overexpressed and accumulated abundantly in the cytoplasm [8]. The cytoplasmic non-genomic function of RARγ can be of oncogenic potential due to its capability of interacting with p85α regulatory subunit of PI3K leading to constitutive activation of AKT and NF-κB [9]. In this regard, targeting to the non-genomic signaling of RARγ can be a promising strategy for novel drug development. Thus, we identified that acacetin, a naturally occurring flavonoid compound isolated from the traditional Chinese medicine *Flos Chrysanthemi Indici*, could bind RARγ and regulate RARγ-dependent balance of AKT and p53 [10]. In this chapter, we present in detail the methodology for identification of acacetin as a non-genomic modulator of RARγ. The methods described here should be able to extend to screen and validate acacetin analogs and other compounds that may target to the non-genomic signaling of RARγ. This chapter should be also a reference for exploring the non-genomic effect and implication of other NRs.

2 Materials

2.1 Cell Lines and Cell Culture Reagents

1. Cell lines: HepG2 cells (ATCC HB-8065™), HEK293T cells (ATCC CRL-3216™).
2. Complete medium: Dulbecco's modified Eagle's medium (DMEM) with 10% fetal bovine serum (FBS) and penicillin/streptomycin sulfate (100×, GIBCO).
3. Reagents: low serum medium (DMEM with 2% FBS without antibiotics), DMEM, 0.25% trypsin/EDTA (GIBCO), Phosphate buffer saline (PBS), Lipofectamine 2000 (Invitrogen).

2.2 Antibodies

Anti-RARγ (C-20, sc-551), anti-p85α (B-9, sc-1637), anti-p53 (sc-126), Goat-anti-Mouse-HRP (Thermo, Rockford, IL), Goat-anti-Rabbit-HRP (Thermo, Rockford, IL), anti-myc (9E10, Santa Cruz), anti-flag (Sigma, M2), anti-β-actin (A5441), anti-p(ser 473)-AKT (cst-4060), anti-AKT1/2/3 (sc-8312), anti-mouse IgG conjugated with Cy3 (Chemicon international), FITC-labeled anti-rabbit IgG (Santa Cruz, CA), normal Rabbit IgG (CST, #2729).

2.3 Plasmid Construction

1. pBind-Gal4-RARγLBD: ligand-binding domain (LBD) of RARγ is fused to the C terminal of Gal4 DNA-binding domain and cloned to pBind vector.
2. pG5*luc*: The pG5*luc* Vector contains five Gal4-binding sites upstream of a minimal TATA box, which in turn is upstream of the firefly luciferase gene.

3. pGL3-RARE: 4× retinoic acid response element (RARE) sequence is cloned to the pGL3-luciferase vector.
4. Renilla luciferase vector pRG Renilla (Promega).
5. pCMV-Myc-RARγ: full length of RARγ cDNA is N-terminally tagged with Myc tag and cloned to pCMV-Myc vector.
6. pEGFP-N1-RARγ: vector with full length of RARγ cDNA fused to the N terminal of enhanced GFP (EGFP) gene.
7. pGEX-RARγ-LBD: expression vector of GST-fused RAR-γ-LBD protein.

2.4 Reporter Assays

1. Dual-Luciferase Assay System Kit (Promega, #E1910).
2. Luminometers: Chameleon V Multifunctional microplate reader.
3. White opaque 96-plates.

2.5 Competitive Ligand-Binding Assays

1. 10 mM Acacetin in DMSO.
2. 10 mM *at*RA in DMSO.
3. [^{3}H]*at*RA.
4. 2.6 mg/mL purified GST-RARγ-LBD: stored at −80 °C.
5. TEGD (pH 7.4): 100 mM Tris, 1.5 mM EDTA, 10% glycerin, 1 mM DTT (added freshly before use).
6. GF/C whatman.
7. PBS (pH 7.4): 0.145 M NaCl, 0.0027 M KCl, 0.0081 M Na_2HPO_4, 0.0015 M KH_2PO_4.
8. Scintillation solution (PE-PerkinElmer).
9. Liquid scintillation counter.

2.6 Immunostaining

1. Gelatin-coated coverslips (8 mm × 8 mm).
2. 1× PBS (0.01 M, pH 7.4): 0.145 M NaCl, 0.0027 M KCl, 0.0081 M Na_2HPO_4, 0.0015 M KH_2PO_4.
3. 4% (w/v) Paraformaldehyde in PBS.
4. Membrane permeable solution: 0.5% Triton X-100 in PBS.
5. PBST: 0.1% (v/v) Tween-20 in PBS.
6. Blocking solution: 1% (w/v) BSA in PBST.
7. Antibody dilution buffer: 1% BSA in PBST.
8. 4,6-Diamidino-2-phenylindole (DAPI) (Sigma).
9. Mounting media (Thermo).
10. Confocal laser scanning microscope: LSM-510 system (Carl-zeiss) or other confocal system.

2.7 Subcellular Fractionation

1. Subcellular fractionation buffer (buffer A): 20 mM Hepes (PH 7.4), 10 mM KCl, 2 mM $MgCl_2$, 1 mM EDTA, 1 mM EGTA, 1 mM DTT (added freshly before use).
2. Nuclei lysis buffer (buffer B): 10 mM Tris–HCl (pH 7.6), 420 mM NaCl, 0.5% (v/v) NP-40, 2 mM $MgCl_2$, 1 mM DTT (added freshly before use).
3. Proteinase inhibitor cocktail (Roche, 11-873-580-001).

2.8 Western Blotting

1. SDS-PAGE running buffer: 0.025 M Tris–HCl (pH 8.3), 0.192 M glycine, 0.1% SDS.
2. Western blot transfer buffer: 0.025 M Tris–HCl, 0.192 M glycine, 20% methanol.
3. SDS loading buffer (5×): 0.3 M Tris–HCl (pH 6.8), 10% SDS, 25% β-mercaptoethanol, 0.1% bromophenol blue, 45% glycerol.
4. TBST: 1.5 M NaCl, 0.1 M Tris–HCl (pH 7.4), 0.05% Tween-20.
5. Blocking buffer: 5% (w/v) skimmed milk in TBST.
6. Antibody dilution buffer: 3–5% BSA in TBST.
7. 30% Acrylamide-bisacrylamide mix (29:1): store at 4 °C in a bottle wrapped with aluminum foil or in amber laboratory bottle.
8. Resolving gel buffer: 1.5 M Tris–HCl, pH 8.8, store at 4 °C.
9. Stacking gel buffer: 0.5 M Tris–HCl, pH 6.8, store at 4 °C.
10. 10% SDS (w/v) in water.
11. 10% (w/v) ammonium persulfate (APS) in water.
12. *N*,*N*,*N*,*N′*-Tetramethyl-ethylenediamine (TEMED).
13. Pre-stained protein standard ladder.
14. PVDF membranes (Millipore).
15. ECL reagents (Thermo).
16. ChemiDoc XRS (Bio Rad).

2.9 Co-immunoprecipitation (Co-IP)

1. Co-IP lysis buffer: 0.5% NP-40, 20 mM Hepes (pH 7.6), 150 mM NaCl, 1 mM EDTA.
2. Protein A/G Plus-Agarose (sc-2003, Santa Cruz).
3. Primary antibody: anti-RARγ.
4. IgG control: nonspecific IgG from the same species and isotype of primary antibody.
5. 2× SDS loading buffer: 0.12 M Tris–HCl (pH 6.8), 4% SDS, 10% β-mercaptoethanol, 0.04% bromophenol blue, 18% glycerol.

2.10 Equipment

1. Luminometers: Chameleon V Multifunctional microplate reader.
2. Liquid scintillation counter.
3. Thermostat incubator (Tokai Hit, INUB-WELS-F1 series) equipped to Leica TCS SP5 laser scanning confocal microscope.
4. Confocal laser scanning microscope: LSM-510 system (Carl-zeiss) or other confocal system.
5. ChemiDoc XRS (Bio Rad).

3 Methods

3.1 Screen and Identify Potential Ligands for RARγ

3.1.1 Luciferase Reporter Systems

A luciferase-based pBind-Gal4-RARγLBD reporter system was developed to screen potential ligands for RARγ. In this system, the yeast Gal4 DNA-binding domain (Gal4-DBD) was fused with RARγ ligand-binding domain (RARγ-LBD) and cloned into pBind vector. The expression of luciferase is under the control of Gal4 Upstream Activator Sequence (UAS) [11]. The binding of a ligand to RARγ-LBD will induce conformational change in the receptor, making it competent for recruitment of co-activators to UAS to transcribe luciferase (Fig. 1a). As a cognate agonist of RARγ, *at*RA was used as a positive control and tool to help identify new ligands. A compound that binds to RARγ-LBD may be a competitor of *at*RA on luciferase transcriptional activity. The candidate ligand was then evaluated for its capability in modulating RARγ-dependent RARE activity by using another luciferase-based pGL3-RARE reporter system that contains 4× RARE sequences cloned to upstream of firefly luciferase gene. By combining these methods, acacetin was identified as a new antagonist of RARγ (Fig. 1b, c).

1. Seed HEK293T cells at a density of 3×10^4 cells per well in 48-well plates and incubate the cells in complete medium at 37 °C with 5% CO_2 overnight (~16 h).
2. Prior to transfection, change the complete medium to low serum medium.
3. Dilute the DNA of pG5*luc* vector (50 ng/well) or pBind-Gal4-RARγLBD expression vector (30 ng/well) in DMEM to a volume of 50 μL in one Eppendorf tube. Dilute 0.5 μL Lipofectamine 2000 in 49.5 μL DMEM in another Eppendorf tube. Mix gently and incubate the tubes at room temperature for 5 min (*see* **Note 1**).
4. Combine the diluted DNA and diluted Lipofectamine 2000 and incubate at room temperature for 20 min to allow the formation of DNA-lipid complexes (*see* **Note 2**).

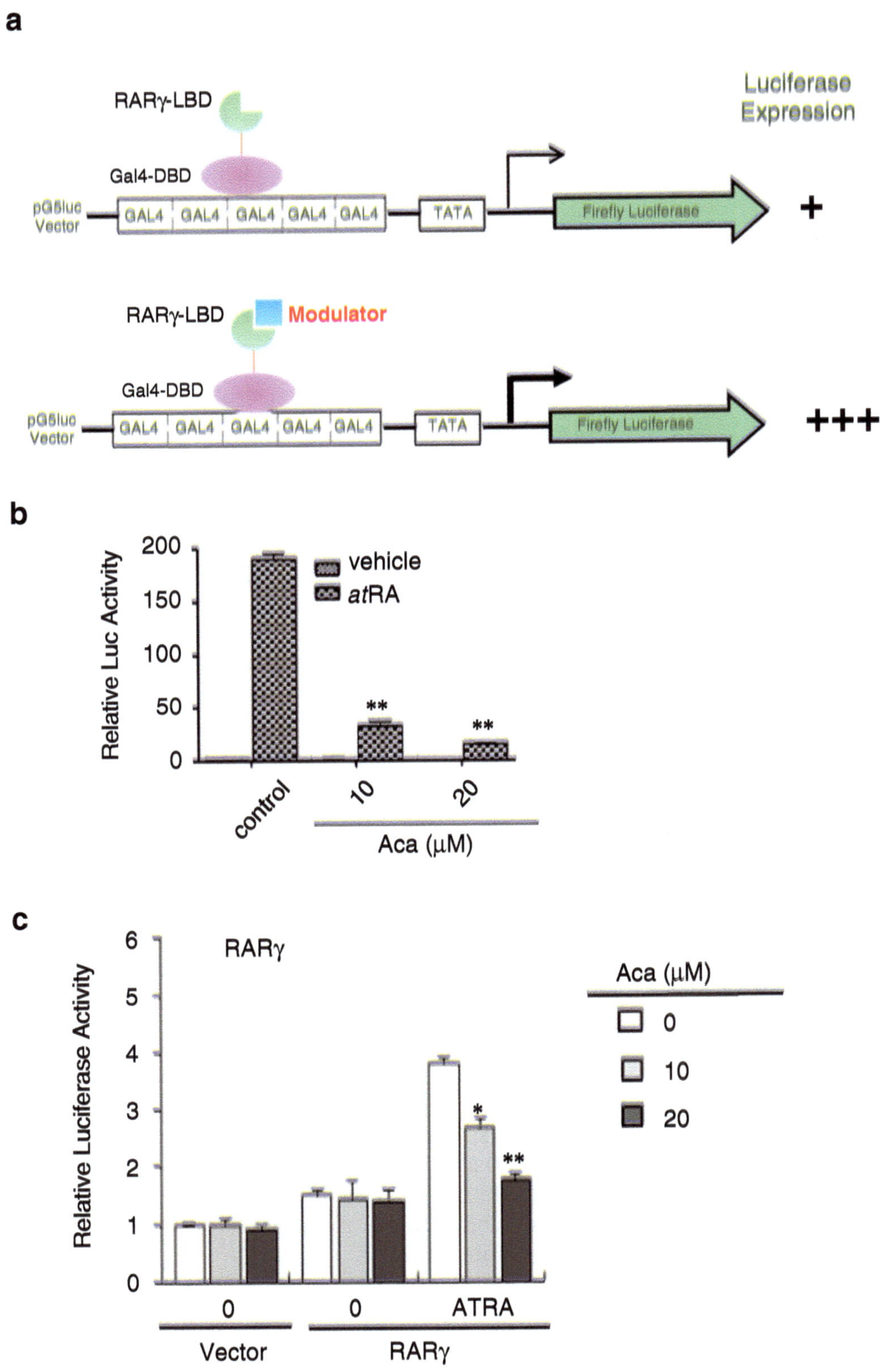

Fig. 1 RARγ ligand identification. (**a**) The construction scheme of pBind-Gal4-RARγ LBD reporter. (**b**) HEK293T cells were co-transfected with pG5*luc* and pBind-Gal4-RARγ LBD constructs and treated with vehicle or

5. Add the mixture to each well, gently rock the plate to mix and return it to 5% CO_2 incubator for 6 h.
6. Replace the medium with complete medium and incubate the transfected cells in 5% CO_2 incubator for 24 h.
7. Replace complete medium with new DMEM (serum free) and treat the cells with vehicle (DMSO) or 10^{-7} M *at*RA in the presence or absence of acacetin at different concentrations for 12 h.
8. Remove the DMEM medium and rinse the cells twice with 1× PBS.
9. Add 100 μL cell lysis buffer to each well and leave the lysates on ice for 20 min (*see* **Note 3**).
10. Pipet cell lysate up and down with 200 μL tip for several times and transfer 50 μL cell lysate to a white opaque 96-well plate.
11. Measure renilla luciferase and Gal4-promoted firefly luciferase activities in each well on luminometers by the guideline of Dual-Luciferase Assay System Kit user manual (*see* **Note 4**).
12. Data analysis: Assays were carried out with triplicate wells at each dose. The relative transcriptional activity was expressed as firefly luciferase light units/Renilla luciferase light units and the results are plotted by GraphPad prim 7 (*see* **Note 5**).

    As an example, *at*RA robustly increased the transcriptional activity of RARγ, while acacetin could dose-dependently antagonize the effect of *at*RA (Fig. 1b, c).

3.1.2 Competitive Ligand-Receptor Binding Assays

The binding activity of acacetin to RARγ was then investigated. The tests were conducted in vitro by incubating the purified GST-RARγ-LBD protein with various concentrations of acacetin in the presence of 1.0 nM [^{3}H]*at*RA. Unlabeled *at*RA was served as positive control (Fig. 2).

1. Prepare TEGD buffer by adding DTT in TEG to a final concentration of 1 mM.
2. Thaw and dilute GST-RARγ-LBD in TEGD buffer to 1.0 mg/mL.
3. Prepare 5× stock solution of [^{3}H]*at*RA (5 nM). The final assay concentration is 1 nM.

Fig. 1 (continued) different concentrations of acacetin for 12 h in the presence or absence of 1.0 nM *at*RA. The cells were lysed and assayed for luciferase and renilla activities. (**c**) HEK293T cells were transfected with pGL3-RARE and pCMV-myc-RARγ or empty vector. The plasmid of pRG Renilla was co-transfected as an internal control. 24 h after transfection, the cells were treated with DMSO or 1.0 nM *at*RA in the absence or presence of different concentration of acacetin for 12 h. Luciferase activities were normalized to renilla activities. Reproduced from *Scientific Reports* 2017;7(1):348 [10]

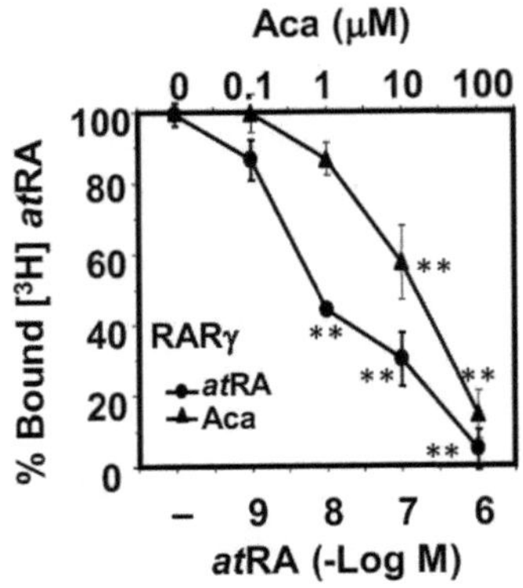

Fig. 2 Acacetin binds RARγ and competitively inhibits *at*RA binding. The purified RARγ-LBD was incubated with 1.0 nM [^{3}H]*at*RA in the presence of either *at*RA or acacetin at an increasing concentrations. The capabilities of unlabeled *at*RA and acacetin to displace the radio-labeled [^{3}H]*at*RA were evaluated by liquid scintillation counting after 12 h treatment. Reproduced from *Scientific Reports* 2017;7(1):348 [10]

4. Prepare 5× assay concentration of unlabeled compounds with ten folds serial dilutions in TEGD buffer:
   *at*RA: 0.05 nM, 0.5 nM, 5 nM, 50 nM, and 500 nM.
   Acacetin: 0.5 μM, 5 μM, 50 μM, 500 μM and 5 mM.
5. Pipet 100 μL TEGD containing 1.0 mg/mL BSA into a 96-well rack of plastic tubes.
6. Add 50 μL [^{3}H]*at*RA and 50 μL tested compound (*at*RA or acacetin) of serial dilutions.
7. Dispense 50 μL GST-RARγ-LBD into each tube. Final total volume of each reaction is 250 μL (*see* **Notes 6** and 7).
8. Incubate the reaction mixture at 4 °C for 14 h in dark.
9. Filtrate these reaction mixtures on GF/C whatman and wash the whatman twice with precool PBS to remove the non-bound [^{3}H]*at*RA.
10. Dry the whatman at 70 °C.
11. Add scintillation solution on the dried whatman and read the CPM value to quantify bound [^{3}H]*at*RA.
12. Calculate the competition rate:

$$\text{Inhibition by } at \text{ RA} = \left[\left(\text{CPM}_{\text{total}} - \text{CPM}_{\text{ra}}\right)/\left(\text{CPM}_{\text{total}} - \text{CPM}_{\text{nonspec}}\right)\right] \times 100\%.$$

$$\text{Inhibition by acacetin} = \left[\left(\text{CPM}_{\text{total}} - \text{CPM}_{\text{aca}}\right)/\left(\text{CPM}_{\text{total}} - \text{CPM}_{\text{nonspec}}\right)\right] \times 100\%.$$

Plot the inhibition rate of acacetin and *at*RA by GraphPad prime 7 software (Fig. 2).

The IC_{50} of inhibition is calculated as: $\lg IC_{50} = X_m - I(P - (3 - P_m - P_n)/4)$, where X_m is Log (Maximal concentration of acacetin), I is the maximal dose/adjacent dose, P is sum of the inhibition rates, P_m is the maximal inhibition rate, and P_n is minimal inhibition rate.

3.1.3 Cell-Based Fluorescence Resonance Energy Transfer (FRET) Assays

The binding activity of acacetin to RARγ was then determined with FRET technique in living cell. Acacetin has fluorescence absorption spectra between 300 and 520 nm, which was shown to be overlapped with fluorescence emission of GFP-RARγ at wave lengths of 500–550 nm. Thus, when acacetin meets and interacts with GFP-RARγ, the fluorescent energy will be transferred from the donor (GFP-RARγ) to the acceptor (acacetin), leading to attenuation of the donor fluorescence intensity (Fig. 3a) [12, 13].

1. Seed HepG2 cells (1×10^5 cells per well) in 6-well plate and incubate the cells in complete medium at 37 °C with 5% CO_2 overnight.
2. Transfect cells with pEGFP-N1-RARγ and pEGFP-N1 expression vectors by Lipofectamine 2000 as described in Subheading 3.1.1.
3. After 24 h of transfection, replace the media with new DMEM (serum free) and treat the cells with 10 μM acacetin or DMSO control.
4. Take the cell culture plate to thermostat incubator (Tokai Hit, INUB-WELS-F1 series) equipped to Leica TCS SP5 laser scanning confocal microscope, select a cell in each well to monitor, and set up the program in Leica TCS SP5 microscope.
5. Measure the GFP fluorescent intensity every 5 min for 120 min (*see* **Note 8**).
6. The fluorescence intensity at zero min was set as basal level; the relative fluorescence intensity at each time point is calculated as the percentage of basal level. Plot the relative fluorescent intensity as indicated in Fig. 3b.

3.2 Non-genomic Activity of RARγ: Nuclear Export and Interaction with Cytoplasmic Protein

RARγ is predominantly nuclear in cells. The non-genomic function of RARγ requires it to be exported to cytoplasm where it can integrate with various signaling cascades. Thus, RARγ translocation and its interaction with p85α were determined.

3.2.1 Immunofluorescence Immunostaining and Confocal Microscopy

Immunofluorescence staining is an excellent method to visualize the localization and translocation of proteins in cells. The expression of RARγ can be detected endogenously in fixed cells with specific antibody or exogenously in live cells transfected with GFP-RARγ. The association of RARγ with p85α can be detected with confocal laser-scanning microscopy by double immunostainings.

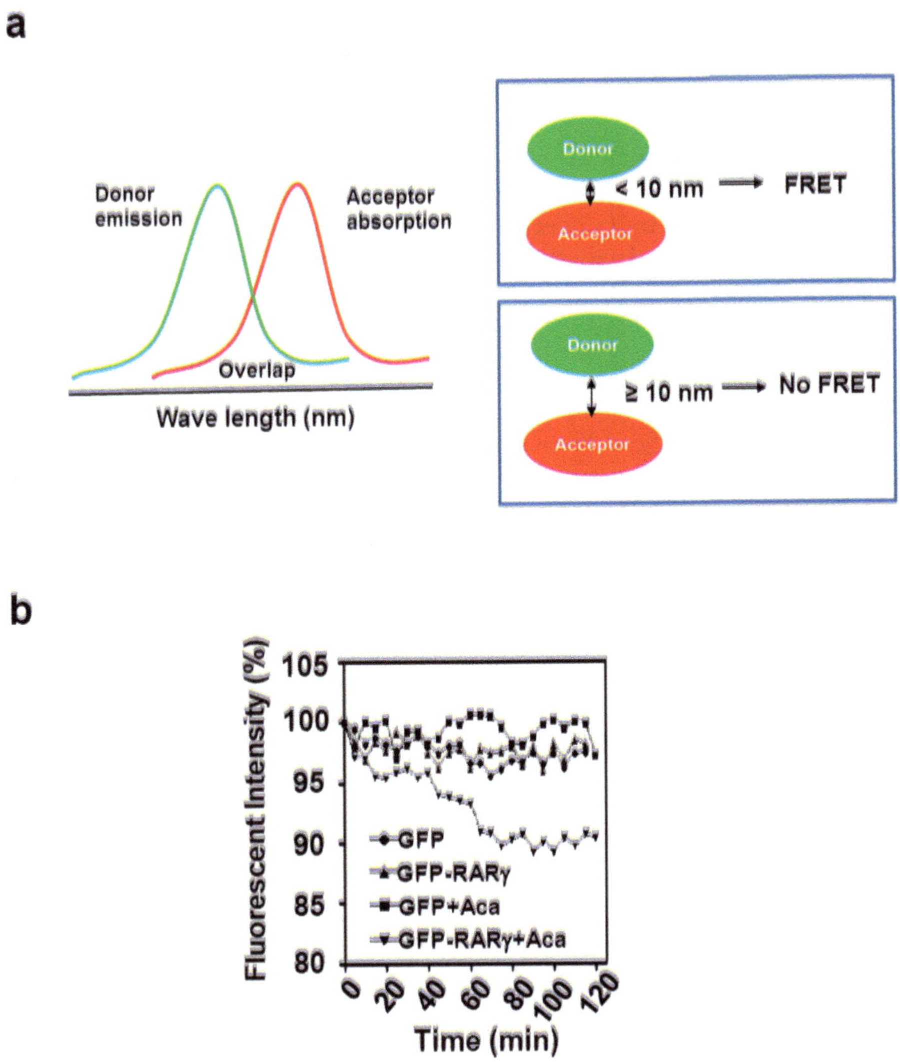

Fig. 3 Acacetin binds RARγ in living cells. (**a**) FRET schematic diagram. (**b**) HepG2 cells were transfected with GFP-RARγ or GFP-vector, and then exposed 10 μM acacetin or vehicle. The plates were moved to 37 °C incubator equipped to Leica TCS SP5 laser scanning confocal microscope. Select one transfected cell and take images every 5 min for 120 min for assaying the fluorescence intensity. Reproduced from *Scientific Reports* 2017;7(1):348 [10]

1. Place sterilized and gelatin-coated coverslips (8 mm × 8 mm) into 24-well plate (*see* **Notes 9** and **10**).
2. Seed and culture HepG2 cells (untransfected and transfected with Myc-RARγ and Flag-p85α) on coverslips at a density of 5000 cells/per well with complete medium overnight.
3. To study the roles of *at*RA and acacetin in regulating the association of RARγ with p85α, replace the complete medium with DMEM (serum free), cells at exponential phase can be treated with 1 nM *at*RA, 10 μM acacetin or vehicle for 12 h.

4. Remove the medium and wash cells twice with 1× PBS.
5. Fix the cells with 4% paraformaldehyde at room temperature for 10 min.
6. Wash the cells with PBS for three times to thoroughly remove the tissue-fixed agent.
7. Permeabilize the cells by adding membrane permeable solution for 15 min on ice.
8. After washing the cells with PBS, add blocking buffer and incubate at room temperature for 1 h.
9. Dilute primary antibodies in dilution buffer. The working titers of antibodies against RARγ and p85 are both at 1:200. As for transfection of Flag-p85α and Myc-RARγ, the antibodies used are anti-Myc (1:1000) and anti-Flag (1:1000), respectively (*see* **Note 11**).
10. Add 200 μL primary antibodies and gently shake the coverslips on a shaking platform at 4 °C for 16–20 h (*see* **Note 12**).
11. Remove the antibodies and wash the cells with PBST three times each for 5 min by softly shaking.
12. Incubate cells with secondary antibody (1:200 in dilution buffer) in a dark chamber at room temperature for 1 h (*see* **Note 13**).
13. The cells are then received three 5 min rinses with PBST and stained with DAPI (1:5000 in PBS) at room temperature for 10 min.
14. Remove DAPI and wash the cells twice with PBS.
15. Invert and mount the coverslips on slides loaded with mounting medium (*see* **Note 14**).
16. Take images using an LSM-510 confocal laser scanning microscope system. As an example, overexpression of RARγ was seen to colocalize with p85α (Fig. 4), which could lead to constitutive activation of PI3K/AKT [9] (*see* **Note 15**).

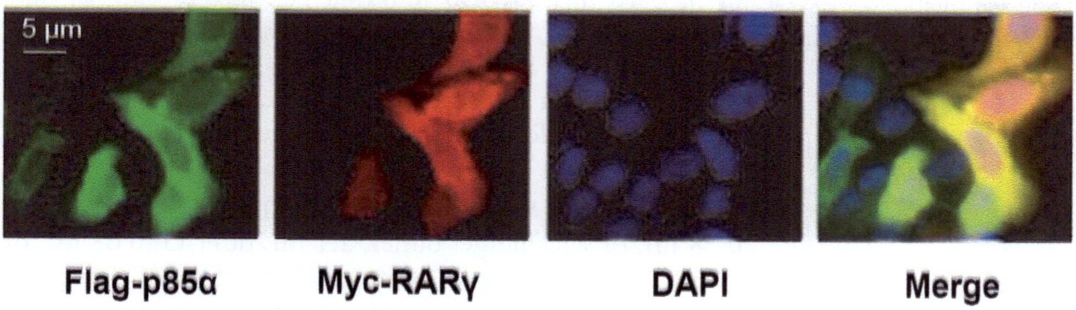

Fig. 4 RARγ interacts with p85α in cytoplasm. HepG2 cells were transfected with Flag-p85α and Myc-RARγ for 48 h and immunostained with anti-Flag and anti-Myc. The images were captured by LSM-510. The scale bar represents 5 μm. Reproduced from *Cancer Research*;70(6):2285–2295 [8]

3.2.2 Subcellular Fractionation

Subcellular fractionation is another useful method to study the protein subcellular localization. Here, we introduce a brief method to divide nuclei and non-nuclei fractions. If it is needed, the non-nuclei fraction can be subdivided to cell membrane, cytoplasm, and mitochondria fractions.

All centrifugations should be done at 4 °C. Samples should be kept on ice throughout the procedure.

1. Seed HepG2 cells in 100 mm dishes and culture the cells to reach 70–90% confluence in 5% CO_2 at 37 °C.
2. Change the complete medium to DMEM (serum free) and treat cells with 10 μM acacetin or DMSO for 12 h.
3. Wash the cells with ice-cold PBS twice.
4. Add 500 μL buffer A with proteinase inhibitor cocktail to each dish and put on ice.
5. Use cell scraper to collect cells and transfer to a 1.5 mL Eppendorf tube.
6. Leave the lysates on ice for 20 min (*see* **Note 16**).
7. Centrifuge samples at 3000 rpm (760 × *g*) at 4 °C for 5 min.
8. Transfer the supernatant (containing cytoplasm, membrane, and mitochondria) to a new Eppendorf tube and store at −80 °C for future use (*see* **Note 17**).
9. Wash the pellet (containing nuclei) with buffer A twice and centrifuge at 3000 rpm (760 × *g*) at 4 °C for 5 min.
10. Remove the supernatant as much as possible and resuspend the pellet in 100 μL buffer B with proteinase inhibitor cocktail. Vortex and incubate on ice for 15 min.
11. Centrifuge the nuclei lysate at 12,000 rpm (16,000 × *g*) for 15 min.
12. Transfer the supernatant (nuclear lysate) to a new Eppendorf tube and store at −80 °C for future use and discard the pellet (*see* **Note 18**).

3.2.3 Western Blotting

1. Prepare 10 mL 8% SDS resolving gel: mix 2.7 mL of 30% Arc-Bis, 3.8 mL of resolving gel buffer, 3.3 mL ddH_2O, 0.1 mL of 10% SDS, 0.1 mL of 10% APS, and 0.006 mL TEMED, mixing sufficiently and cast gel within a gel cassette and allow space for stacking gel, then gently overlay with iso-butanol or water.
2. Prepare 3 mL 5% SDS stacking gel: mix 0.5 mL of 30% Arc-Bis, 0.38 mL of stacking gel buffer, 2.1 mL ddH_2O, 0.03 mL of 10% SDS, 0.03 mL APS, and 0.003 mL TEMED. Add the mixture overlay the resolving gel. Insert a 10-well (or 15-well) gel comb immediately without introducing air bubbles.

3. Add 5× SDS loading buffer to the cell lysates (pre-stored at −80 °C) and boil at 95 °C for 5 min (*see* **Note 19**).
4. Load 20 μL of boiled sample per well. Each gel is also loaded with 2 μL pre-stained protein standard ladder in one well.
5. Run the gel with SDS running buffer at 80 V for 30 min and then at 120 V for 80 min.
6. Transfer protein from gel to PVDF membrane in transfer buffer at 4 °C for 80 min at 260 mA (*see* **Note 20**).
7. Block the PVDF membrane with blocking buffer (5% skim milk) at room temperature for 1 h or at 4 °C overnight.
8. Wash PVDF membrane with TBST for three times each for 5 min at room temperature with shaking.
9. Incubate PVDF membrane with primary antibody (diluted in antibody dilution buffer at appropriate concentration) at 4 °C overnight or at room temperature for 1 h, with softly shaking.
10. Remove primary antibody and wash membrane with TBST for three times each for 5 min at room temperature with shaking (*see* **Note 21**).
11. Incubate PVDF membrane with HRP-conjugated second antibody with appropriate dilution in antibody dilution buffer at room temperature for 1 h.
12. Remove second antibody, and wash PVDF membrane with TBST for three times each for 5 min at room temperature with shaking.
13. Remove TBST as much as possible, mix ECL reagent A and reagent B (1:1), add the mixture to PVDF membrane for 1 or 2 min, and scan the chemiluminescence at ChemiDoc XRS (*see* **Notes 22** and **23**).

3.2.4 Co-immunoprecipitation (Co-IP) Assays

RARγ can interact with other proteins in cytoplasm to perform its non-genomic functions. Co-immunoprecipitation (Co-IP) is a common method to detect protein–protein interaction in cells. In this method, the Co-IP lysis buffer is critical for results, the concentration of NaCl and NP-40 is often lower than 300 mM and 1% (v/v). Since *at*RA and acacetin were shown to promote and disrupt the assembling between RARγ and p85α within the cytoplasm, respectively, they can serve as probes to explore the non-genomic signaling of RARγ.

1. Seed HepG2 cells in 100 mm dishes and culture the cells to reach 70–90% confluence in 5% CO_2 at 37 °C overnight.
2. Change the complete medium to DMEM (serum free) and treat the cells with 1.0 nM *at*RA, 15 μM acacetin, or vehicle for 12 h.
3. Remove the medium and wash the cells with ice-cold PBS.

4. Add 1 mL PBS, scratch and transfer the cells to 1.5 mL Eppendorf tubes.
5. Centrifuge at 500 × *g* at 4 °C for 5 min. Remove the supernatant as much as possible.
6. Add 500 μL Co-IP lysis buffer with proteinase inhibitor cocktail to resuspend the cell pellet. Leave the lysates on ice for 30 min.
7. Centrifuge at 12,000 × *g* at 4 °C for 15 min, remove the supernatants into new tubes, and discard the pellets. The supernatant samples can be directly used to co-immunoprecipitate or stored at −80 °C until use. Each sample is divided into three aliquots: 10–50 μL for input loading, 300 μL for immunoprecipitation, and 150 μL with normal IgG as a control (*see* **Notes 24** and **25**).
8. Incubate the samples with anti-RARγ monoclone antibody (1 μg in 300 μL of each sample) or with normal rabbit IgG (1 μg in 300 μL of sample from untreated and treated cells). Agitate the reaction tubes on rotator at 4 °C overnight.
9. Prepare the beads: wash protein A/G agarose with 1 mL Co-IP lysis buffer, centrifuge at 500 × *g* at 4 °C for 3 min and discard the supernatant.
10. Add 30 μL prewashed protein A/G agarose to the samples from **step 8** and rotate on rotator at 4 °C for 2 h (*see* **Note 26**).
11. Centrifuge at 500 × *g* at 4 °C for 3 min, carefully remove and discard the supernatants.
12. Wash the beads with Co-IP lysis buffer at 4 °C, centrifuge at 500 × *g* at 4 °C for 3 min, and discard the supernatants.
13. Repeat **step 12** for two times.
14. Remove supernatant as much as possible, add 30 μL 2× SDS loading buffer to resuspend the beads, and boil at 95 °C for 10 min.
15. Spin down the agarose and take the supernatant to new Eppendorf tubes for Western blotting analysis as detailed in Subheading 3.2.3. As examples, the interaction between RARγ and p85α could be enhanced by *at*RA but disrupted by acacetin (Fig. 5).

4 Notes

1. In pGL3-RARE reporter system, cells were co-transfected with pGL3-RARE (40 ng) and Renilla (30 ng), in combination with pCMV-Myc-RARγ (30 ng) or pCMV-Myc (30 ng) vector.

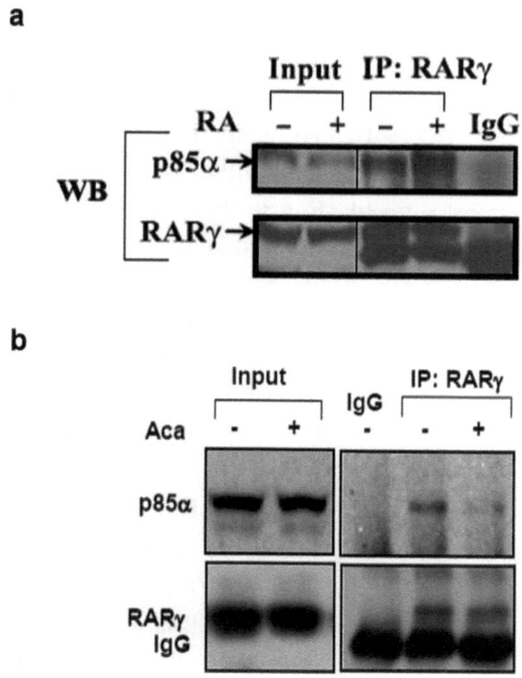

Fig. 5 Effects of *at*RA and acacetin on the interaction between RARγ and p85α. HepG2 cells were treated with 1.0 nM *at*RA (**a**), 15 μM acacetin (**b**), or vehicle for 12 h. The lysates were incubated with anti-RARγ antibody and blotted with anti-RARγ and anti-p85α antibodies. Rabbit IgG was served as a control. Reproduced from *Cancer Research*; 70 (6):2285–2295 and *Scientific Reports* 2017;7(1):348 [8, 10]

2. Pipet softly the mixture of DNA and Lipofectamine 2000 for 6–8 times to mix the complex thoroughly.
3. Check for complete cell lysis under a light microscope. If lysis is incomplete, shake the plate for additional 20 min.
4. Keep this operation away from excessive light and perform it as quickly as possible.
5. Renilla expression is used as an internal control value to which expression of the experimental firefly luciferase reporter gene may be normalized.
6. A nonspecific binding control should be set as 50 μL [^{3}H]*at*RA in 200 μL TEGD.
7. When working with radioisotopes, the proper safety rules for work with radioactive materials must be followed during the entire experimental procedure and waste must be disposed properly.
8. This cell-based fluorescence intensity should be measured from the same cell.

9. Many types of cultured cells do not adhere well to glass coverslips. Thus, coverslips should be coated with gelatin (or poly-L-lysine) to enhance the cell adhesion before use.
10. To avoid the coverslip moved in the well, lid a drop of PBS before placing the coverslip into the plate.
11. For double staining, the two primary antibodies must be from different species so that the co-immunostaining features can be detected and visualized with different fluorescent-labeled secondary antibodies.
12. Not necessary to rinse the cells with PBST before adding antibodies.
13. From this step, the next operations must be performed in dark place.
14. Lay down the coverslip slowly to avoid causing bubbles.
15. The coverslips should not be exposed to laser radiation for long time to avoid fluorescence quenching.
16. Pipet the lysate for several times to avoid cell precipitating during the 20 min incubation.
17. Don't draw up pellet to supernatant.
18. The fractions should be assayed for protein concentrations.
19. The boiled samples can be detected by Western blotting or stored at −80 °C for future use.
20. The PVDF membrane must be pre-incubated with methanol for more than 15 s. Ensure no dry areas on the membrane for a complete protein transfer.
21. Recycle these primary antibodies as they can be reused for 10–15 subsequent experiments over several months.
22. Take pictures from short (a few seconds) and long (a few minutes) exposures of PVDF membrane to X-ray film so that strong and weak expressions can be clearly detected.
23. The membranes can be stripped with Western Blot Stripping Buffer and reprobed with another primary antibody.
24. Prior to this step, you may perform a pre-clearing step to reduce noise background as follows: incubate the samples with 30 μL protein A/G agarose at 4 °C for 30–60 min, spin samples using a microcentrifuge for 10 min at 2500 × *g*, 4 °C, transfer the supernatants to new tubes for immunoprecipitation, and discard pellet.
25. Isotype IgG (Rabbit IgG) is used as the nonspecific binding control.
26. Use a cut micropipette tip to draw up the pre-prepared beads to prevent them from being shattered or damaged.

References

1. De Mendonca RL, Bouton D, Bertin B, Escriva H, Noel C, Vanacker JM, Cornette J, Laudet V, Pierce RJ (2002) A functionally conserved member of the FTZ-F1 nuclear receptor family from *Schistosoma mansoni*. Eur J Biochem 269(22):5700–5711
2. Altucci L, Leibowitz MD, Ogilvie KM, de Lera AR, Gronemeyer H (2007) RAR and RXR modulation in cancer and metabolic disease. Nat Rev Drug Discov 6(10):793–810
3. Germain P, Chambon P, Eichele G, Evans RM, Lazar MA, Leid M, De Lera AR, Lotan R, Mangelsdorf DJ, Gronemeyer H (2006) International Union of Pharmacology. LX. Retinoic acid receptors. Pharmacol Rev 58(4):712–725
4. Gutierrez-Mazariegos J, Schubert M, Laudet V (2014) Evolution of retinoic acid receptors and retinoic acid signaling. Subcell Biochem 70:55–73
5. Piskunov A, Rochette-Egly C (2012) A retinoic acid receptor RARalpha pool present in membrane lipid rafts forms complexes with G protein alphaQ to activate p38MAPK. Oncogene 31(28):3333–3345
6. Dey N, De PK, Wang M, Zhang H, Dobrota EA, Robertson KA, Durden DL (2007) CSK controls retinoic acid receptor (RAR) signaling: a RAR-c-SRC signaling axis is required for neuritogenic differentiation. Mol Cell Biol 27(11):4179–4197
7. Han YH, Zhou H, Kim JH, Yan TD, Lee KH, Wu H, Lin F, Lu N, Liu J, Zeng JZ, Zhang XK (2009) A unique cytoplasmic localization of retinoic acid receptor-gamma and its regulations. J Biol Chem 284(27):18,503–18,514
8. Yan TD, Wu H, Zhang HP, Lu N, Ye P, Yu FH, Zhou H, Li WG, Cao X, Lin YY, He JY, Gao WW, Zhao Y, Xie L, Chen JB, Zhang XK, Zeng JZ (2010) Oncogenic potential of retinoic acid receptor-gamma in hepatocellular carcinoma. Cancer Res 70(6):2285–2295
9. Huang GL, Luo Q, Rui G, Zhang W, Zhang QY, Chen QX, Shen DY (2013) Oncogenic activity of retinoic acid receptor gamma is exhibited through activation of the Akt/NF-kappaB and Wnt/beta-catenin pathways in cholangiocarcinoma. Mol Cell Biol 33 (17):3416–3425
10. Zeng W, Zhang C, Cheng H, Wu YL, Liu J, Chen Z, Huang JG, Ericksen RE, Chen L, Zhang H, Wong AS, Zhang XK, Han W, Zeng JZ (2017) Targeting to the non-genomic activity of retinoic acid receptor-gamma by acacetin in hepatocellular carcinoma. Sci Rep 7(1):348
11. Tanago A, Ikeuchi T (2014) Stable reporter gene assay based on Gal4-vitamin D receptor beta fusion proteins in medaka (*Oryzias latipes*), and its transactivational properties. Zool Sci 31(4):195–201
12. Li HY, Ng EK, Lee SM, Kotaka M, Tsui SK, Lee CY, Fung KP, Waye MM (2001) Protein-protein interaction of FHL3 with FHL2 and visualization of their interaction by green fluorescent proteins (GFP) two-fusion fluorescence resonance energy transfer (FRET). J Cell Biochem 80(3):293–303
13. Nishi M, Tanaka M, Matsuda K, Sunaguchi M, Kawata M (2004) Visualization of glucocorticoid receptor and mineralocorticoid receptor interactions in living cells with GFP-based fluorescence resonance energy transfer. J Neurosci 24(21):4918–4927

Chapter 3

Lentiviral-Mediated shRNA Approaches: Applications in Cellular Differentiation and Autophagy

Nina Orfali, Jennie N. Jeyapalan, Corinne L. Woodcock, Tracey R. O'Donovan, Dalyia Benjamin, Mary Cahill, Sharon McKenna, Lorraine J. Gudas, and Nigel P. Mongan

Abstract

Acute myeloid leukemia (AML) is characterized by the accumulation of immature white blood cell precursors in the bone marrow and peripheral circulation. In essence, leukemic cells fail to differentiate and are stalled at a particular step of hematopoietic maturation and are unable to complete their development into functional blood cells with a finite life cycle. They are thus said to possess a "differentiation block." Pharmacological override of this block is one attractive avenue of therapy, termed "differentiation therapy." The most successful example of this therapeutic strategy is the use of the physiologic retinoid all-*trans*-retinoic acid (ATRA) in the treatment of acute promyelocytic leukemia (APL). In this chapter, we will outline the methods used to characterize the mechanisms mobilized by retinoid signaling and will use the activation of a key regulator of autophagy, ATG7, as an example of the functional characterization of a retinoid regulated gene during differentiation. We will discuss how lentiviral delivery of shRNA constructs into cultured APL cells, such as NB4, can be used to functionally deplete key proteins. We will also describe how the effect of protein knockdown on ATRA-induced differentiation and autophagy can be assessed using quantitative PCR, Western blotting, and flow cytometry.

Key words Differentiation, Leukemia, Malignancy, Flow cytometry, Cancer

1 Introduction

The importance of retinoid signaling in the regulation of embryonic development and stem cell differentiation is well established [1–4]. A role for retinoids in the regulation of hematopoiesis has also emerged and have revealed important roles for retinoids in hematopoietic stem cells (HSC) homeostasis [5, 6] and in myeloid [7] and lymphoid lineages [8, 9]. Acute myeloid leukemia (AML) describes a heterogenous group of clonal malignancies arising in hematopoietic stem or progenitor cells. Leukemic cells are characterized by an increased rate of proliferation, a resistance to apoptosis, and an inability to differentiate into functionally mature

Swapan K. Ray (ed.), *Retinoid and Rexinoid Signaling: Methods and Protocols*, Methods in Molecular Biology, vol. 2019, https://doi.org/10.1007/978-1-4939-9585-1_3,

myeloid cells with a finite life cycle. Consequently, a lethal accumulation of malignant hematopoietic precursors occurs in the bone marrow and peripheral blood of affected individuals with variable compromise of normal hematopoiesis [10]. Acute promyelocytic leukemia (APL) represents approximately 10% of all cases of AML. APL is distinguished by a clinical coagulation defect, characteristic morphology, and chromosomal translocations that lead to oncogenic fusions of the retinoic acid receptor alpha (RARα) locus on chromosome 17q21 with partner genes. In over 95% of cases, this fusion is with the promyelocytic leukemia (PML) gene [11]. The resultant PML-RARα fusion protein binds to retinoic acid response elements (RAREs) with high affinity and recruits repressive complexes, including histone-deacetylases (HDACs) and DNA methyltransferases (DNMTs), that block the transcription of RARα target genes necessary for myeloid differentiation [12, 13].

Retinoids are an important treatment for APL [14, 15]. Consistent with this, exogenous retinoid treatment of myeloid cells (Figs. 1 and 2) in culture leads to changes in the gene expression of known myeloid associated transcription factors, some of which harbor RAREs in their promoter regions, such as CCAAT enhancer binding protein ε (CEBPε) and others, which likely represent secondary response genes, such as PU.1 [16, 17]. The specificity of all *trans* retinoic acid (ATRA) for APL relates to its effect on the PML-RARα oncoprotein. Pharmacologic doses of ATRA cause a conformational change in PML-RARα, resulting in the dissociation of HDACs and DNMTs [13]. This enables the recruitment of coactivator complexes which re-open chromatin structure and restores the transcription of retinoid target genes, including those involved in hematopoietic differentiation [18]. Pharmacological concentrations of ATRA have the secondary effect of inducing the degradation of the PML-RARα oncoprotein through several non-overlapping co-operating proteolytic mechanisms [19, 20]. Nasr and colleagues recently reported that synthetic retinoids capable of reactivating transcription in PML-RARα-expressing leukemic cells but with no effect on PML-RAR protein successfully induce differentiation but fail to eliminate the leukemia-initiating activity of transplanted clones in vivo [21]. This establishes two distinct mechanisms of therapeutic action exerted by ATRA and emphasizes the importance of oncoprotein proteolysis by autophagy pathways in achieving long-term cures. Autophagy is a ubiquitous cellular pathway involved in protein turnover. Unlike the ubiquitin-proteasome protein degradation system, autophagy is not sterically limited and is capable of catabolizing large protein complexes and organelles [22]. Crucially, mechanistic links between retinoid signaling and autophagy have emerged [23–25].

The in vitro study of ATRA-induced differentiation is possible due to the availability of appropriate cell line models such as NB4. NB4 cells can be induced to differentiate by treatment with

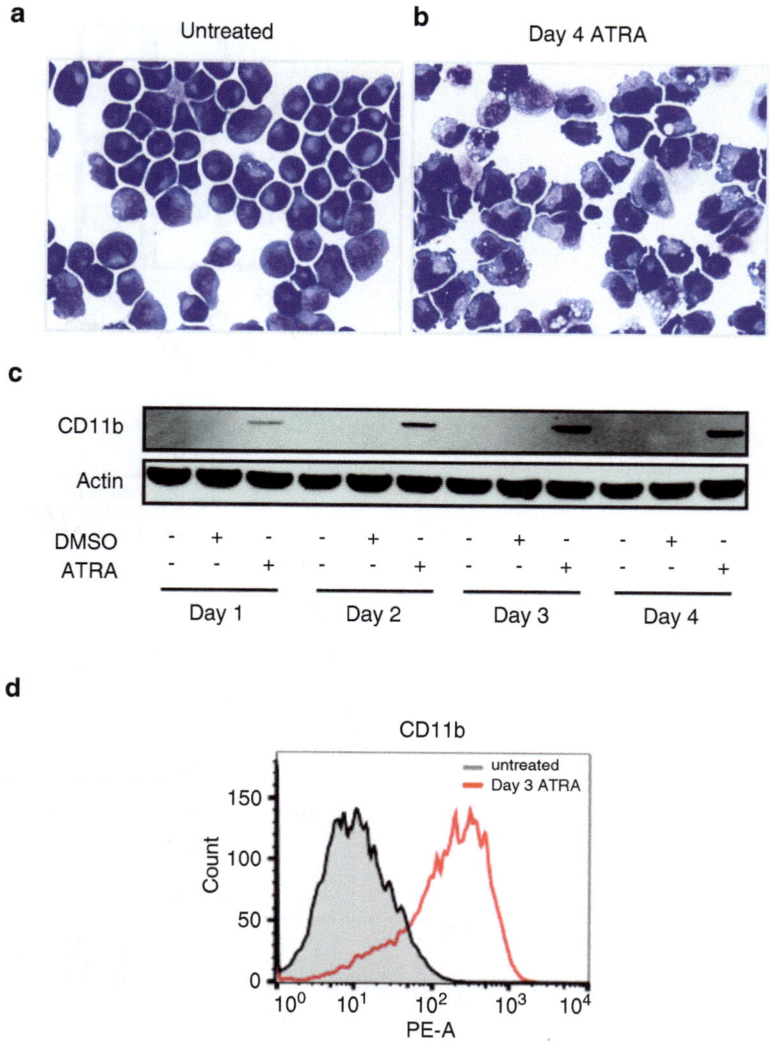

Fig. 1 Measurement of all-*trans*-retinoic acid (ATRA)-induced differentiation in acute promyelocytic leukemia (APL) cells. NB4 cells were seeded at 2×10^5 cells/mL and treated with 1 μmol/L ATRA for 4 days to induce granulocytic differentiation. (**a**) Morphologic analysis at day 4 revealed increased nuclear lobulation and increased cytoplasmic volume with a decreased nuclear-to-cytoplasmic (N:C) ratio (40×). (**b**) Total CD11b protein expression, as a measure of granulocytic differentiation, was measured in daily whole-cell extracts by Western blot (Abcam). Results were normalized to a β-actin loading control ($n = 3$). A single representative blot is shown. (**c**) Surface CD11b was measured daily by flow cytometry in fresh unfixed cells (eBioscience). A representative result at day 3 is shown ($n = 3$). DMSO = dimethyl sulfoxide. Adapted and reproduced, with permission from Elsevier, from Orfali and colleagues [24]

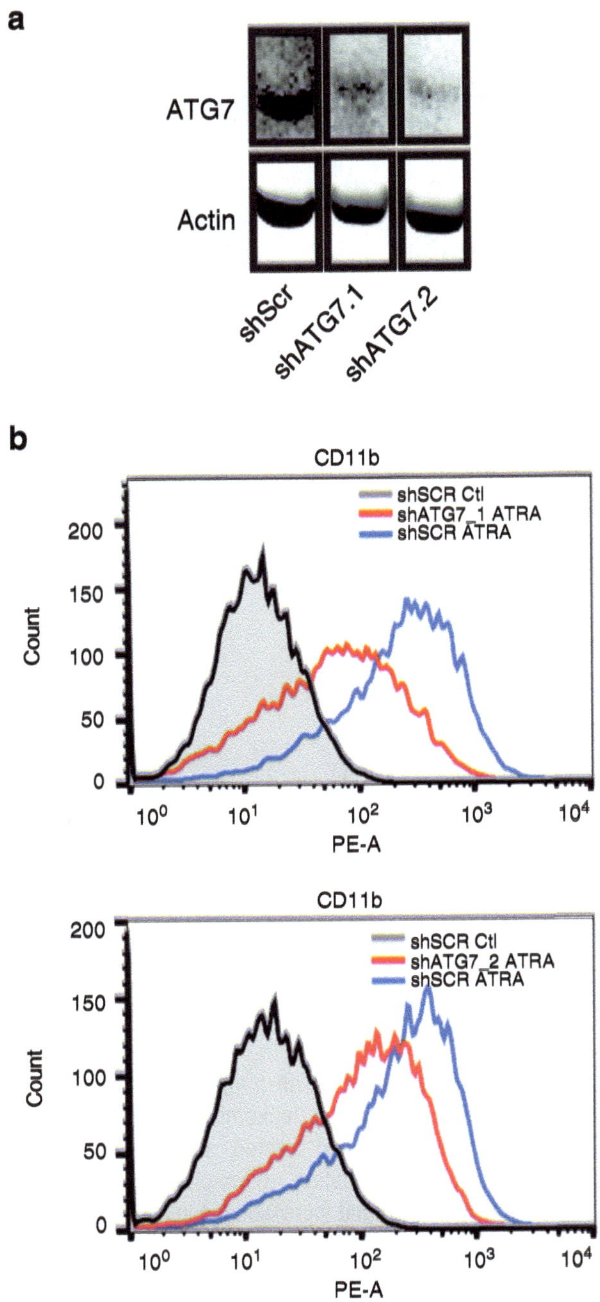

Fig. 2 Effects of genetic ATG7 knockdown on all-trans-retinoic acid (ATRA)-induced acute promyelocytic leukemia (APL) cell differentiation. ATG7 was knocked down in NB4 cells using lentiviral transduction of target-specific short hairpin (sh)RNAs (shATAG7.1, shATG7.2). (**a**) Successful knockdown was confirmed by measurement of ATG7 protein levels by western blot in untreated and treated cells (Cell Signalling Technology), using β-actin as a loading control. (**b**) Two knockdown NB4 clones and scrambled controls were seeded at 2×10^5 cells/mL and treated with ATRA (1 μmol/L) for 3 days. Differentiation was assessed at day 3 by measurement of surface CD11b expression (eBioscience) ($n = 2$). Gray filled histograms are untreated SCR controls, blue histograms are SCR treated with ATRA, and red histograms reveal reduced differentiation in the ATG7 knockdown clones. (**c**) Morphologic analysis by light microscopy at day 3 in ATG7 knockdown cell lines as assessed by nuclear shape and nuclear-to-cytoplasmic (N:C) ratio (40×)

pharmacological doses of ATRA (1 μM) (Fig. 1). ATRA-treated NB4 cells have increased nuclear lobulation and a decreased nuclear-to-cytoplasmic ratio characteristic of maturing granulocytes. Differentiation is confirmed by the detection of the granulocyte surface marker CD11b by western blotting and/or flow cytometry. Autophagy is characterized by the accumulation of double-membraned vesicles within the cell cytoplasm. Subsequent fusions events with endosomes and lysosomes can often result in a vesicular morphology, detectable with light microscopy. In order to assess autophagy, levels of the established autophagy markers can be examined. LC3B-I becomes conjugated to phosphatidylethanolamine (PE) to form LC3B-II, which is incorporated into autophagosome membranes. Other autophagy proteins, GABARAP and GATE-16/GABARAPL2, can also be incorporated into autophagosomes [26]. Autophagic activity can also be assessed by flow cytometry using the Cyto-ID assay, which fluorescently labels autophagosomes [27].

Here we will outline key methods used to decipher the role of autophagy in ATRA-induced differentiation of APL cells. By using shRNA to deplete a key component of the autophagy pathway, such as ATG7 (Fig. 2), it is possible to dissect the functional relationship between ATRA-induced differentiation and induction of autophagy by examining the expression of differentiation and autophagy markers by quantitative PCR, western blotting, and flow cytometry [24].

2 Materials

All solutions should be prepared in ultra-pure reverse osmosis grade 18.2 MΩ water or similar using analytical or cell culture grade reagents as appropriate. Local health and safety, risk assessment requirements, chemical hazard/MSDS sheets and environmental and waste disposal rules should be consulted prior to beginning the method.

2.1 Cell Culture

1. Cell lines such as NB4 APL cells (DSMZ cell culture collection) and HEK293T (ATCC) should be available.
2. NB4 Media: RPMI 1640 (Sigma) supplemented with 10% heat-inactivated fetal calf serum and 1% penicillin/streptomycin.
3. HEK293T media: DMEM (Sigma) supplemented with 10% fetal calf serum.
4. Tissue culture dishes (10 cm) and multi-well plates (Thermo Fisher, Nunc).
5. Transfection reagent, such as Fugene (Promega).

6. ATRA (Sigma) is maintained short-term at 1 mM stock in ethanol at −20 °C and used at a final concentration of 1 μM and used in low light conditions.
7. Hank's balanced salt solution (Thermo Fisher).
8. Rapi-Diff (Atom Scientific) modified Giemsa stains.
9. Phosphate-buffered saline (PBS. 1×): 8 g NaCl, 0.2 g KCl, 1.44 g Na_2HPO_4, 0.24 g KH_2PO_4, pH 7.4 in 1 L of reverse osmosis grade water.
10. HistoMount mounting solution (Thermo Fisher).
11. Methanol (Fisher Scientific, 10511732).
12. Ethylenediamine tetra-acetic acid (EDTA) (Sigma).
13. Cyto-ID Assay (ENZO-Life Science).
14. PE-conjugated anti CD11b antibody—FACS analysis (ThermoFisher Scientific).

2.2 Lentiviral shRNA-Mediated Knockdown

1. Plasmids expressing validated (sh)RNA for the human and mouse genomes generated by The RNAi Consortium (TRC) and non-gene targeting negative controls are available from Sigma (https://www.sigmaaldrich.com/life-science/functional-genomics-and-rnai/shrna.html).
2. pLKO.1 plasmid vectors expressing shRNAs targeting human ATG7 were purchased from Sigma (TRCN0000007584 = shATG7_1 and TRCN0000007587 = shATG7_2).
3. A non-targeting scramble control pLKO.1 shRNA plasmid (Addgene).
4. Lentiviral packaging "helper" plasmids pCMVΔ8.9 or pCMV-dR8.2 and pCMV-VSVG are available from Addgene (www.addgene.org) respectively.
5. Transfection reagent such as Lipofectamine 2000 (ThermoFisher).

2.3 Protein Electrophoresis, Western Blotting, and Visualization

1. Modified radioimmunoprecipitation assay "RIPA" buffer (50 mM Tris/HCl (pH 7.4), 150 mM NaCl, 0.25% sodium deoxycholate, 1% Igepal, 1 mM EDTA, 1× Pefabloc, 1× protease inhibitor cocktail, 1 mM Na_3VO_4, 1 mM NaF).
2. Bio-Rad Bradford Reagent (ThermoFisher Scientific) for protein quantification.
3. Precision Plus protein standard (Bio-Rad).
4. NuPAGE 4–12%, Bis-Tris gels (ThermoFisher).
5. NuPAGE MES SDS Running Buffer (20×) (ThermoFisher) diluted to 1× with water.
6. PVDF membranes (ThermoFisher).
7. Primary rabbit polyclonal antibodies were: anti-CD11b (Abcam), anti-LC3A/B (MBL), anti-LC3B (Novus

Biologicals), anti-GABARAP (Abgent), anti-GABARAPL2/GATE-16 (Abgent), and anti-ATG7 (Cell Signalling).

8. IR-Dye 800 secondary antibody (LiCOR Biosciences).

2.4 RNA Isolation

1. Trizol (ThermoFisher) or TriReagent (Sigma).
2. Chloroform (Fisher Scientific).
3. Isopropanol/2-propopanol (Fisher Scientific).
4. Ethanol (Fisher Scientific).
5. Diethylpyrocarbonate-treated RNAse-free water (Thermo Fisher).

2.5 Reverse Transcription and Quantitative PCR

1. RNA can be reverse transcribed to cDNA using master-mixes, which include reverse transcriptase enzyme, dNTPs, and buffers, such as OneStep RT-PCR Kit (Qiagen) or qScript One-Script (Quanta) (*see* **Note 3**).
2. Validated quantitative PCR (qPCR) assays for human and mouse genes using hydrolysis probe assays, such as TaqMan®, assays (ThermoFisher). Gene Expression Assays for CEBPE (HHs00357657_m1) and CSF3R (Hs00167918_m1) with HMBS (Hs00609296_g1) used as an internal control (ThermoFisher).
3. StepOnePlus qPCR system (Applied Biosystems, Zug, Switzerland).

2.6 Equipment and Software

1. Category II tissue culture laminar flow hood.
2. Humidified tissue culture incubator at 37 °C with 5% CO_2.
3. Cytospin 4 centrifuge (Thermo Scientific).
4. BD-LSRII flow cytometer.
5. Western apparatus, such as mini-Protean II (BioRad).
6. Microcentrifuge.
7. Vortex.
8. Histopathology slides, coverslips, and Coplin staining jars.
9. Light microscope such as from Leica, Nikon, or Zeiss.
10. Nanodrop Microvolume UV-Visible Spectrophotometer (ThermoFisher).
11. Quantitative PCR machine such as a LightCyclerII (Roche) or StepOnePlus qPCR system (Applied Biosystems, Zug, Switzerland).
12. Odyssey IR imaging system (Li-Cor).
13. Flow cytometer and FlowJo flow cytometry software (FlowJo LLC).
14. NC200 automatic cell counter (Chemometec).

15. XCell SureLock Mini-Cell western blot analysis (Thermo Fisher) or Mini-Protean western and transfer apparatus (BioRad).

3 Methods

3.1 Cell Lines and Tissue Culture Reagents

1. NB4 cells are maintained in suspension cultures at a density of between 0.5×10^6 and 1×10^6 /mL in 10 cm tissue culture dishes containing 10 mL media.
2. HEK293T cells are maintained in monolayer cultures and maintained in 75 cm^2 dishes and subcultured at a ratio of 1:3–1:8 every 2–3 days.
3. Cells are counted using a hemocytometer with light microscope, Coulter particle counter (Beckman), or NC200 automatic cell counter (Chemometec).

3.2 Lentiviral shRNA-Mediated Knockdown

The RNAi consortium has built a library of shRNAs targeting human and mouse genes utilizing the pLKO.1 cloning vector [28]. The pLKO.1 vector can be introduced into cells via direct transfection or via lentiviral particles. The puromycin resistance marker within the pLKO.1 vector allows for stable selection. In this protocol pLKO.1 plasmid vectors expressing shRNAs targeting human ATG7 are available from Sigma (TRCN0000007584 = shATG7_1 and TRCN0000007587 = shATG7_2), along with a non-target shRNA control vector (*see* **Note 1**).

3.2.1 Lentiviral Production in HEK293T Cells

1. Day 1—Seed a T25 flask with 5×10^5 HEK293T cells per shRNA vector to be transfected in 5 mL DMEM +10% FBS, without antibiotics.
2. Day 2—In a polypropylene microfuge tube prepare the transfection mix for each shRNA as stated below. Tube 1 will contain the DNA in Opti-MEM, consisting of 1 μg pLKO.1 shRNA, 750 ng pCMVΔ8.9 (packaging vector), and 250 ng pVSVG (envelope vector) and make up to 40 μL with Opti-MEM, incubate at room temperature for 5 min.
3. Tube 2 will contain the transfection reagent, add 10 μL of Fugene to 150 μL of Opti-MEM, mix well, and incubate at room temperature for 5 min. Transfer tube 2 to tube 1, mix, and incubate at room temperature for 20 min.
4. Add the 200 μL transfection mix dropwise to the HEK293T cells and rock the flask gently to mix gently as HEK293T do not adhere strongly to the plastic surface and incubate for 12–15 h at 37 °C, 5% CO_2.
5. Day 3—Replenish cells with fresh growth media (5 mL) following overnight recovery. Incubate for a further 48 h for lentiviral production.

6. Day 5—Lentivirus will be released into the medium by the HEK293T cells. Harvest the HEK293T culture supernatant and filter through a 0.45 μm filter or centrifuge the supernatant at 1200 rpm for 5 min to remove any cell debris that is present. Viral particles can be stored at 4 °C overnight or at −80 °C prior to use.

3.2.2 Determining the Optimal Puromycin Concentration

The protocol to identify the optimal puromycin selection concentration is on NB4 cell line at relative densities used in subsequent transduction procedures.

1. Day 1—Seed 1×10^4 cells in 200 μL medium per well of 96 well plate, 24 wells in total.
2. Day 2—Replace growth medium with 200 μL of selection medium for 3 wells per puromycin concentration. Puromycin concentrations used in this protocol are 0, 0.5, 1, 2, 4, 6, 8, and 10 μg/mL. Include untreated control cells containing growth medium only.
3. Days 4–15—Replace medium every other day with fresh selection medium. Monitor cells daily using a microscope and observe the percentage of surviving cells. The minimum puromycin concentration to use is the lowest concentration that kills 100% of cells in 3–5 days from the start of selection.

3.2.3 Lentiviral Transduction of NB4 Cells

1. Day 1: Plate NB4 cells at 1×10^6 cells/mL in 10 cm dishes containing 4 mL growth medium.
2. Day 2: Cells should be around 70% confluent. Transduce NB4 cells with 1:1 ratio with 2× growth medium supplemented with no antibiotics and 8 μg/mL of polybrene transfection reagent. Incubate overnight at 37 °C, 5% CO_2.
3. Day 3: Change to fresh growth medium (4 mL) for 48 h.
4. Day 5: After 48 h change to medium containing 2 μg/mL of puromycin to select for transduced cells. Change selection medium every other day for 10 days to select puromycin-resistant cells stably harboring the pLKO1 plasmid.

3.3 Western Protein Analysis

3.3.1 Protein Extraction

1. All aspects of protein separation, including centrifuging, to be done at 4 °C.
2. Remove cell pellets from −80 °C freezer and place immediately on ice. Do not allow to thaw.
3. Add 15–30 μL of freshly prepared RIPA buffer to each pellet. (1× Pefabloc, 1× protease inhibitor cocktail, 1 mM Na_3VO_4, and 1 mM NaF are added freshly to the RIPA buffer just before use).

4. Exact volume depends on the size of the pellet but volume of RIPA should be kept to a minimum to allow for maximum protein concentration.
5. Incubate on ice for 30 min, vortexing every 10 min.
6. Centrifuge for 20 mins at max speed (14,000 rpm).
7. Remove the supernatant from each tube and transfer to fresh, prechilled 1.5 mL tubes. Discard pellet containing cell components. Proteins samples can be stored long term at −80 °C.

3.3.2 Western Blotting

In this protocol, precast polyacrylamide NuPAGE 4–12%, Bis-Tris gels are used and electrophoresis performed using XCell SureLock Mini-Cell.

1. Assemble the NuPAGE 4–12% Bis-Tris precast gel into the unit following manufacturer's protocol.
2. Fill tank with 1× electrophoresis buffer.
3. In a microfuge tube prepare 30–50 μg of protein in a maximum volume of 15–20 μL, which includes loading buffer (1×).
4. Add samples to heating block and incubate at 100 °C for 5 min.
5. Load protein samples in the required wells. In 1 well, load a pre-stained protein marker, e.g., Precision Plus protein standard (Bio-Rad, 161-0374). Fill any empty well with 1× loading buffer.
6. Assemble the lid to the Mini-Cell unit, filling the space between gels, ensuring the tops of both gels are covered. Run the gel at 100 V until the samples align and the dye front has moved a few mm through the gel—increasing to 130 V for the duration of the run. Ensure that the blue loading dye reaches bottom of the gel.
7. Remove gel from apparatus and place in deionized water while preparing the transfer pack.
8. Remove cassette from Turbo blot transfer unit and place the Trans-Blot turbo Transfer pack, bottom section containing filter paper and the PVDF membrane into the bottom cassette electrode (anode). Place the gel onto the membrane and cover with the remaining transfer pack filter paper, using a roller to remove any bubbles. Add the top cassette electrode (cathode).
9. Place cassette into the Turbo blot unit and set to semi-dry transfer for 30 min at 25 V.
10. When transfer is complete remove cassette from unit and disassemble semi-dry sandwich, keeping the membrane.

3.3.3 Antibody Incubations and Imaging

1. Place membrane into a 50 mL tube containing 5 mL blocking buffer. Blocking buffer chosen depends on the buffer recommended for the primary antibody. 5% skim milk powder in 1×

TBS, 0.1% Tween® 20, Odyssey buffer (Li-COR, 927-40000) or 5% BSA, 1× TBS, 0.1% Tween® 20.

2. Place onto a rotator and incubate at room temperature for 1 h to block membrane.
3. Discard blocking solution and replace with primary antibody in relevant blocking buffer, e.g., 5 μL of anti-ATG7 primary antibody (1:1000) and place on rotator.
4. Incubate overnight at 40 °C.
5. Discard primary antibody and wash membrane with 10 mL of 1× TBS, 0.1% TWEEN® 20 for 5 min.
6. Repeat washes 3×.
7. Membranes are then incubated with the relevant infrared fluorescent dyes—IR-Dye 800 or IR-Dye 700 (anti-rabbit or anti-mouse, etc.) secondary antibody (Li-COR). Place on rotator and incubate at room temperature for 1 h.
8. Discard secondary antibody and wash blot with 10 mL of 1× TBS, 0.1% TWEEN® 20 for 5 min.
9. Repeat washes 3×.
10. Remove membrane and place on the flat surface or the Odyssey IR imaging system (Li-COR) (*see* **Note 2**).

3.4 RNA Isolation

Isolate RNA with a procedure of your choice, e.g., TRI-reagent® or RNA extraction kit following manufacturer's suggestions. We describe the purification of total RNA from human APL cell line, NB4, using TRI-reagent® protocol.

1. Harvest cells in a 1.5 mL microcentrifuge tube, for 5–10 × 10^6 mammalian cells add 1 mL of TRI-reagent® to lyse cells.
2. Add 200 μL chloroform per 1 mL of TRI-reagent®, close tube tightly, and shake vigorously for 15 s until cloudy.
3. Incubate sample at room temperature for 5 min.
4. Centrifuge the sample at 12,000 × *g*, 4 °C for 15 min.
5. At this point there should be 3 layers, the clear aqueous upper layer contains total RNA. Remove the aqueous phase layer using a pipette taking care not to disrupt the interphase, and transfer to a clean, sterile microcentrifuge tube.
6. Add 500 μL of isopropanol per 1 mL of TRI-reagent®, mix by inverting tube. Incubate at room temperature for 5 min.
7. Centrifuge the sample at 12,000 × *g*, 4 °C for 10 min.
8. The RNA-containing pellet should now be present but may not be visible. Discard the supernatant gently using a pipette taking care not to disturb the pellet.

9. Wash the pellet in 1 mL of 75% ethanol per 1 mL of TRI-reagent®
10. Briefly vortex the sample and centrifuge at 12,000 × *g* for 5 min at 4 °C.
11. Gently discard the supernatant using a pipette taking care not to disturb the pellet.
12. Air dry the pellet at room temperature for 5–10 min.
13. Resuspend pellet in 50 μL of RNAse-free water and store at −80 °C.
14. Quantify RNA using spectrophotometer, e.g., Nanodrop, a fluorescent dye-based method or using the Agilent 2100 Bioanalyzer.

3.5 Reverse Transcription and qPCR

3.5.1 Reverse Transcription

RNA (1 μg) is reverse transcribed using qScript reverse transcriptase (Quanta Biosciences) in a final reaction volume of 20 μL.

1. Thaw all frozen components. Mix thoroughly, and briefly centrifuge to collect contents before use. Place all components, including qScript RT on ice.
2. Make up the following in a 0.2 mL tube:
   (a) 1 μg RNA (variable volume)
   (b) 4 μL of qScript Reaction Mix (5×)
   (c) 1 μL qScript RT
   (d) Nuclease-free water, variable volume to yield a final reaction volume of 20 μL.
3. Mix by flicking tube and then centrifuge for 10 s to collect contents.
4. Place tube(s) in a thermal cycler programmed as follows:
   (a) 22 °C, 5 min
   (b) 42 °C, 30 min
   (c) 85 °C, 5 min
   (d) 4 °C hold.

After completion of cDNA synthesis, the resulting cDNAs are diluted 1:10 in nuclease-free water and stored at −20 °C.

3.5.2 Quantitative PCR Using SYBR Green and Taqman Assays

1. Quantitative PCR (qPCR) analysis of CEBPε and GSF3R mRNA is performed using "TaqMan®" hydrolysis probe assays and the StepOnePlus qPCR system (Applied Biosystems, Zug, Switzerland).
2. In general, standard qPCR protocol of
   Step 1: 94 °C, 5 min × 1 cycle
   Step 2: 94 °C, 30 s

Step 3: 70 °C, 20 s

Step 4: Fluorescence measurement, go to step 2 × 39 cycle

Step 5: 12 °C hold

3. Raw C_t values were normalized to HMBS and to the untreated control samples using the Pfaffl method [29], where the equation below is used to determine relative expression of the target gene relative to the reference housekeeping gene:

$$R = \frac{(E_{\text{target}}) = \Delta\text{CpTarget (control-sample)}}{(E_{\text{ref}}) = \Delta\text{CpReference (control-sample)}}$$

3.6 Cellular Phenotyping: Morphological Assessment

1. Cultured NB4 cells are grown in suspension so can be retrieved by centrifugation at 1000 × *g* for 5 min and counted using a hemocytometer or cell/particle counter. Cells are resuspended at 5 × 10^5 cells/mL in PBS.
2. The cell suspension (100 μL) is placed in the chamber of labeled cytospin slides, which are spun at 800 rpm for 5 min at room temperature in a balanced Cytospin 4 centrifuge.
3. The slide is allowed to air dry prior to fixing in Rapi-Diff solution A (thiazine dye in methanol) for 10 s and staining with Rapi-Diff solution B (eosin Y in phosphate buffer). Excess stain is removed by rinsing in tap water and blotting onto absorbent paper. The slide is then counter-stained with Rapid-Diff solution C (polychromed methylene blue in phosphate buffer), washed with tap water and air dried.
4. Differentiation morphology based on nucleo-cytoplasmic ratio, nuclear morphology, the presence of azurophilic granules, and vesiculation can be assessed using a light microscope at 40× magnification.

3.6.1 Detection of Differentiation

1. Cultured NB4 cells can be retrieved by centrifugation at 1000 rpm for 4 min, counted as described earlier and washed with ice-cold PBS prior to being resuspended at a concentration of 1 × 10^6 cells/mL.
2. A 1% BSA solution in PBS is prepared by diluting 1 g in 100 mL PBS.
3. Each marker analyzed uses 1.5 mL (equivalent to 1.5 × 10^6 cells).
4. For differentiation analysis, cells are incubated with 5 μL PE-conjugated anti-CD11b antibody in 100 μL 1% BSA/PBS solution/sample, for 30 min at 4 °C, in the dark.
5. A negative control is prepared where cells are incubated with PE-labeled nonspecific IgG 2.5 μL in 100 μL 1% BSA/PBS for 30 min at 4 °C in the dark.

6. Cells are then washed twice with 500 μL of 1% BSA/PBS, pelleted by centrifugation at 1000 rpm for 4 min and resuspended in 400 μL PBS for analysis on the BD-LSRII flow cytometer (BD Biosciences).

3.6.2 Detection of Autophagy

1. The Cyto-ID assay kit is used to assess the levels of autophagy in NB4 cells (Enzo Life Sciences). Cells (1.5×10^6 cells) are prepared by centrifugation at 1000 rpm for 4 min, washed once in PBS and resuspended in 1× assay buffer. Cells are again centrifuged and resuspended in 200 μL phenol red free RPMI 1640 media supplemented with 5% FCS—sample is mixed thoroughly to ensure uniform cell suspension.
2. Cyto-ID reagent is prepared in phenol red free medium (5% FCS) 1:1000 dilution/200 μL medium per sample.
3. 200 μL of CytoID® reagent is added to the resuspended cell solution, giving a total volume of 400 μL in each sample and incubated at 37 °C for 30 min, protected from light.
4. The cells are then pelleted by centrifugation at 1000 rpm for 4 min and washed twice with 1× assay buffer and the supernatant discarded.
5. The cell pellet is resuspended in 1× assay buffer and cells can be analyzed using either wide-field fluorescence microscopy or flow cytometry.

3.6.3 Flow Cytometry

1. Cells for differentiation detection are analyzed through the PE filter on the LSRII (filter range 574–589 nm), excited with the yellow/green laser (561 nm).
2. Cells for autophagy detection are analyzed through the 488 2 filter (filter range 500–550 nm), excited with the blue laser (488 nm). All data analysis and histogram overlays are performed using FlowJo software.

4 Notes

1. The generation and use of lentiviral mediate shRNAs are governed by local health and safety and genetic modification regulations. It will be necessary to become familiar with the necessary biohazard precautions and to prepare robust risk assessments and standard operating procedures consistent with local rules before initiating such experiments. It is important to test two or more shRNAs as knockdown can vary between targets and cell lines. It is also important to include a non-gene targeting scramble control as the process of lentiviral transduction can influence intra-cellular signaling pathways including autophagy. The protocol followed is based on the

method provided by Addgene/TRC consortium for the pLKO.1 vector. Cells should be usually grown with antibiotics, but during transfection process antibiotics will also pass into the cells and kill them. Transfections should be carried late afternoon, so that the cells are not in transfection reagent for >15 h. Freeze thawing stocks greatly reduces the virus viability; therefore, using the virus immediately after production is recommended. The puromycin working concentration range for mammalian cells is 0.5–10 μg/mL.

2. PVDF membrane is hydrophobic; therefore, aqueous buffers will not penetrate the membrane. Rinsing in methanol prior to transfer activates the membrane, allowing the penetration of aqueous solutions and more efficient binding of proteins. Chemiluminescence is an alternative detection strategy for western blots. For chemiluminescence, add fresh 5 mL of 5% w/v BSA, 1× TBS, 0.1% Tween® 20, plus 2.5 μL of Anti-rabbit IgG, hHRP-linked secondary antibody. Place on rotator and incubate at room temperature for 1 h. Discard secondary antibody and wash blot with 10 mL of 1× TBS, 0.1% TWEEN® 20 for 5 min. Repeat washes 3×. Remove membrane and place on flat surface. Prepare enhanced chemiluminescence reagent (ECL) by combining 0.5 mL of solution A to 0.5 mL of solution B in a microfuge tube (ECLTM Prime Western Blotting System, Thermo Fisher). Mix and add 1 mL ECL solution to membrane distributing it evenly across membrane. Incubate at room temperature for 5 min. Image using chemiluminescent detection system, e.g., ChemiDoc system (BioRad) at varying exposures.

3. The Chomczynski RNA purification method utilizing phenol-guanidine thiocyanate and chloroform, with subsequent 2-propanol, ethanol precipitation wash steps, provides a cost-effective and well-established approach to isolate RNA [29].

References

1. Gudas LJ (1994) Retinoids and vertebrate development. J Biol Chem 269:15399–15402
2. Laursen KB, Wong PM, Gudas LJ (2012) Epigenetic regulation by RARalpha maintains ligand-independent transcriptional activity. Nucleic Acids Res 40:102–115
3. Kashyap V, Gudas LJ (2010) Epigenetic regulatory mechanisms distinguish retinoic acid-mediated transcriptional responses in stem cells and fibroblasts. J Biol Chem 285:14534–14548
4. Kashyap V, Gudas LJ, Brenet F, Funk P, Viale A, Scandura JM (2011) Epigenomic reorganization of the clustered Hox genes in embryonic stem cells induced by retinoic acid. J Biol Chem 286:3250–3260
5. Purton LE, Bernstein ID, Collins SJ (2000) All-trans retinoic acid enhances the long-term repopulating activity of cultured hematopoietic stem cells. Blood 95:470–477
6. Purton LE, Dworkin S, Olsen GH, Walkley CR, Fabb SA, Collins SJ, Chambon P (2006) RARgamma is critical for maintaining a balance between hematopoietic stem cell self-renewal and differentiation. J Exp Med 203:1283–1293
7. Ghatpande S, Ghatpande A, Sher J, Zile MH, Evans T (2002) Retinoid signaling regulates

primitive (yolk sac) hematopoiesis. Blood 99:2379–2386

8. Ross AC (2012) Vitamin A and retinoic acid in T cell-related immunity. Am J Clin Nutr 96:1166S–1172S
9. Mucida D, Park Y, Kim G, Turovskaya O, Scott I, Kronenberg M, Cheroutre H (2007) Reciprocal TH17 and regulatory T cell differentiation mediated by retinoic acid. Science 317:256–260
10. Caceres-Cortes JR (2013) Blastic leukaemias (AML): a biologist's view. Cell Biochem Biophys 66:13–22
11. Lengfelder E, Hofmann WK, Nolte F (2013) Management of elderly patients with acute promyelocytic leukemia: progress and problems. Ann Hematol 92:1181–1188
12. Grignani F, De Matteis S, Nervi C, Tomassoni L, Gelmetti V, Cioce M, Fanelli M, Ruthardt M, Ferrara FF, Zamir I, Seiser C, Grignani F, Lazar MA, Minucci S, Pelicci PG (1998) Fusion proteins of the retinoic acid receptor-alpha recruit histone deacetylase in promyelocytic leukaemia. Nature 391:815–818
13. Di Croce L, Raker VA, Corsaro M, Fazi F, Fanelli M, Faretta M, Fuks F, Lo Coco F, Kouzarides T, Nervi C, Minucci S, Pelicci PG (2002) Methyltransferase recruitment and DNA hypermethylation of target promoters by an oncogenic transcription factor. Science 295:1079–1082
14. Kayser S, Krzykalla J, Elliott MA, Norsworthy K, Gonzales P, Hills RK, Baer MR, Racil Z, Mayer J, Novak J, Zak P, Szotkowski T, Grimwade D, Russell NH, Walter RB, Estey EH, Westermann J, Gorner M, Benner A, Kramer A, Smith BD, Burnett AK, Thiede C, Rollig C, Ho AD, Ehninger G, Schlenk RF, Tallman MS, Levis MJ, Platzbecker U (2017) Characteristics and outcome of patients with therapy-related acute promyelocytic leukemia front-line treated with or without arsenic trioxide. Leukemia 31:2347–2354
15. Lo-Coco F, Avvisati G, Vignetti M, Thiede C, Orlando SM, Iacobelli S, Ferrara F, Fazi P, Cicconi L, Di Bona E, Specchia G, Sica S, Divona M, Levis A, Fiedler W, Cerqui E, Breccia M, Fioritoni G, Salih HR, Cazzola M, Melillo L, Carella AM, Brandts CH, Morra E, von Lilienfeld-Toal M, Hertenstein B, Wattad M, Lubbert M, Hanel M, Schmitz N, Link H, Kropp MG, Rambaldi A, La Nasa G, Luppi M, Ciceri F, Finizio O, Venditti A, Fabbiano F, Dohner K, Sauer M, Ganser A, Amadori S, Mandelli F, Dohner H, Ehninger G, Schlenk RF, Platzbecker U, Gruppo Italiano Malattie Ematologiche dell'Adulto, German-Austrian Acute Myeloid Leukemia Study Group, Study Alliance Leukemia (2013) Retinoic acid and arsenic trioxide for acute promyelocytic leukemia. N Engl J Med 369:111–121
16. Park DJ, Chumakov AM, Vuong PT, Chih DY, Gombart AF, Miller WH Jr, Koeffler HP (1999) CCAAT/enhancer binding protein epsilon is a potential retinoid target gene in acute promyelocytic leukemia treatment. J Clin Invest 103:1399–1408
17. Mueller BU, Pabst T, Fos J, Petkovic V, Fey MF, Asou N, Buergi U, Tenen DG (2006) ATRA resolves the differentiation block in t(15;17) acute myeloid leukemia by restoring PU.1 expression. Blood 107:3330–3338
18. Tomita A, Kiyoi H, Naoe T (2013) Mechanisms of action and resistance to all-trans retinoic acid (ATRA) and arsenic trioxide (As2O 3) in acute promyelocytic leukemia. Int J Hematol 97:717–725
19. Nervi C, Ferrara FF, Fanelli M, Rippo MR, Tomassini B, Ferrucci PF, Ruthardt M, Gelmetti V, Gambacorti-Passerini C, Diverio D, Grignani F, Pelicci PG, Testi R (1998) Caspases mediate retinoic acid-induced degradation of the acute promyelocytic leukemia PML/RARalpha fusion protein. Blood 92:2244–2251
20. vom Baur E, Zechel C, Heery D, Heine MJ, Garnier JM, Vivat V, Le Douarin B, Gronemeyer H, Chambon P, Losson R (1996) Differential ligand-dependent interactions between the AF-2 activating domain of nuclear receptors and the putative transcriptional intermediary factors mSUG1 and TIF1. EMBO J 15:110–124
21. Ablain J, Leiva M, Peres L, Fonsart J, Anthony E, de The H (2013) Uncoupling RARA transcriptional activation and degradation clarifies the bases for APL response to therapies. J Exp Med 210:647–653
22. Mizushima N, Klionsky DJ (2007) Protein turnover via autophagy: implications for metabolism. Annu Rev Nutr 27:19–40
23. Orfali N, McKenna SL, Cahill MR, Gudas LJ, Mongan NP (2014) Retinoid receptor signaling and autophagy in acute promyelocytic leukemia. Exp Cell Res 324:1–12
24. Orfali N, O'Donovan TR, Nyhan MJ, Britschgi A, Tschan MP, Cahill MR, Mongan NP, Gudas LJ, McKenna SL (2015) Induction of autophagy is a key component of all-trans-retinoic acid-induced differentiation in leukemia cells and a potential target for pharmacologic modulation. Exp Hematol 43:781–93 e2

25. Chen ZH, Wang WT, Huang W, Fang K, Sun YM, Liu SR, Luo XQ, Chen YQ (2017) The lncRNA HOTAIRM1 regulates the degradation of PML-RARA oncoprotein and myeloid cell differentiation by enhancing the autophagy pathway. Cell Death Differ 24:212–224
26. Weidberg H, Shvets E, Shpilka T, Shimron F, Shinder V, Elazar Z (2010) LC3 and GATE-16/GABARAP subfamilies are both essential yet act differently in autophagosome biogenesis. EMBO J 29:1792–1802
27. Lihuan D, Jingcun Z, Ning J, Guozeng W, Yiwei C, Wei L, Jing Q, Yuanfang Z, Gang C (2014) Photodynamic therapy with the novel photosensitizer chlorophyllin f induces apoptosis and autophagy in human bladder cancer cells. Lasers Surg Med 46:319–334
28. Moffat J, Grueneberg DA, Yang X, Kim SY, Kloepfer AM, Hinkle G, Piqani B, Eisenhaure TM, Luo B, Grenier JK, Carpenter AE, Foo SY, Stewart SA, Stockwell BR, Hacohen N, Hahn WC, Lander ES, Sabatini DM, Root DE (2006) A lentiviral RNAi library for human and mouse genes applied to an arrayed viral high-content screen. Cell 124:1283–1298
29. Chomczynski P, Sacchi N (1987) Single-step method of RNA isolation by acid guanidinium thiocyanate-phenolchloroform extraction. Anal Biochem 162(1):156–159

Chapter 4

Ligand Design for Modulation of RXR Functions

Claudio Martínez, José A. Souto, and Angel R. de Lera

Abstract

Retinoid X receptors (RXRs) are promiscuous partners of heterodimeric associations with other members of the Nuclear Receptor (NR) superfamily. RXR ligands ("rexinoids") either transcriptionally activate the "permissive" subclass of heterodimers or synergize with partner ligands in the "nonpermissive" subclass of heterodimers. The rationale for rexinoid design with a wide structural diversity going from the structures of existing complexes with RXR determined by X-Ray, to natural products and other ligands discovered by high-throughput screening (HTS), mere serendipity, and rationally designed based on Molecular Modeling, will be described. Included is the new generation of ligands that modulate the structure of specific receptor surfaces that serve to communicate with other regulators. The panel of the known RXR agonists, partial (ant)agonists, and/or heterodimer-selective rexinoids require the exploration of their therapeutic potential in order to overcome some of the current limitations of rexinoids in therapy.

Key words Rexinoids, RXR, Agonists, Antagonists, RXR heterodimers, (Virtual) ligand design, Computer-assisted RXR ligand discovery

1 Introduction

The central biological role of retinoid X receptor (RXRα, NR2B1; RXRβ, NR2B2; and RXRγ, NR2B3), members of the nuclear receptor (NR) superfamily of transcriptional regulators [1, 2], derives from its presence as common heterodimerization partner of several NRs [3]. In these heterodimers [4], RXR can be an active or a silent transcription partner [5]. In the first case, also known as permissive, they are activated by the RXR ligand or by the partner's ligand, and synergistic effects are observed upon double ligand binding pocket (LBP) occupancy. These heterodimers respond to dietary lipids (fatty acids, steroids, etc.), and play essential roles in nutrient acquisition (farnesoid X receptor, FXR, liver X receptor, LXR, peroxisome proliferator-activated receptor, PPAR) and clearance (FXR, constitutive androstane receptor, CAR), thus controlling the dynamics of energy flow and organ communication [5]. The second class or nonpermissive, formed mostly with receptors for hormones including thyroid hormone receptors (TRs),

Swapan K. Ray (ed.), *Retinoid and Rexinoid Signaling: Methods and Protocols*, Methods in Molecular Biology, vol. 2019, https://doi.org/10.1007/978-1-4939-9585-1_4,

vitamin D receptors (VDRs), and retinoic acid receptors (RARs), are activated by the partner's ligand whereas ligand-bound RXR is silent, the so-called "RXR subordination," i.e., the heterodimer does not respond to rexinoids alone [6]. A challenge for medicinal chemists is the discovery of selective RXR ligands that also exhibit selectivity for a particular heterodimer.

The ligand-binding domain (LBD) of this NR is a single protein domain with a primarily helical scaffold of twelve α-helices (Hx) arranged in three layers termed "anti-parallel alpha-helical sandwich," and a short S1-S2 β-turn. Agonist binding triggers the exchange of co-repressor (CoR) protein complexes by coactivator (CoA) complexes and also of chromatin-remodeling enzymes, which shift the condensed transcriptionally inactive to the activated and accessible active state leading ultimately to gene transcription [7–14].

A recent search on the Protein Data Bank (PDB) database revealed about 57 X-Ray crystal structures of RXR with different ligands (13 of which contain 9-*cis*-retinoic acid). Comprehensive revisions of the tridimensional structures of the *C*-terminal LBD and the central DNA-binding domain (DBD) in the presence and absence of various agonists or antagonists and cognate DNA response elements have recently been reported [15].

Ligand-free full-length RXR also showed a tendency to associate into homotetramers both in solution and when bound to DNA, which could represent an inactive form of the receptor. In fact, recent crystalographic studies have shown that the autorepressed RXRα tetramer is in equilibrium with the unliganded repressed RXRαLBD-CoR (SMRT) tetramer [16]. Some RXR antagonists appear to displace SMRT for AF-2 binding instead of helping SMRT recruitment stabilizing the inactive RXR tetramer with an enlarged tetrameric interface [16].

Despite the maturity of the field, only two rexinoids have been approved by the Food and Drug Administration (FDA) for clinical applications: Alitretinoin (9-*cis*-retinoic acid **1.1**, Fig. 1) and Targretin (LGD1069 or Bexarotene **3.1**, Fig. 3).

2 Materials: RXR Ligands

2.1 Structural Binding to RXR Receptors

The CoR- or the CoA-bound states of the receptor are exquisitively sensitive to ligand action (agonists, antagonists, inverse agonists), and therefore to their structure [12–15, 17–22]. In RXR-*agonist* complexes [12–15] residue L436 in H11 (human RXRα numbering) keeps a highly conserved conformation inducing a sharp turn on the LBP volume, and therefore these are ligands either conformationally or configurationally twisted or sufficiently flexible to twist around single bonds.

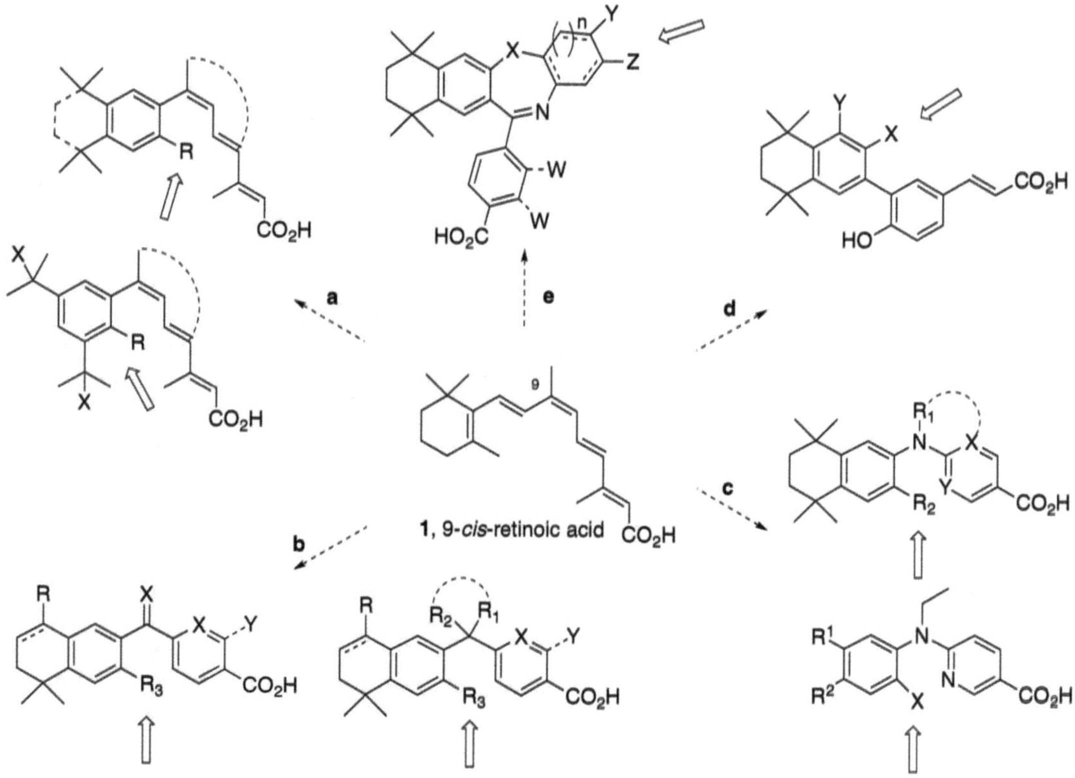

Fig. 1 Structures of the rationally designed rexinoid classes organized according to the subdivisions shown below (series **2–6** in the text). The structural diversity of rexinoids is summarized in these building blocks with hydrophobic, linkers and polar motifs. Antagonist feature groups attached to the arrowed positions indicated

More than 500 RXR agonists and antagonists (with IC/IC_{50} lower than 50 μM) are annotated in European Molecular Biology Laboratory Chemical Database (ChEMBL), the database dependent on the European Bioinformatics Institute (EBI). The rationale for the structure-based design of the most important classes of RXR modulators [12–14], going from naturally occurring compounds, those discovered by high-throughput screening or mere serendipity, will be described, and the structures of the ligands along with relevant biological activities will be presented in order to summarize our recent revision in this topic [14]. In addition, computer-aided drug design (CADD) applied to RXR receptors, as well as the discovery of compounds that modulate the activities of RXR in a non-canonical way, will be described, including those that bind to the receptor surface and others that attach to the coregulatory binding surface (coactivator binding site and dynamic protein interfaces).

2.2 Endogenous RXR Ligands

Except in pancreatic β-cells and recently in human serum in very low concentrations [23] 9-*cis*-retinoic acid **1.1** (Fig. 1) has not been conclusively shown to be endogenously present. Either its concentration is under the detection limit, it is present transiently, or it accumulates in local cell populations [24].

The endogenous presence of 9-*cis*-13,14-dihydroretinoic acid (structure **1.1** saturated at the C13-C14 double bond) and its all-*trans* isomer has recently been proven in several mice organs (liver, serum, brain) by HPLC-MS-MS and UV and quantified (quantitative binding affinity of K_d 90 ± 20 nM; cf. 20 ± 10 nM for 9-*cis*-retinoic acid **1.1** by fluorescence quenching assays) to be sufficient to maintain RXR activities [25]. Crystal structure studies of the *R* enantiomer bound to hRXRα ligand-binding domain

(LBD) and a 13-residue peptide comprising the nuclear receptor-binding surface NR2 of NCoA2 confirmed binding to the canonical binding pocket similarly to 9-*cis*-retinoic acid **1.1**. This ligand and other dihydroretinoids are presumably produced by retinol saturase (RetSat) [26].

2.3 Dietary Compounds

Our recent review [14] provided a thorough description of the collection of flexible unsaturated natural fatty acids acquired in the diet which have been reported to bind to and activate RXR. The promiscuity of RXR on binding several metabolites indicates a putative role for this receptor as an intracellular sensor of the cell metabolic status, binding either the natural ligand or fatty acids, depending on their supply and local concentration.

2.4 Rational Design of RXR Modulators. Structural Similarity to 9-cis-RA and Ring-Locked Analogs

2.4.1 LBP Architectures and Ligand Binding Pockets (LBP) Determinants

Recent revisions of the crystal structures of RXR-ligand complexes [12, 15] clearly confirmed that selective RXR ligands should feature a bent structure or a twisted polyene side-chain for optimizing binding to the L-shaped RXR ligand-binding pocket (LBP). Such a twist is structurally unattainable by the elongated fully conjugated polyenes, such as all-*trans*-retinoic acid, although not for partially unsaturated analogs. Only highly flexible ligands, such as 9-*cis*-retinoic acid, 9-*cis*-13,14-dihydroretinoic acid, and certain unsaturated fatty acids will show pan-RAR/RXR agonistic properties.

2.4.2 Rationally Designed Rexinoid Scaffolds

Three structural units could be noted in most of the rexinoid structures: a hydrophobic region (usually a carbo- or a heterocyclic ring), a terminal polar group (usually, a carboxylic acid), and a linker fragment. In most of the cases the latter contains the functionality responsible for the permanent or induced skeletal twist responsible for the optimal adaptation to (and filling of) the L-shaped RXR LBP [27]. Only the most common scaffolds have been selected to discuss the systematic structure–activity relationship (SAR) studies on "rationally designed" rexinoids. Figure 1 summarizes the classification of the most important rexinoid scaffolds, for which sufficient structural and functional information is available in order to extract SAR lessons. According to the nature and functionality of the linker, grouping of most of the reported

RXR ligands can be proposed, including: (a) polyenoic acids and 9-*cis*-locked analogs, (b) aryl rings connected by Csp^2 or Csp^3 atoms, (c) diarylamines, (d) biphenyls with *ortho* substituents, and (e) fully condensed rings (Fig. 1) [14].

For the rational design of rexinoid scaffolds and taking into consideration the chemical instability of the natural ligands, (part of) the polyunsaturated side chain of the parent system has been replaced by aryl and heteroaryl rings, giving rise to the so-called (hetero)arotinoids ((hetero)aromatic retinoids). These ligands show greater stability and are easier to synthesize than the parent polyenes, while partly retaining, or in some cases improving, binding affinity to the receptor and potency relative to the native ligands. The more rigid (hetero)arotinoid skeletons are moreover easier to functionalize at defined positions, and synthetic planning allows to place particular substituents at strategic positions of the receptor LBP if desired. In particular, these substituents are primarily responsible for the ligand-dependent functional modulation of the receptors [14]. For example, canonical RXR antagonists feature long or bulky substituents attached to the central linker region (see arrows in Fig. 1), which disrupt the hydrophobic interactions that stabilize the agonist conformation of H12 [27]. The functional transition from agonists to partial agonists/antagonists within a family of rexinoids is usually modulated by the length or bulk of these substituents.

2.4.3 Structure–Activity Relationship Studies of RXR Ligands

Subtype-Selective Rexinoids

RXRs are expressed in virtually every tissue of the body: RXRβ is ubiquitously expressed, but at highest level in the central nervous system (CNS); RXRα is mainly expressed in liver, lung, muscle, kidney, epidermis, and intestine and is the major subtype in skin; and RXRγ is found in brain (regions of the CNS such as the olfactory bulb and the pituitary gland), cardiac and skeletal muscle [2]. The expression levels on tissues vary, however, with cell type and cell differentiation status. A central question in rexinoid biology and pharmacology research is the role and functions of the different subtypes, and whether subtype-selective ligands are able to regulate specific biological pathways. This issue remains unsolved, due to the lack of subtype-specific rexinoids, since in contrast to RARs, all residues that constitute the LBP of RXR (H3, H5, H7, H11, and the β-turn) are highly conserved in all three RXR subtypes (α, β, and γ). As a consequence, the discovery of subtype-selective RXR ligands (with about 2log difference in potency) remains to be solved [14].

An alternative approach to obtain subtype selectivity would be the exploitation of the differences in aminoacid composition of the second layer away from the ligand [28]. Whether this is the structural basis for some of the reported moderate selectivities when the three subtypes were used in the assays or instead is the result of allosteric effects remains to be clarified.

Rexinoid Antagonists

The recent structural and functional studies on rexinoid antagonists have revealed a variety of modulation modes that adds to the classical mechanism of antagonistic action. In contrast to agonists, antagonists (and partial agonists) fail to shift the equilibrium of multiple RXRα H12 conformations observed in the unliganded state to a more compact form. Antagonists generally feature a bulky side chain at the level of the C19 methyl group of 9-*cis*-retinoic acid. This additional fragment was intended to fill the void space in this region revealed by the structure of RXRα-9-*cis*-retinoic acid complex, sterically preventing helix H12 from adopting the active conformation and disrupting the interaction surface with coactivators. These principles are maintained in the heterodimeric crystal structures solved [15, 17, 29], in which ligand substituents were found oriented toward H11, which adopts an antagonist conformation displacing H12 toward the solvent and preventing CoA peptide recruitment to this site of the heterodimer. The agonist to antagonist transition within a family of analogs is a function of the length or bulk of the substituent. These findings are consistent with the re-positioning of H12 upon increase of the length of the protruding arm of the synthetic rexinoid (see below) [12–15].

Partial Rexinoid Agonists/ Antagonists

The comparison of a series of X-Ray structures of compounds with small variations on the substituents at key positions, characterized as partial agonists, with that of RXRα LBD bound to the full agonist, revealed some residue reorientations, and a more dynamic AF2 helix [30, 31]. Partial agonists therefore "sense" intracellular coregulator levels and act as context-dependent cell-selective modulators with agonist or antagonist properties. The therapeutic potential of RXR partial (ant)agonists, although promising, remains unexplored.

Dual RAR/RXR Modulators

Compounds that bind both RAR and RXR as agonists are rare, given the different structural constraints of their LBPs. 9-*cis*-retinoic acid and similarly flexible analogs bind both receptors as a result of their conformational adaptation to the different binding pockets of RAR and RXR, but the same profile in conformationally constrained analogs is more unusual. Less common is the binding with contrasting activities, such as agonist of one receptor and antagonist of the other, although some examples have been reported [12–15].

Heterodimer-Selective Rexinoids

The so-called selective RXR modulators are an important class of rexinoids that exhibit heterodimer selectivity. In addition, the rexinoid might act as agonist or antagonist of heterodimers with different partners in preference to others or to the RXR/RXR homodimers. As the functional consequences of rexinoid binding will be primarily determined by the ligand, different agonistic and antagonistic structures should result in different actions on coregulator interactions and RXR function [12–15]. In addition, RXR heterodimer-selective antagonists can inhibit specifically the synergistic activity of an RXR agonist on permissive NR/RXR heterodimers.

Selective modulators of heterodimers have been developed that separate the physiological activities of the RXR and its partner receptors and might be considered part of the so-called Specific NR Modulators (SNuRMs) group. This important class of rexinoids might act as agonist (or synergists if they allosterically increase the potencies of the partner ligand) or antagonist of RXR heterodimers with different partners, sometimes by subtle variations in their structures. Thus, different (ant)agonistic scaffolds should result in different actions on coregulator interactions and RXR function. However, despite the advances in this area, no precise structural principles have yet emerged to guide chemical engineering of heterodimer-specific ligands as chemical probes or candidates for drug development [12–15].

2.4.4 Methods. Selected RXR Ligands

A selection of the most common and potent rexinoids grouped by structural similarity as described in Fig. 1 is shown below, together with the described activities as ligands of RXR and some hetero/homodimers. The general structures contain substituents of different chemical nature and length in order to summarize the main structural contributions to rexinoid activity (primarily as agonists or antagonists) [12–15].

Obviously, it is not possible, within the aim of the Series articles, to describe the preparation of these compounds, since the synthetic schemes are highly dependent upon their structure and therefore it is not possible to provide general methods and reaction conditions for their preparation. Thus, those interested on the synthesis will have to consult the references provided for each case.

(Poly)enylcarboxylic Acids

The (poly)enylcarboxylic acids are structurally similar to the native retinoids due to the preservation of part or the unsaturated chain and moreover incorporate other hydrophobic or (hetero)aryl rings replacing some double bonds of the polyene (Fig. 2).

Compound **2.1** (LG100567) is a dual RAR and RXR agonist, ca. 25 times more potent than 9-*cis*-retinoic acid, but analogs with a substituent at C2 of the 3,5-di-alkylbenzene scaffold (**2.1–2.4**) are selective RXR ligands [32, 33]. With shorter side chains (**2.3**, LG101506) they are full agonists of the RXR/RXR homodimer and PPARγ/RXR heterodimer, binding specifically to RXR with high affinity (K_i values of 3.0 ± 0.8 nM, 9.0 ± 1.2 nM, and 11.0 ± 3.6 nM for RXRα, β, and γ, respectively), and with longer side chains, such as heptyl ether **2.4** (LG101208) [34], they function as homodimer antagonists. Analogs with benzofused

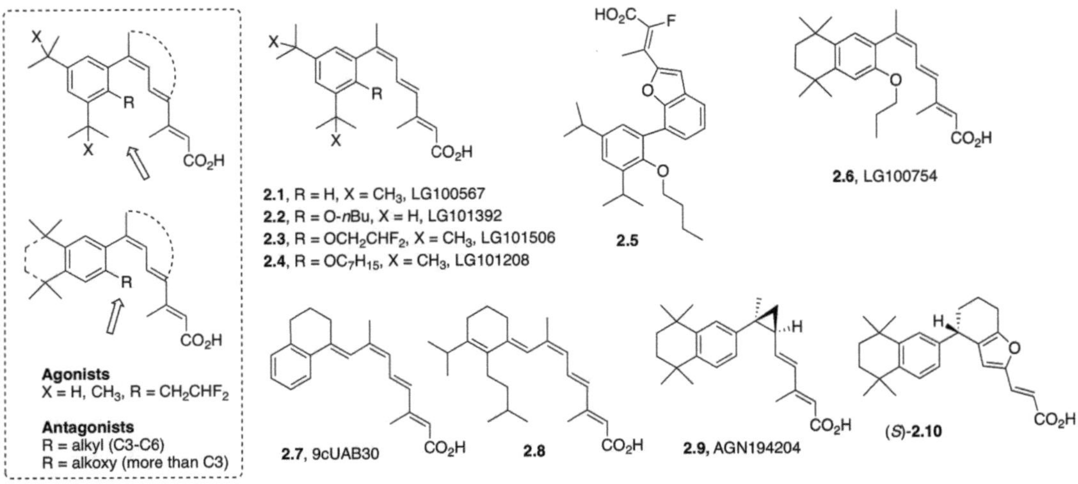

Fig. 2 (Poly)enylcarboxylic acids, including 9-*cis*-locked and acrylic (crotonic) acid side chains

heteroaryl rings also showed the entire panel of activities upon variation of the aryl ether chain, and the incorporation of fluorine atoms at the acrylic or crotonic acid termini as in the full antagonist **2.5** increased the binding affinity and co-transfection activity [35].

Derivatives of 5,6,7,8-tetrahydro-5,5,8,8-tetramethylnaphthalene with small alkoxy groups at the *ortho* position (C3) and thus vicinal to the trienoic acid side chain exhibited the entire range of RXR ligand binding activities (with weak or no activity on RARs). The aryl-propyl ether **2.6** (LG100754) retained high affinity for RXR, acted as an agonist for RXR heterodimers and antagonist of RXR homodimers [36]. Acting as transcriptional activator through RAR/RXR in a cellular context, LG100754 is called "the phantom ligand." This is a more general behavior of certain ligands in permissive nuclear receptor heterodimers that have the ability to activate the LBD of unliganded heterodimeric partners, possibly due to ligand-induced allosteric signal transmission [37]. In the case of LG100754 **2.6** this effect was shown to be due to the direct binding of this ligand to RAR and stabilization of coactivator interaction, combined with the inhibition of the latter at the RXR side [17].

Rexinoid 9cUAB30 **2.7** [38] showed induction of apoptosis, reduction of proliferation, and prevention of mammary cancer in rodent models [39, 40], and is undergoing NCI clinical trials as cancer chemopreventive agent after animal studies revealed no significant toxicity [41] (although hepatomegaly was observed at higher doses) [42]. Methyl-substituted derivatives of this scaffold have been prepared at several positions of the ring [43, 44]. Other analogs have more recently been studied for prevention of breast cancer without toxicity, with **2.8** as the most effective in in vivo rat models alone or in combination with tamoxifen [45], without observing increase in triglyceride levels.

The (*S*,*S*)-9-*cis*-locked retinoid with a cyclopropane at C6 and C7 **2.9** (AGN194204) is the most potent and specific RXR agonist described [46], with EC_{50} values ranging from 0.08 nM (RXRγ) to 0.8 nM (RXRβ), and a potent hypoglycemic agent in the *db/db* mouse model (genetically defective in leptin signaling). Rexinoid (*S*)-**2.10** with the skeleton constrained up to C-12 by a (tetrahydro) benzofused heterocyclic compounds (furan, thiophene, isoquinoline), tetrahydrobenzofuranylpropenoic was reported to also be highly potent (EC_{50} of 4 nM) [47].

Rexinoids with geminal diarylethenes, heteroatom-containing derivatives, and geminal diaryl groups attached to a C_{sp3}, and gem di(hetero)aryl carbo- and heterocyclic benzoic acids

Ligands (Fig. 3) based on aryl groups with substituents at the *ortho* position (alkyl, halogen) and connected by a Csp2 (alkene, ketone, oxime…) exhibited greater affinity for RXR than for RAR. LGD1069 **3.1** [48–50] (K_d values of 14 ± 3; 21 ± 4; 29 ± 7 nM for RXR subtypes) is the rexinoid drug (bexarotene) approved for the treatment of cutaneous T-cell lymphoma (CTCL). It is also undergoing clinical trials for the treatment of diseases caused by uncontrolled cell proliferation, although so far with less-promising

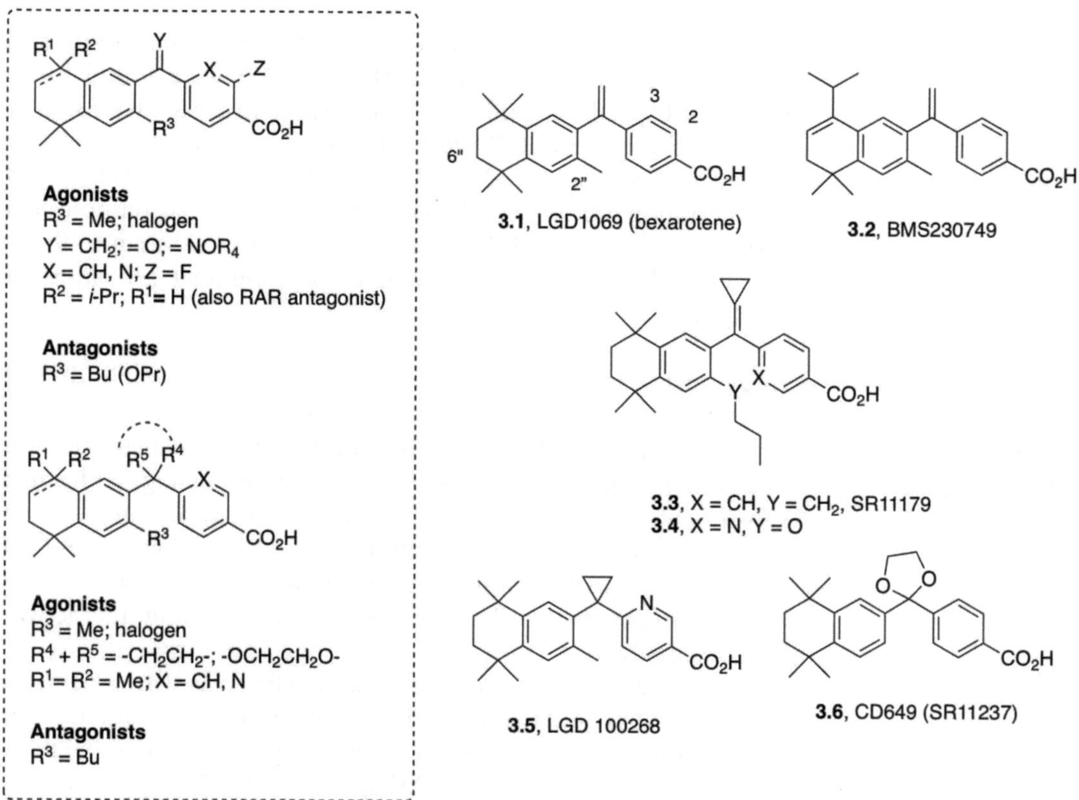

Fig. 3 Rexinoids with geminal diarylethenes, heteroatom-containing derivatives, and geminal diaryl groups attached to a C_{sp3}, and gem di(hetero)aryl carbo- and heterocyclic benzoic acids

results due to the induction of hypothyroidism, hyperlipidemia, and cutaneous toxicity as a result of residual RAR agonism [51, 52]. SAR studies on the skeleton proved that the reduced polarity ($C{=}CH_2$ > C=O) and greater lipophilicity at the Csp2 connector ($C{=}CMe_2$ > $C{=}CH_2$) combined with the presence of a C3-methyl group and a benzoic acid or dienoic acid terminus were synergistically required for good RXR selectivity. Analogs with alkyl groups (size larger than methyl), and heteroatom substituents (halogens, hydroxy / alkoxy groups) replacing the C2″-methyl group of **3.1** proved to be less potent, with the larger alkyl derivatives behaving as antagonists [50].

Diarylethylene **3.2** (BMS230749) exhibited the peculiar profile of rexinoid with RAR antagonistic activity, which was used to support the discovery of a novel rexinoid-dependent apoptotic pathway through permissive PPARγ/RXR heterodimers and production of apoptogenic NO [53].

Internal olefins substituted at one terminus by aryl rings and on the other by alkyl or alkoxy substituents likewise exhibited RXR selectivity as was the case of the alkylidenecyclopropyl-based analogs [54]. Increasing the size of the alkyl or alkoxyl groups at the C3' position in this series rendered RXR antagonists such as **3.3** and **3.4**.

Rexinoids with a C_{sp3} replacing the olefin to connect both aryl rings such as nicotinic acid LGD100268 **3.5** were found to be more RXR specific (>1000 for RXR relative to RAR), with similar binding affinity than LGD1069 **3.1** (dissociation constant of 3 nM; cf. 9-*cis*-retinoic acid, 9–12 nM) and 50% maximal activation at a concentration at least tenfold lower than that of 9-*cis*-retinoic acid [50]. Dioxolane SR11237 **3.6** at 10^{-7} M was equipotent to 9-*cis*-RA **1** in the TRE-pal assay, and caused minimal activation of RARβ [54].

Diaryl(heteroaryl)amines with a Tetrahydrotetramethylnaphthalene Skeleton and Analogs

In general, compounds with the basic diarylamine scaffold [55] were shown to be RXR agonists with highly potent retinoid synergistic activity, a duality that can be modulated by the incorporation of a *N*-substituent of moderate length and size. Analogs with small substituents did not discriminate among retinoid receptors, those with medium-size alkyl groups (in particular if branched at C3–C4) caused loss of RAR agonistic activity, and derivatives with longer or bulkier *N*-substituents showed reduced RXR binding and activation profiles [56].

N-Methylamino-pyrimidine-5-carboxylic acids with a pentoxy (**4.1**) or hexoxy (**4.2**) groups (Fig. 4) are selective RXR antagonists that inhibit retinoid synergism by RXR agonists in combination with low concentration of a RAR agonist in RAR/RXR heterodimers [57]. Heavily fluorinated *N*-benzylderivative **4.3** was able to

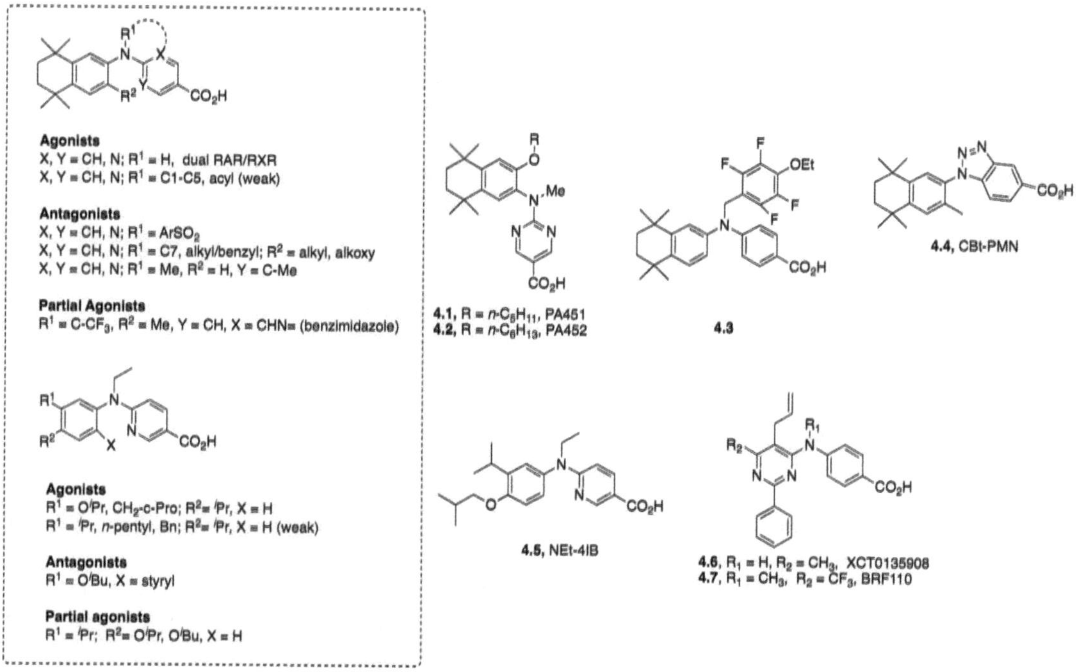

Fig. 4 Diaryl(heteroaryl)amines with a tetrahydrotetramethylnaphthalene skeleton and analogs

dose-dependently inhibit the differentiation of HL-60 cells in the presence of RAR agonist, and was characterized as antagonist of the RAR/RXR heterodimer acting at the RAR site [55].

The reduction of conformational flexibility by incorporation of the nitrogen into a heterocyclic ring fused to the benzoic acid (benzotriazole or benzimidazole) afforded partial (ant)agonists, such as benzotriazole CBt-PMN **4.4** (EC_{50} = 15 nM) [58], which showed in vitro selective activation of the PPARγ/RXRα and LXRα/RXRα heterodimers. In vivo evaluation of CBt-PMN **4.4** revealed beneficial effects including potent glucose-lowering, improvement in insulin secretion, and glucose tolerance in type 2 diabetes model mice.

RXR partial agonist NEt-4IB **4.5** exhibited also a low EC_{50} value (EC_{50} = 169 nM) [59] and 50–60% efficacy toward all RXR subtypes in luciferase reporter gene assay [60]. Oral administration of **4.5** to mice resulted in antidiabetic effects without the undesired side effects of full agonists [59].

Aminopyrimidine derivative **4.6** (XCT0135908, discovered after screening of a chemical library) is a rexinoid that selectively activated the Nurr1/RXR heterodimers in African green monkey CV-1 cells [61]. Although in itself is a silent receptor, the orphan nuclear receptor Nurr1 is an essential heterodimer partner in RXR-ligand-induced neuronal survival, and the activators may be useful in the treatment of neurodegenerative diseases (such as Parkinson's disease). Further structure–activity relationship (SAR) studies

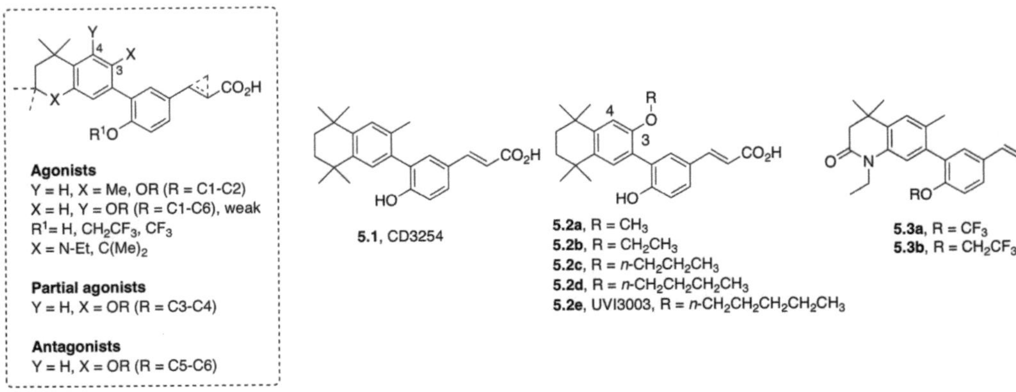

Fig. 5 *o,o′*-Disubstituted biaryl cinnamic acids

afforded BRF110 **4.7** as an in vivo active Nurrl (nuclear receptor related 1)-RXRα-selective ligand that prevents dopaminergic (DAErgic) neuron demise and striatal DAErgic denervation in vivo against PD-causing toxins in a Nurr-1-dependent manner [62].

o,o′-Disubstituted Biaryl Cinnamic Acids

Sterically hindered biaryls with a cinnamic acid at the connecting *meta* position such as CD3254 **5.1** [31] are selective RXR agonists, and the modulation of their activities to partial agonists-antagonists **5.2a-e** (Fig. 5) could be achieved through *O*-alkyl substituents at C3, with UVI3003 **5.2e** being characterized as a potent RXR antagonist [30]. The structural basis for the functional transition along the series of these analogs was deduced from the crystal structures of the partial agonists bound to RXRα LBD and coactivator peptide transcriptional mediators/intermediary factor 2 (TIF2) NR2 [30, 31]. Compounds with a small alkoxy side chains (**5.2a–5.2c**) induced conformational changes and alteration on the orientation of L436 and thus on helix H12, through a water molecule or a steric clash.

Analogs based on the more polar dihydro-[1*H*]-quinolin-2-ones (**5.3**) showed nanomolar potency and high selectivity as activators of LXR/RXR in preference to PPARγ/RXR heterodimers, in particular cinnamic acid **5.3a** ($EC_{50} = 3.6$ nM) [63, 64]. The 2,2,2-trifluoroethyloxy derivative **5.3b** was characterized as a potent activator of the expression of ABCA1 in the human monocytic cell line THP-1.

Benzofused Locked Poly(hetero)cyclic Benzoic Acids

Diaryl-diazepinylbenzoic acids behave in general as RAR/RXR synergists and this activity was potentiated by the substitution pattern of the diazepine ring (Fig. 6). HX600 **6.1** [65], which was inactive by itself in a human promyelocytic leukemia HL-60 differentiation assay, showed a dual function in the regulation of retinoid activities, acting as RXR synergyst (rexinoid) or as RAR

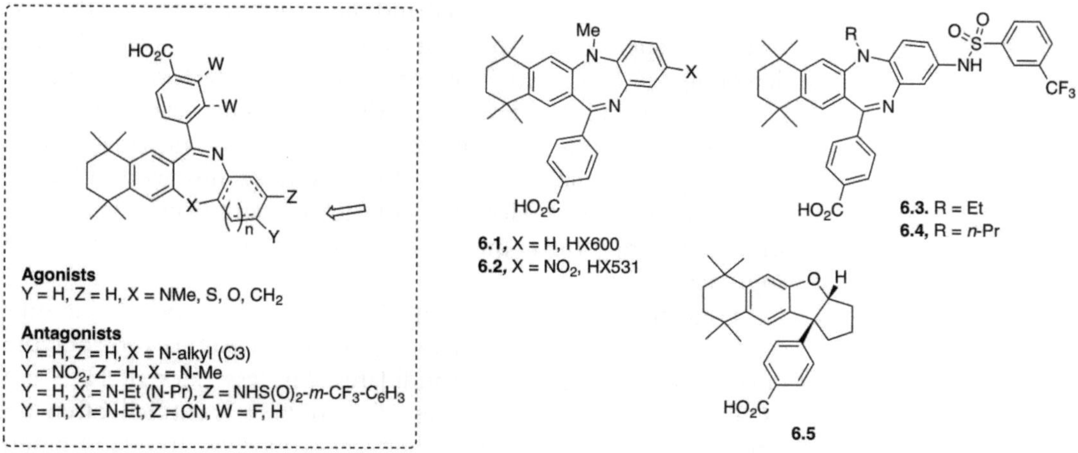

Fig. 6 Dibenzodiazepinyl rexinoids

antagonist at low and high concentrations, respectively. It was able to selectively activate the nerve growth factor receptor (NGFI-B) / RXR (inducing allosteric effects) and the Nurrl/RXR (both orphan heterodimer co-receptors) [66].

RXR antagonists were designed by modifications of the skeleton with *N*-alkyl substituents of size greater than C3 or groups (arylsulfonamide or nitro) at the *para* position to the formal imine of the dibenzodiazepine [67]. Sulfonamides **6.3** and **6.4** with antagonistic activity showed IC_{50} values below 0.1 μM for RXR. Pharmacokinetic profiles were measured for the latter after intravenous (1 mg/Kg bolus) and peroral (3 mg/Kg as suspension) administration in rats and found to be appropriate for therapeutical indications in diabetes and obesity [68, 69].

Bexarotene **3.1** activates Nurrl/RXR heterodimers (Nuclear Receptor Related 1 Protein), rescues dopamine neurons, and restores behavioral functions in a rat model of Parkinson's disease. Chiral dihydrobenzofurans have been developed to also function as Nurrl/RXR dimer activators [70], with sterically constrained enantiomer **6.5** being the most potent of the series.

2.5 Computer-Aided Discovery of RXR Ligands. Chemical Hit Derivatization

The above approach to rexinoid drug discovery has been called "chemocentric approach" following the principle that "structural similarity should reflect functional similarity" [71]. In addition, the combination of computational similarity search and de novo design has been a viable strategy for chemical scaffold hopping and rexinoid drug discovery of some of the compounds selected in Figs. 2, 3, 4, 5, and 6. The design potential of Computational Medicinal Chemistry and advances in the use of software in drug discovery after target identification for industrial applications have been recently reviewed [71, 72].

The ligand similarity-based and the structure-based are the two common approaches to drug discovery using Virtual Ligand Screening (VLS), with the first generally performing better [71, 73]. VLS on the target's binding site using large libraries of selected collections of molecules by docking or pharmacophore modeling are usually the tools for new drug discovery. Ligand- and structure-based Molecular Modeling is the most successful strategy through the combination of sequential, parallel, or hybrid approaches.

High-throughput Screening (HTS) allows compound identification and Computer-aided Drug Design (CADD) provides the prioritization of screening candidates. Success rates of HTS are usually very low and the series of candidates identified often do not show optimized physichochemical properties. In ligand-based CADD the chemical structures of small molecules known to interact with the target (for which comprehensive information such as binding affinities and physicochemical properties should be known) are further modified in order to discover more potent compounds. Chemical similarity and information on already described variations of activity with structural modifications are key consideration for the success of this approach. Moreover, Quantitative Structure--Activity Relationships (QSAR) models allow enquiring about the dependence of the target response with the ligand structures.

In structure-based CADD (somehow analogous to HTS), Molecular Dynamics simulations based on Newtonian mechanics using a force field that relate trajectories of conformations as a function of time provides highly useful information on pharmacophores for ligand design based on ligand docking, including the prediction of the preferred orientations of ligands to the receptor. Pharmacophore modeling provides a selection of the steric and electronic features of the ligands that might suggest the synthetic modifications that would provide optimal occupancy of the LBP. The combination of these two VLS modeling tools has allowed the identification of new RXR ligands.

After a computational hit structural identification has been completed, lead compounds are found using fragment-based drug discovery (FBDD). The limitation of this strategy is the use of commercial collections of compounds of limited structural diversity covering a rather small chemical space [71].

Advances in automated drug discovery with the development of programs such as SPIDER (Self-organizing map-based prediction of drug equivalence relationships) for target-prediction based on physicochemical properties and two-dimensional pharmacophoric feature distributions and TIGER (Target Interference Generator) [74] to cluster the collection of reference compounds into sets of locally similar molecules have been recently developed. DOGS (ligand-based de novo design) generates ligands with virtual organic synthesis by optimizing the topological similarity between newly generated candidates and templates [74]. Self-organizing maps (SOM) algorithm in combination with pharmacophore feature representations have recently been used for ligand discovery.

Merging information from programs for target prediction and 2D pharmacophoric similarity to the known RXR ligands, such as Chemically Advanced Template Search (CATS) [75], allow ranking of the screening compounds, which can be completed with computational metrics to cover the fundamental aspects of the molecular structure as well as 2D and 3D pharmacophores.

2.5.1 Computational Methods

- Screening of libraries and reference compounds after protonation at pH7.
- Distance calculations and ranking taking as reference some of the compounds described in Figs. 2, 3, 4, 5, and 6.
- Macromolecular target prediction for the entire screening library, and selection of the compounds placed in the top positions.
- Virtual screening for agonists/antagonists using programs descriptors and fingerprints after pre-filtering for reference compounds, computation of the Euclidean distances to the selected compounds, as well as Tanimoto similarity for fingerprints. Ranking of the compounds according to their distance/similarity to the query compounds and selection.
- De novo design using several RXR agonists or natural products, and selection of candidates after considerations of design frequency of the structures, individual ranks, and building block availability.
- Scaffold-diversity analysis, with flexible structure alignment.

2.5.2 Molecular Docking

- Selection of human RXR structures (α, β, γ) and preparation in MOE2016.08 with protonation of molecular structures at pH 7, and correction of structural features (addition of hydrogen and capping ends).
- Deletion of water molecules further away than about 5 Å from the receptor or ligand.
- Restrainment of the receptor atoms, with those located at distances higher than 8 Å from the ligand being fixed, except hydrogen atoms close to the ligand.
- Minimization of the structures in the AMBER10:EHT (Extended Hückel Theory) forcefield (with a termination value of 0.1 kcal mol^{-1} $Å^{-1}$).
- Addition of hydrogens to all ligands, protonation of ligands at pH 7, and further minimization.

- Docking of ligands in MOE2016.08 using the integrated GOLD docking program, generation of 100 poses of the ligand, further refinement of the poses using the "Induced Fit" method and scoring leading to a selection of the best 10 final poses for each ligand.

2.5.3 Rexinoids by Computer-Aided Discovery

Structure-based virtual screening of RXRα LBP by Glide of 1280 drugs (downloaded from Drugbank 2.0) approved by FDA in different conformations into the LBP of the agonist- and antagonist-bound conformations of RXRα afforded several potential ligands, in what is another example of the repositioning of these drugs for new indications. According to their high score, statin drugs (inhibitors of sterol biosynthesis) were predicted to be RXRα antagonists, including pitavastatin **7.1**, fluvastatin **7.2**, and analogs [76], which showed antagonistic effects in RXRα (Fig. 7). Surface Plasmon Resonance (SPR) experiments indicated binding of these compounds to the purified RXRα-LBP protein although rates of association and dissociation were found to be very fast. Molecular Dynamics simulations suggested that these antagonists bind to the classical LBP and the hydroxyl groups do not make any contact with residues of the LBD [76]. The bis-dihydrated derivative of pitavastatin (compound **7.3**) was also characterized as RXRα antagonist with greater affinity than parent **7.1**.

Although the discovery of these new RXR ligands resulted from the evaluation of structurally diverse collections of natural products

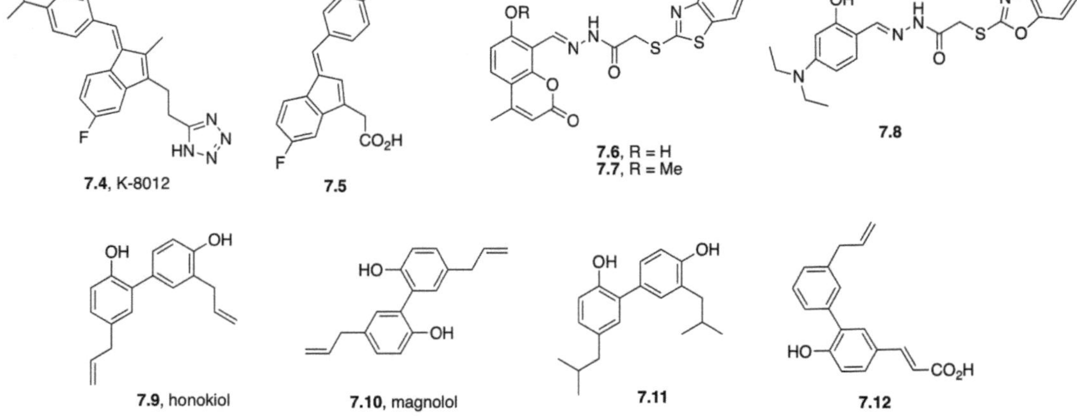

Fig. 7 RXR modulators from VLS and computational-based drug discovery

and their metabolites, as well as commercial libraries, their relatively small structural variability, easy adaptation to the LBP, and, in most of the cases, their smaller size result in suboptimal occupancy of the ligand binding cavity. Their dissimilar structures (see Fig. 7) are just an indication of the conformational adaptability of RXR LBP to bind compounds with a variety of skeletons [14]. However, synthetic and medicinal chemists can further develop these scaffolds in order to improve potency and selectivity.

2.6 Non-canonical Modulation of RXR

Truncated RXRα (tRXRα) resides in the cytoplasm, where it interacts with the p85 subunit of the phosphatidylinositol-3-OH kinase (PI3K) leading to activation of PI3K/AKT signaling in cancer cells. Sulindac, a non-steroidal anti-inflammatory drug (NSAID) and analogs, such as compound K-8012 **7.4**, have been reported as tRXRα ligands (Fig. 7). A different binding mode targeting the receptor surface was proposed for analog **7.5** based on time-resolved fluorescence resonance energy transfer (FRET) RXR-α-coactivator peptide competition assay, due to the observation of hydrophobic interactions with the carboxylate oriented away from Arg316. In addition, ethyltetrazole K-8012 **7.4** showed improved activity in the inhibition of the tRXRα-mediated PI3K/AKT signaling pathway. These compounds were characterized as antagonists of RXRα in co-transfection assays [77]. The crystal structure of ligand K-8012 **7.4** bound to RXRα revealed the presence of tetrameric oligomers with two homodimers packed in a bottom-to-bottom manner. Moreover, ligand **7.4** (1:2 ligand/protein stoichiometry) was located in a hydrophobic region of LBD near the entrance and the edge of the LBP that does not overlap with the binding site of 9-*cis*-retinoic acid. The tetrazole unit instead occupied a region away from Arg316 sitting atop the *N*-terminus of H11 and stabilized by charge-helix dipole interactions [77].

A VLS campaign using the coactivator-binding site of the crystal structure of RXRα-CD3254-CoA peptide led to the discovery of antagonist **7.6** [78] (Fig. 7) with a hydrazide bond that exhibited some selectivity for RXRα over other NRs, as well as for the RXR heterodimers. Competition experiments with tritiated 9-*cis*-retinoic acid allowed to discard binding of **7.6** to the LBP and supported interaction to the coregulatory binding-site. Similarly to Sulindac analogs, **7.6** inhibited TNFα-activated PI3K/protein kinase B (AKT)-dependent activation in several cancer cell lines in the absence or presence of TNFα, and this effect might be related to its ability to induce apoptosis through inhibition of the interaction of tRXRα with p85α [77]. Antagonist **7.6** appears to be the first RXR antagonist that targets the coregulatory binding site surface, not the LBD, to regulate its non-genomic actions. More recent studies have proven that the compound occupies a coactivator binding groove [79], since methyl ether **7.7**, with even lower binding to RXRα due to the lack of the hydroxyl group, appears to interact with Glu453, located at H12. Modifications of the somewhat unstable 2-oxo-2*H*-chromen-8-yl fragment and replacement by substituted benzenes led to **7.8**, with a 4-diethylamino-2-hydroxy imine group, which inhibited RXRα with greater potency in the mediation of the TNFα-induced NF-κB signaling pathway.

Other RXR modulators lack the carboxylic acid, such as natural product honokiol **7.9** (Fig. 7), and positional isomer magnolol **7.10**, which induced the transactivation of the PPARγ/RXRα heterodimer, but not that of the RXRα/RXRα homodimer. The crystal structure of the ligand-bound RXRα revealed that the allyl phenol moieties of ligand **7.9** occupy the L-shaped arms of the LBD and one of the phenol hydroxyl groups forms a hydrogen bond with Asn306 on N-terminal helix H5. More recent Molecular Modeling and NMR spectroscopy concluded that honokiol targets RXR at both sides of the interface, acting on the coactivator side of the dynamic activation function (AF2) or alternatively switches from one to the other side of the interface [80]. Rational design allowed to split the dual-binding properties of the natural product honokiol at the dynamic nuclear receptor interface AF2, and led to compound **7.12**, which stabilizes AF2 through binding at the LBP (as confirmed by the X-Ray structure) and analog **7.11**, which destabilizes AF2 through binding at the coactivator side of the interface [80]. Compound **7.12** is the first-in-class modulator reported to inhibit coactivator binding at the solvent-exposed side of AF2.

Acknowledgments

This work was supported by grants from the Spanish MINECO (SAF2016-77620-R-FEDER), Xunta de Galicia (Consolidación GRC 2017/61 from DXPCTSUG; ED-431G/02-FEDER "Unha maneira de facer Europa" to CINBIO, a Galician research center 2016-2019); Juan de la Cierva Contract to J. A. S.).

References

1. Laudet V, Gronemeyer H (2002) The nuclear receptor facts book. Academic Press, San Diego
2. Germain P, Chambon P, Eichele G, Evans RM, Lazar MA, Leid M et al (2006) The pharmacology and classification of the nuclear receptor superfamily. RETINOID X RECEPTORS (RXRs). Pharmacol Rev 58:760–772
3. Mangelsdorf DA, Evans RM (1995) The RXR heterodimers and orphan receptors. Cell 83:841–850
4. Kojetin DJ, Matta-Camacho E, Hughes TS, Srinivasan S, Nwachukwu JC, Cavett V et al (2015) Structural mechanism for signal transduction in RXR nuclear receptor heterodimers. Nat Commun 6:8013
5. Evans RM, Mangelsdorf DJ (2014) Nuclear receptors, RXR, and the big bang. Cell 157:255–266
6. Germain P, Iyer J, Zechel C, Gronemeyer H (2002) Co-regulator recruitment and the mechanism of retinoic acid receptor synergy. Nature 415:187–192
7. Gronemeyer H, Gustafsson J-A, Laudet V (2004) Principles for modulation of the nuclear receptor superfamily. Nature Rev Drug Disc 3:950–964

8. Hermanson O, Glass CK, Rosenfeld MG (2002) Nuclear receptor coregulators: multiple modes of modifications. Trends Endocrinol Metab 13:55–60
9. Jepsen K, Hermanson O, Onami TM, Gleiberman AS, Lunyak V, McEvilly RJ et al (2000) Combinatorial roles of the nuclear receptor corepressor in transcription and development. Cell 102:753–763
10. Smith CL, O'Malley BW (2004) Coregulator function: a key to understanding tissue specificity of selective receptor modulators. Endocr Rev 25:45–71
11. Dollé P, Niederreither K (2015) The retinoids: biology, biochemistry and disease. John Wiley & Sons, Hoboken, NJ
12. Bourguet W, Moras D (2015) Retinoid receptors: protein structure, DNA recognition and structure-function relationships. In: Dollé P, Niederreither K (eds) The retinoids: biology, biochemistry, and disease. John Wiley & Sons, Hoboken, NJ, pp 131–149
13. Mendoza-Parra MA, Bourguet W, de Lera AR, Gronemeyer H (2015) Retinoid receptor-selective modulators: chemistry, 3D structures and systems biology. In: Dollé P, Neiderreither K (eds) The retinoids: biology, biochemistry, and disease. Wiley-Blackwell, Hoboken, NJ, pp 165–192
14. Dominguez M, Alvarez S, de Lera AR (2017) Natural and structure-based RXR ligand scaffolds and their functions. Curr Top Med Chem 17:631–662
15. Huang P, Chandra V, Rastinejad F (2014) Retinoic acid actions through mammalian nuclear receptors. Chem Rev 114:233–254
16. Zhang H, Chen L, Chen J, Jiang H, Shen X (2011) Structural basis for retinoic X receptor repression on the tetramer. J Biol Chem 286:24593–24598
17. Sato Y, Ramalanjaona N, Huet T, Potier N, Osz J, Antony P et al (2010) The "Phantom Effect" of the rexinoid LG100754: structural and functional insights. PLoS One 5:e15119
18. Germain P, Gaudon C, Pogenberg V, Sanglier S, Potier N, Van Dorsselaer A et al (2009) Differential action on coregulator interaction defines inverse retinoid agonists and neutral antagonists. Chem Biol 16:479–489
19. le Maire A, Teyssier C, Erb C, Grimaldi M, Alvarez S, de Lera AR et al (2010) A unique secondary-structure switch controls constitutive gene repression by retinoic acid receptor. Nat Struct Mol Biol 17:801–807
20. Bourguet W, de Lera AR, Gronemeyer H (2010) Inverse agonists and antagonists of retinoid receptors. In: Conn PM (ed) Meth Enzymol. Academic Press, New York, pp 161–195
21. Pérez E, Bourguet W, Gronemeyer H, de Lera AR (2012) Modulation of RXR function through ligand design. Biochim Biophys Acta 1821:57–69
22. le Maire A, Alvarez S, Shankaranarayanan P, de Lera AR, Bourguet W, Gronemeyer H (2012) Retinoid receptors and therapeutic applications of RAR/RXR modulators. Curr Top Med Chem 12:505–527
23. Arnold SLM, Amory JK, Walsh TJ, Isoherranen N (2012) A sensitive and specific method for measurement of multiple retinoids in human serum with UHPLC-MS/MS. J Lipid Res 53:587–598
24. Kane MA (2012) Analysis, occurrence, and function of 9-*cis*-retinoic acid. Biochim Biophys Acta 1821:10–20
25. Rühl R, Krzyzosiak A, Niewiadomska-Cimicka A, Rochel N, Szeles L, Bn V et al (2015) 9-*cis*-13,14-Dihydroretinoic acid is an endogenous retinoid acting as RXR ligand in mice. PLoS Genet 11:e1005213
26. Moise AR, Alvarez S, Domínguez M, Alvarez R, Golczak M, Lobo GP et al (2009) Activation of retinoic acid receptors by dihydroretinoids. Mol Pharmacol 76:1228–1237
27. Egea PF, Mitschler A, Rochel N, Ruff M, Chambon P, Moras D (2000) Crystal structure of the human RXRα ligand-binding domain bound to its natural ligand: 9-*cis*-retinoic acid. EMBO J 19:2592–2601
28. Egea PF, Mitschler A, Moras D (2002) Molecular recognition of agonist ligands by RXRs. Mol Endocrinol 16:987–997
29. Bourguet W, Vivat V, Wurtz J-M, Chambon P, Gronemeyer H, Moras D (2000) Crystal structure of a heterodimeric complex of RAR and RXR ligand-binding domains. Mol Cell 5:289–298
30. Nahoum V, Pérez E, Germain P, Rodríguez-Barrios F, Manzo F, Kammerer S et al (2007) Modulators of the structural dynamics of RXR to reveal receptor function. Proc Natl Acad Sci U S A 104:17323–17328
31. Pérez-Santín E, Germain P, Quillard F, Khanwalkar H, Rodríguez-Barrios F, Gronemeyer H et al (2009) Modulating retinoid X receptor with a series of (*E*)-3-[4-hydroxy-3-(3-alkoxy-5,5,8,8-tetramethyl-5,6,7,8-tetrahydronaphthalen-2-yl)phenyl]acrylic acids and their 4-alkoxy isomers. J Med Chem 52:3150–3158
32. Zhang L, Nadzan AM, Heyman RA, Love DL, Mais DE, Croston G et al (1996) Discovery of

novel retinoic acid receptor agonists having potent antiproliferative activity in cervical cancer cells. J Med Chem 39:2659–2663
33. Bennani JL, Marron KS, Mais DE, Flatten K, Nadzan AM, Boehm MF (1998) Synthesis and characterization of a highly potent and selective isotopically labeled retinoic acid receptor ligand, ALRT1550. J Org Chem 63:543–550
34. Canan-Koch SS, Dardashti LJ, Cesario RM, Croston GE, Boehm MF, Heyman RA et al (1999) Synthesis of retinoid X receptor-specific ligands that are potent inducers of adipogenesis in 3T3-L1 cells. J Med Chem 42:742–750
35. Michellys PY, D'Arrigo J, Grese TA, Karanewsky DS, Leibowitz MD, Mais DA et al (2004) Design, synthesis and structure-activity relationship of novel RXR-selective modulators. Bioorg Med Chem Lett 14:1593–1598
36. Canan Koch SS, Dardashti LJ, Hebert JJ, White SK, Croston GE, Flatten KS et al (1996) Identification of the first retinoid X receptor homodimer antagonist. J Med Chem 39:3229–3234
37. Schulman IG, Li C, Schwabe JW, Evans RM (1997) The phantom ligand effect: allosteric control of transcription by the retinoid X receptor. Genes Dev 11:299–308
38. Muccio DD, Brouillette WJ, Breitman TR, Taimi M, Emanuel PD, X-k Z et al (1998) Conformationally defined retinoid acid analogues. 4. Potential new agents for acute promielocytic and juvenile myelomonocytic leukemias. J Med Chem 41:1679–1687
39. Atigadda VR, Vines KK, Grubbs CJ, Hill DL, Beenken SL, Bland KI et al (2003) Conformationally defined retinoic acid analogues. 5. Large-scale synthesis and mammary cancer chemopreventive activity for (2*E*,4*E*,6*Z*,8*E*)-8-(3′,4′-Dihydro-1′(2′*H*)-naphathalen-1′yliden e)-3,7-dimethyl-2,4,6-octatrienoic acid. J Med Chem 46:3766–3769
40. Grubbs CJ, Lubet RA, Atigadda VR, Christov K, Deshpande AM, Tirmal V et al (2006) Efficacy of new retinoids in the prevention of mammary cancers and correlations with short-term biomarkers. Carcinogenesis 27:1232–1239
41. Kolesar JM, Hoel R, Pomplun M, Havighurst T, Stublaski J, Wollmer B et al (2010) A pilot, first-in-human, pharmacokinetic study of 9cUAB30 in healthy volunteers. Cancer Prev Res 3:1565–1570
42. Kapetanovic IM, Horn TL, Johnson WD, Cwik MJ, Detrisac CJ, McCormick DL (2010) Murine oncogenicity and pharmacokinetics studies of 9-*cis*-UAB30, an RXR agonist, for breast cancer chemoprevention. Int J Toxicol 29:157–164
43. Desphande A, Xia G, Boerma LJ, Vines KK, Atigadda VR, Lobo-Ruppert S et al (2014) Methyl-substituted conformationally constrained rexinoid agonists for the retinoid X receptors demonstrate improved efficacy for cancer therapy and prevention. Bioorg Med Chem 22:178–185
44. Atigadda VR, Xia G, Desphande A, Boerma LJ, Lobo-Ruppert S, Grubbs CJ et al (2014) Methyl substitution of a rexinoid agonist improves potency and reveals site of lipid toxicity. J Med Chem 57(12):5370–5380
45. Atigadda VR, Xia G, Deshpande A, Wu L, Kedishvili N, Smith CD et al (2015) Conformationally defined rexinoids and their efficacy in the prevention of mammary cancers. J Med Chem 58:7763–7774
46. Vuligonda V, Lin Y, Chandraratna RAS (1996) Synthesis of highly potent RXR-specific retinoids: The use of a cyclopropyl group as a double bond isostere. Bioorg Med Chem Lett 6:213–218
47. Haffner CD, Lenhard JM, Miller AB, McDougald DL, Dwornik K, Ittoop OR et al (2004) Structure-based design of potent retinoid X receptor a agonists. J Med Chem 47:2010–2029
48. Lehmann J, Jong L, Fanjul A, Cameron J, Lu X, Haefner P et al (1992) Retinoids selective for retinoid X receptor response pathways. Science 258:1944–1946
49. Boehm MF, Zhang L, Badea BA, White SK, Mais DE, Berger E et al (1994) Synthesis and structure-activity relationships of novel retinoid X receptor-selective retinoids. J Med Chem 37:2930–2941
50. Boehm MF, Zhang L, Zhi L, McClurg MR, Berger E, Wagoner M et al (1995) Design and synthesis of potent retinoid X receptor selective ligands that induce apoptosis in leukemia cells. J Med Chem 38:3146–3155
51. Tanaka T, De Luca LM (2009) Therapeutic potential of "rexinoids" in cancer prrevention and treatment. Cancer Res 69:4945–4947
52. Sherman SI, Gopal J, Haugen BR, Chiu AC, Whaley K, Nowlakha P et al (1999) Central hypothyroidism associated with retinoid X receptor-selective ligands. New Engl J Med 340:1075–1079
53. Shankaranarayanan P, Rossin A, Khanwalkar H, Alvarez S, Alvarez R, Jacobson A et al (2009) Growth factor-antagonized rexinoid apoptosis involves permissive PPARγ/RXR heterodimers to activate the intrinsic death pathway by NO. Cancer Cell 16:220–231

54. Dawson MI, Jong L, Hobbs PD, Cameron JF, Chao W-R, Pfahl M et al (1995) Conformational effects on retinoid receptor selectivity. 2. Effects of retinoid bridging group on retinoid X receptor activity and selectivity. J Med Chem 38:3368–3383
55. Ohta K, Kawachi E, Fukasawa H, Shudo K, Kagechika H (2011) Diphenylamine-based retinoid antagonists: Regulation of RAR and RXR function depending on the N-substituent. Bioorg Med Chem 19:2501–2507
56. Ohta K, Tsuji M, Kawachi E, Fukasawa H, Hashimoto Y, Shudo K et al (1998) Potent retinoid synergists with a diphenylamine skeleton. Biol Pharm Bull 21:544–546
57. Takahashi B, Ohta K, Kawachi E, Fukasawa H, Hashimoto Y, Kagechika H (2002) Novel retinoid X receptor antagonists: specific inhibition of retinoid synergism in RXR-RAR heterodimers. J Med Chem 45:3327–3329
58. Kakuta H, Yakushiji N, Shinozaki R, Ohsawa F, Yamada S, Ohta Y et al (2012) RXR partial agonist CBt-PMN exerts therapeutic effects on type 2 diabetes without the side effects of RXR full agonists. ACS Med Chem Lett 3:427–432
59. Ohsawa F, Yamada S, Yakushiji N, Shinozaki R, Nakayama M, Kawata K et al (2013) Mechanism of retinoid X receptor partial agonistic action of 1-(3,5,5,8,8-pentamethyl-5,6,7,8-tetrahydro-2-naphthyl)-1H-benzotriazole-5-carboxylic acid and structural development to increase potency. J Med Chem 56:1865–1877
60. Kawata K, Morishita K-i, Nakayama M, Yamada S, Kobayashi T, Furusawa Y et al (2015) RXR partial agonist produced by side chain repositioning of alkoxy RXR full agonist retains antitype 2 diabetes activity without the adverse effects. J Med Chem 58:912–926
61. Wallén-Mackenzie Ö, Mata de Urquiza A, Petersson S, Rodriguez FJ, Friling S, Wagner J et al (2003) Nurr1-RXR heterodimers mediate RXR ligand-induced signaling in neuronal cells. Genes Dev 17:3036–3047
62. Spathis AD, Asvos X, Ziavra D, Karampelas T, Topouzis S, Cournia Z et al (2017) Nurr1: RXRα heterodimer activation as monotherapy for Parkinson's disease. Proc Natl Acad Sci U S A 114:3999–4004
63. Lagu B, Pio B, Lebedev R, Yang M, Pelton PD (2007) RXR-LXR heterodimer modulators for the potential treatment of dyslipidemia. Bioorg Med Chem Lett 17:3497–3503
64. Lagu B, Lebedev R, Pio B, Yang M, Pelton PD (2007) Dihydro-[1*H*]-quinolin-2-ones as retinoid X receptor (RXR) agonists for potential treatment of dyslipidemia. Bioorg Med Chem Lett 17:3491–3496
65. Umemiya H, Fukasawa H, Ebisawa M, Eyrolles L, Kawachi E, Eisenmann G et al (1997) Regulation of retinoidal actions by diazepinylbenzoic acids. Retinoid synergists activate the RXR-RAR heterodimers. J Med Chem 40:4222–4234
66. Morita K, Kawana K, Sodeyama M, Shimomura I, Kagechika H, Makishima M (2005) Selective allosteric ligand activation of the retinoid X receptor heterodimers of NGFI-B and Nurr1. Biochem Pharmacol 71: 98–107
67. Kagechika H, Shudo K (2005) Synthetic retinoids: recent developments concerning structure and clinical utility. J Med Chem 48:5875–5883
68. Sakaki J, Kishida M, Konishi K, Gunji H, Toyao A, Matsumoto Y et al (2007) Synthesis and structure-activity relationship of novel RXR antagonists: orally active anti-diabetic and anti-obesity agents. Bioorg Med Chem Lett 17:4804–4807
69. Sakaki J, Konishi K, Kishida M, Gunji H, Kanazawa T, Uchiyama H et al (2007) Synthesis and structure-activity relationship of RXR antagonists based on the diazepinylbenzoic acid structure. Bioorg Med Chem Lett 17:4808–4811
70. Sundén H, Schäfer A, Scheepstra M, Leysen S, Malo M, Ma J-N et al (2016) Chiral dihydrobenzofuran acids show potent retinoid X receptor–nuclear receptor related 1 protein dimer activation. J Med Chem 59:1232–1238
71. Katsila T, Spyroulias GA, Patrinos GP, Matsoukas M-T (2016) Computational approaches in target identification and drug discovery. Comput Struct Biotechnol J 14:177–184
72. Schneider G (2017) Automating drug discovery. Nature Rev Drug Disc 17:97–113
73. Sliwoski G, Kothiwale S, Meiler J, Lowe EW (2014) Computational methods in drug discovery. Pharmacol Rev 66:334–395
74. Reker D, Rodrigues T, Schneider P, Schneider G (2014) Identifying the macromolecular targets of de novo-designed chemical entities through self-organizing map consensus. Proc Natl Acad Sci U S A 111:4067
75. Reutlinger M, Koch Christian P, Reker D, Todoroff N, Schneider P, Rodrigues T et al (2013) Chemically advanced template search (CATS) for scaffold-hopping and prospective target prediction for 'orphan' molecules. Mol Inf 32:133–138
76. Xu D, Cai L, Guo S, Xie L, Yin M, Chen Z et al (2017) Virtual screening and experimental validation identify novel modulators of nuclear receptor RXRα from Drugbank database. Bioorg Med Chem Lett 27:1055–1061

77. Chen L, Wang Z-G, Aleshin AE, Chen F, Chen J, Jiang F et al (2014) Sulindac-derived RXRα modulators inhibit cancer cell growth by binding to a novel site. Chem Biol 21: 596–607
78. Chen F, Liu J, Huang M, Hu M, Su Y, X-k Z (2014) Identification of a new RXRα antagonist targeting the coregulator-binding site. ACS Med Chem Lett 5:736–741
79. Xu D, Guo S, Chen Z, Bao Y, Huang F, Xu D et al (2016) Binding characterization, synthesis and biological evaluation of RXRα antagonists targeting the coactivator binding site. Bioorg Med Chem Lett 26:3846–3849
80. Scheepstra M, Nieto L, Hirsch AKH, Fuchs S, Leysen S, Lam CV et al (2014) A natural-product switch for a dynamic protein interface. Angew Chem Int Ed 53:6443–6448

Chapter 5

In Vitro Models to Study the Regulatory Roles of Retinoids in Angiogenesis

Ayman Al Haj Zen

Abstract

Retinoids are reported to regulate vascular growth and remodeling during embryonic development and wound healing. A better understanding of angiogenic mechanisms of retinoids has clinical implication in pathological conditions such as chronic nonhealing wounds. Here, we describe in vitro angiogenesis assays that can be a useful tool to study the role of retinoids and retinoic acid signaling in the regulation of different features of angiogenesis such as tubulogenesis and branching.

Key words Retinoids, Endothelial cells, Angiogenesis, Wound healing, Coculture, Tube formation

1 Introduction

Angiogenesis is defined as the formation of new capillaries from preexisting blood vessels and is a critical component for a proper tissue repair [1]. Insufficient angiogenesis can result in a delay in wound healing and chronic wound formation [2]. The interaction between fibroblasts and microvascular endothelial cells in the granulation tissue of wound is a requisite element of an established healing response. Fibroblasts drive the angiogenic process through their secreted extracellular matrix and their local release of vascular growth factors [3]. The deficiency of the formation of a well-vascularized granulation tissue impairs the process of wound healing in diabetic patients [4]. Many studies have shown an involvement of retinoic acid (RA) signaling in angiogenesis and vascular morphogenesis. RA plays a crucial role in vascular morphogenesis during embryogenesis. The lack of Raldh2, an important enzyme for RA synthesis, results in a significant defect in the early vascular development associated with embryonic death [5]. It has been suggested that RA influences the formation of primitive vascular plexus because it regulates endothelial cell growth and vascular remodeling via controlling the expression of genes involved in

Swapan K. Ray (ed.), *Retinoid and Rexinoid Signaling: Methods and Protocols*, Methods in Molecular Biology, vol. 2019,
https://doi.org/10.1007/978-1-4939-9585-1_5, © Springer Science+Business Media, LLC, part of Springer Nature 2019

vascular patterning and morphogenesis such as Vascular Endothelial Growth Factor (VEGF), Fibroblast Growth Factor (FGF), Transforming Growth Factor-beta (TGF-β), and Hedgehog [6]. Exogenous retinoids can increase proangiogenic behavior of bovine aortic endothelial cells and it can induce angiogenesis in two different animal models: in vivo Matrigel plug assay and Chicken Chorioallantoic Membrane Assay [7]. Retinoids induces the expression of vascular growth factors such as VEGF, Hepatocyte Growth Factor (HGF), and Placental Growth Factor (PGF). Tazarotene, a retinoid with selective activity toward the retinoic acid receptors, RARβ and RARγ, triggers skin wound regeneration, which is associated with an increase of neovascularization [8, 9].

In this chapter, we describe two phenotypic assays to examine the biological role of RA signaling pathway and retinoids in angiogenesis. One assay evaluates the effect of retinoids on the formation of capillary-like tubes by endothelial cells seeded on a basement membrane matrix (*see* Fig. 1a). Endothelial tube formation assay recapitulates many steps occurring in angiogenesis as cell adhesion,

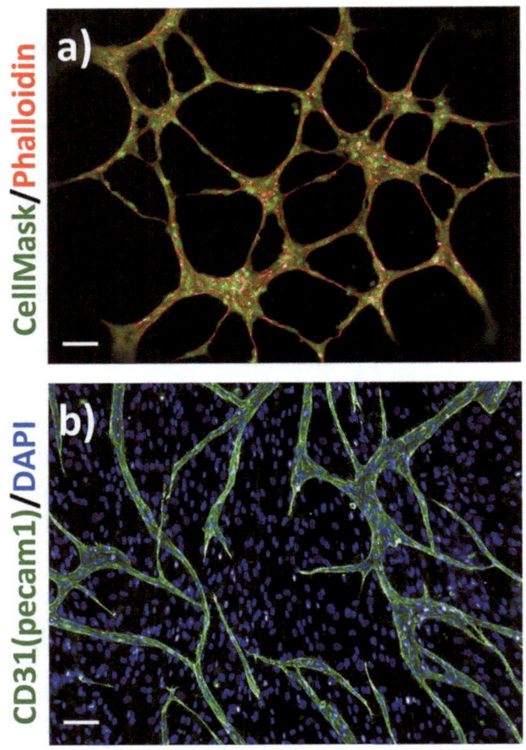

Fig. 1 In vitro cell models for angiogenesis. (**a**) *Endothelial tube formation assay:* The tubes are fixed after 8 hours and stained with phalloidin-alexa568 (red) and Cell Mask (green). (**b**) *Endothelial fibroblast coculture angiogenesis assay:* Cocultures were fixed at 5 days after plating and stained with an antibody against CD31 (endothelial cell marker, green) and DAPI (Nuclei, blue). Scale bar 100 μm

motility, tubulogenesis, and branching morphogenesis [10]. This assay is a rapid, quantitative, yields highly reproducible results, and can be easily adapted for high throughput use [11]. To investigate the angiogenic potential of retinoids in more complex context of wound repair, we employ a second cell-based assay where endothelial cells are cocultured with dermal fibroblasts (*see* Fig. 1b). In contrary to the tube formation assay, this assay is a long-term assay where the extracellular matrix is secreted and conditioned by fibroblasts [12, 13]. Previous reports showed that endothelial cell tube-like structures were surrounded by a basement membrane and contained a stable lumen for up to 5 weeks in culture [14]. The primary advantage of this model is that it allows examining the role of retinoids in the paracrine interactions between fibroblasts and endothelial cells, which modulate the formation of granulation tissue during wound healing. RA diffuses from its cell of origin, acting in a paracrine fashion to influence neighboring cell behaviors [15]. This model could also be useful to study this paracrine mechanism of endogenous retinoic acid signal.

Both assays have been proven to be robust tools to elucidate the signaling pathways including RA signaling and retinoids as they relate to angiogenesis. We have used these assays in a combination of relevant selective small molecules to demonstrate that, Tazarotene enhanced in vitro angiogenesis by promoting branching morphogenesis and tubule remodeling. The proangiogenic phenotype is mediated by RAR but not Retinoic X Receptor (RXR) activation. Moreover, blockade of retinoic acid degradation, which leads to increase the intracellular level of retinoic acid, induces the angiogenic behavior of endothelial cells similar to the effect of Tazarotene [9].

2 Materials

2.1 Endothelial Tube Formation Assay

1. Standard Matrigel® matrix basement membrane (9–10 mg/mL, Corning) should be thawed on cold ice at 4 °C overnight and then aliquoted into single use vials, and stored at −20 °C.
2. Primary Human Umbilical Vein Endothelial Cells (HUVEC) pooled (Lonza).
3. Endothelial cell growth medium-2 (EGM-2) contains Endothelial Basal Medium-2 (EBM-2) supplemented with EGM-2 SingleQuot Kit (Lonza).
4. Accutase® solution consisting of a mixture of proteolytic and collagenolytic enzymes (Sigma).
5. Hematocytometer.
6. Microplates 96-well tissue culture-treated with black walls and clear thin bottom suitable for immunofluorescent studies and

microscopic viewing (e.g., CellCarrier-96 Black, Perkin Elmer).

7. Pipetting reagent reservoirs (25 mL capacity, sterile, and disposable).
8. Fixation solution: 4% Paraformaldehyde, PBS (pH 7.4–7.6).
9. Permeabilization solution: 0.1% Triton x-100, PBS (pH 7.4–7.6).
10. Staining dyes (ThermoFisher Scientific): Alexa Fluor™ 568 Phalloidin and HCS CellMask™ Green Stain.
11. Washing solution: 0.05% TWEEN-20, (pH 7.4–7.6).

2.2 Endothelial Cell Fibroblast Coculture Angiogenesis Assay

1. Microplates 96-well tissue culture-treated with black walls and clear bottom suitable for immunofluorescent studies and microscopic viewing.
2. Primary adult dermal fibroblasts (NHDF) (Lonza).
3. Accutase® solution consisting of a mixture of proteolytic and collagenolytic enzymes (Sigma).
4. Hematocytometer.
5. Pipetting reagent reservoirs (25 mL capacity, sterile, and disposable).
6. Medium 106 (Gibco™, Thermofisher) supplemented with Low Serum Growth Supplement (LSGS) contains 2% fetal bovine serum, basic fibroblast growth factor, heparin, hydrocortisone, and epidermal growth factor (Gibco™, Thermofisher).
7. Primary Human Umbilical Vein Endothelial Cells (HUVEC) pooled (Lonza).
8. Endothelial cell growth medium-2 (EGM-2) contains Endothelial Basal Medium-2 (EBM-2) supplemented with EGM-2 SingleQuot Kit (Lonza).
9. Fixation solution: 4% Paraformaldehyde, PBS (pH 7.4–7.6).
10. Blocking and permeabilization solution: 3% Bovine Serum Albumin (BSA) diluted in 0.03% TWEEN-20, PBS buffer.
11. Mouse anti-human CD31/PECAM-1 antibody, Clone # 9G11 (R&D systems).
12. Goat anti-mouse IgG, biotin conjugated secondary antibody (Santa Cruz).
13. Streptavidin, Alexa Fluor™ 488 conjugate (ThermoFisher).
14. DAPI solution (1 mg/mL), (ThermoFisher).

3 Methods

3.1 Endothelial Tube Formation Assay (See Fig. 2a)

3.1.1 Assay Preparation

1. Calculate the total volume of Matrigel® required for the experiment. Use 50 μL of Matrigel® solution for each well of the 96-well plate (e.g., 60 wells will require 3 mL of Matrigel® solution) (*see* **Note 1**).
2. Place the frozen Matrigel® solution on ice and defrost at 4 °C for overnight. Matrigel® solidifies when warm, so it must be kept on ice all time until use (*see* **Notes 2** and **3**).

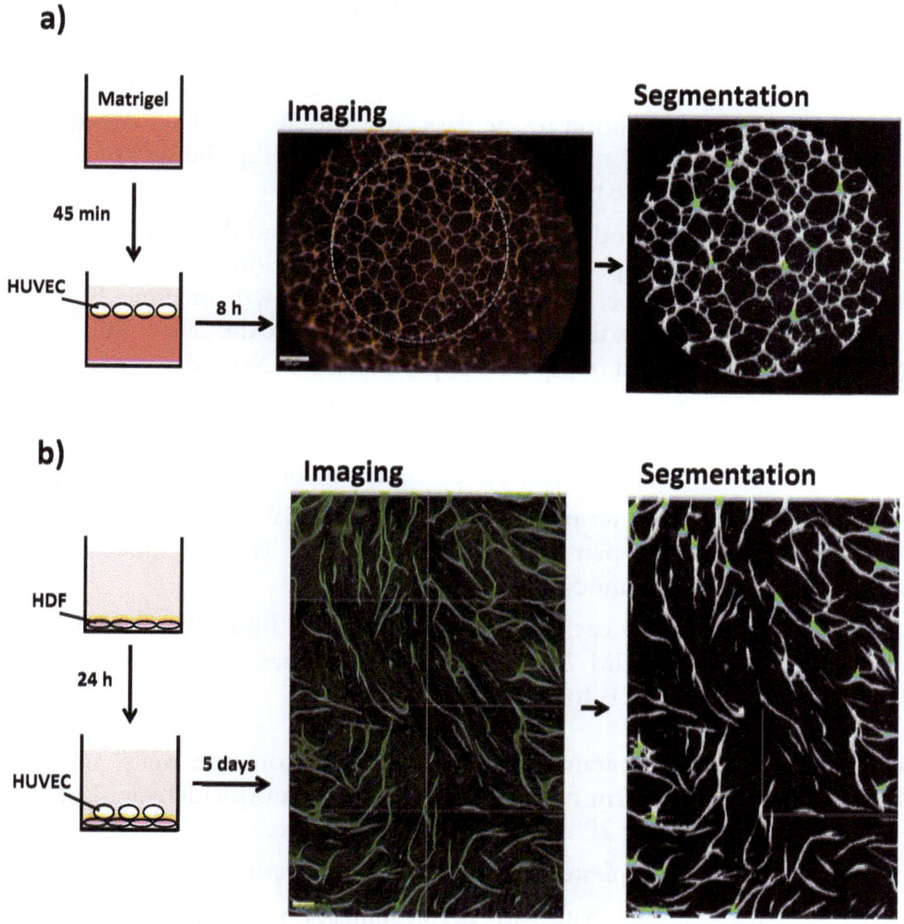

Fig. 2 Schematic representation of procedures for using in vitro angiogenesis assays. (**a**) *Endothelial tube formation assay:* HUVEC are seeded on top of Matrigel® and incubated at 37 °C/%5 CO_2 for 8 h. The tubes are stained with phalloidin-alexa568 (orange) and the whole well is imaged using ×2 objective lens. The Region-Of-Interest is cropped, segmented and quantified using MetaMorph® angiogenesis tube formation app. Tubes (white mask); cell aggregates (green mask), scale bar 500 μm. (**b**) *Endothelial fibroblast coculture angiogenesis assay*: HUVEC are plated onto a confluent human dermal fibroblast layer. Cocultures were fixed at 5 days after plating and stained with an antibody against CD31 (green). Nine fields are imaged using ×10 objective lens. The tubes are segmented and quantified using MetaMorph® angiogenesis tube formation app. Scale bar 200 μm

3. Gently, so as not to introduce bubbles, pipet 50 μL of cold Matrigel® solution into each well of 96-well plate, which is placed on ice (*see* **Note 4**).
4. Place the plate containing the Matrigel® into a tissue culture incubator at 37 °C, 5% CO_2 for at least 45 min to allow polymerization of Matrigel®.
5. Meanwhile, dilute the retinoid compounds to be tested and DMSO (vehicle) in the EGM-2 media.
6. Gently add 50 μL of compound solution prepared on top of the gelled Matrigel® and return back the plate into the tissue culture incubator (*see* **Note 5**).
7. Remove and discard the media from culture flask with 80% confluent HUVECs and rinse cells with PBS. Add Accutase® solution to the flask, swirl briefly, and incubate at 37 °C for a few minutes to release the cells. Tap the side of the flask to be sure that the cells are detaching.
8. Gently pipette the solution up and down to make a single cell suspension. Transfer cell suspension in a sterile 50 mL conical tube. Add equal amount of growth medium EGM-2, gently mix the suspension and centrifuge the cells at 300 × *g* for 5 min and remove the supernatant (*see* **Note 6**).
9. Aspirate supernatant and resuspend the cell pellet in pre-warmed EGM-2 media. The cells should be gently pipetted up and down a few times to obtain a single cell suspension.
10. Determine cell number and viability by mixing 5 μL of cell suspension with 5 μL of Trypan blue and using a hemocytometer.
11. To each well, carefully add 15,000 cells diluted in 50 μL of EGM-2 on top of media surface and incubate at 37 °C, 5% CO_2 for 8–16 h (*see* **Note 7**).

3.1.2 Fixation and Staining

1. Aspirate carefully the media from the wells, add 100 μL of warm fixative solution (4% paraformaldehyde, PBS). Incubate for 20 min at room temperature.
2. Aspirate fixative solution and wash with PBS twice.
3. Incubate with permeabilization solution (0.1% Triton x-100, PBS) for 10 min at room temperature.
4. Remove the permeabilization solution and incubate with staining buffer containing: Alexa Fluor-568 Phalloidin (1/200) and HCS CellMask green (1/5000) in washing buffer for 45 min at room temperature.
5. Wash twice with washing solution for 20 min each at room temperature.

6. Add PBS and keep at 4 °C for 1 week to reduce the background noise.
7. Wash once with PBS and the plate is now ready for viewing and imaging using inverted fluorescent microscope.

3.1.3 Imaging and Quantification

1. Acquire the images of tubes 2× objective lens using a high content screening system (e.g., Operetta "Perkin Elmer," InCell 6000 "GE Healthcare") or a fluorescent inverted microscope equipped with a wide working distance ×2 objective lens and CCD camera.
2. On high content screening system, determine the height of the focal plan, which is usually between 1200 and 1400 μm for 96-well plate.
3. Use the red channel of phalloidin alexa-568 to acquire images at exposure times of 400 ms for quantifying the tube formation.
4. Export the image tiff files with 16-bit depth to analyze the images using image analysis software (MetaMorph®, Molecular Device), (*see* **Note 8**).
5. In MetaMorph®, using "create region" crop the 16-bit depth images to extract the region-of-interest from the well images and to remove the out-of-focus objects in the border area of the well. Use "angiogenesis tube formation application" to segment and analyze the cropped images to quantify the morphological features of the vascular-like network. Set the pixel intensity threshold to distinguish actin-labeled endothelial tubes from local background along with a minimum and maximum tube width and applied for analyzing all images. MetaMorph® can measure more than 15 quantitative parameters depicting the tube network such as total tube length, branching point numbers, tube thickness, and tube area.

3.2 Endothelial Fibroblast Coculture Angiogenesis Assay (See Fig. 2b)

3.2.1 Assay Preparation

1. ***Day 1:*** Prepare single cell suspension of cultured human skin fibroblasts in complete medium-106 using Accutase® solution as described in Subheading 3.1.1 (**steps** 7 and **8**). Count cells in a hemocytometer.
2. Add 100 μL (15,000 cells) per well of the single cell suspension to a 96-well plate (*see* **Note 9**) and transfer the plate into a tissue culture incubator 37 °C, 5% CO2 for 24 h.
3. ***Day 2*:** Check if cultured fibroblasts are 100% confluent. Harvest the HUVEC using Accutase® solution and prepare a single cell suspension in a full-growth factor supplemented medium consisting of a 1:1 mixture of medium-106: EGM-2 medium (*see* **Notes 10–12**). Count cells in a hemocytometer.

4. Add 100 μL of HUVEC (5000 cells) per well onto the over-confluent monolayer of fibroblasts and incubate the plate at 37 °C, 5% CO_2 for 1 h.
5. Remove the culture media and add compounds (retinoids) and DMSO (vehicle) to be tested which are diluted in culture media mixture.
6. ***Days 3 and 5:*** Aspirate the culture media containing compounds and replace with fresh culture media mixture and prepared compounds (*see* **Note 13**).
7. ***Day 7:*** Stop the assay at this point.

3.2.2 Fixation and Staining

1. Aspirate the culture media from the wells and wash with PBS. Add 100 μL of fixative solution (4% paraformaldehyde, PBS). Incubate for 20 min at room temperature.
2. Aspirate fixative solution and wash with PBS twice
3. Add 100 μL of blocking and permeabilization solution, and incubate for 20 min at room temperature.
4. Incubate with 50 μL of monoclonal mouse anti-human CD31/PECAM antibody (1:100) diluted in PBS at 4 °C for overnight.
5. Wash twice with PBS for 5 min each.
6. Incubate with goat anti-mouse IgG, biotin conjugated secondary antibody (1/200) diluted in PBS for 45 min at room temperature.
7. Wash twice with PBS for 5 min each.
8. Incubate with Streptavidin, AlexaFluor-488 conjugate (1/200) and DAPI solution (1/500) diluted in PBS for 30 min at room temperature.
9. Wash twice with PBS for 10 min each.
10. Add PBS into wells and store the plate at 4 °C for imaging later.

3.2.3 Imaging and Quantification

1. Observe and acquire the images of endothelial tubes using a high content screening system such as Operetta "Perkin Elmer," InCell 6000 "GE Healthcare," or an available fluorescent inverted microscope equipped with ×10 objective lens.
2. Acquire nine identically positioned fields from each well of 96-well plate.
3. Export the image tiff files with 16-bit depth to analyze the images using image analysis software (MetaMorph®, Molecular Device).
4. In MetaMorph®, use "angiogenesis tube formation application" to segment and analyze the images to extract the morphological features of the vascular-like network. Set the pixel

intensity threshold to distinguish CD31-labeled endothelial tubes from local background along with a minimum and maximum tube width and applied for analyzing all images. We found that the following quantitative parameters of the tube networks: total tube length, branching point numbers, tube thickness and total node area (cell aggregates), are most suitable for further multiparametric analysis of the vascular-like network (e.g., hierarchical clustering and principal component analysis). This analysis could be useful to delineate the retinoid mode of action from the acquired image-based phenotypic fingerprint.

4 Notes

1. Because the results from peripheral wells show edge effect and this subsequently causes inter-plate variations, therefore, avoid the use of the peripheral wells on 96-well plates to eliminate potential edge effects but fill them with PBS solution.
2. Rapid defrost of Matrigel® may result in increasing of Matrigel® due to an inconsistency of thermal gradients in the Matrigel® solution in the preparation tube and cause premature polymerization of parts of Matrigel®.
3. Where possible, the same lot of Matrigel® should be used for the entire study. The concentration of growth factors and basement membrane matrix components in Matrigel® is highly variable from lot to lot even within the same vendor. This may affect the quality of the assay.
4. Dispose the Matrigel® in the same location of the well for each well to reduce the variability.
5. While adding media containing compounds, care must be taken not to damage the gels. When pipetting, slowly allow each drop of liquid to spread over the surface of the gel rather than drop from a distance.
6. It is a critical step to obtain a single cell suspension and avoid cell aggregates. A disposable transfer pipet (5-mL capacity) works well for resuspending the cells in Accutase® solution before adding complete EGM-2 media and centrifugation.
7. The time of the assay depends on several factors such as endothelial cell type, passage number, and growth factors in medium. When Matrigel® with growth factors is used, HUVEC form capillary-like tubes 4 h after seeding. The tubes mature by 6–16 h. After 18–24 h, the tubes disintegrate and cells undergo apoptosis. We have found that the optimal time to stop the assay is 8–10 h.

8. The quantification of the tube formation may be performed by other image analysis software such as open-source Image J software with the Angiogenesis Analyzer plugin (Carpentier G, 2012), which has capability to measure many angiogenic parameters: tube length, branching points, number of meshes, and number of tubes.
9. It is recommendable to use an early passage fibroblasts and HUVEC. HUVEC must be not seeded until fibroblasts are confluent.
10. Microvascular endothelial cells isolated from skin can be used in this assay.
11. This means that the growth factor additives (VEGF-A, FGF-2, IGF-1, and epidermal growth factor) are present in the medium at half the amount intended for the complete EGM-2 medium. Alternatively, only EGM2 media can be used in the assay in order to enhance its dynamic range for investigating the anti-angiogenic factors. The choice of medium should be based on the experiment aims.
12. If endothelial cells grow as cluster of cells rather than tubes, this may occur due to the conditions of coculture: fibroblasts are not confluent, the seeding density of endothelial cells is more than 1.8×10^4 per cm^2 and/or the vascular growth factors concentration in the culture media is suboptimal.
13. Using the current settings of this assay, the formation of established endothelial tubes will be from 5 days following plating of endothelial cells. Therefore, the effect of tested compounds on tube formation must be assessed from day 5 after endothelial cell plating.

Acknowledgments

This work is supported from British Heart Foundation Centre of Research Excellence (Oxford), RE/08/004/23915 and RE/13/1/30181.

References

1. Potente M, Gerhardt H, Carmeliet P (2011) Basic and therapeutic aspects of angiogenesis. Cell 146(6):873–887. https://doi.org/10.1016/j.cell.2011.08.039
2. Okonkwo UA, DiPietro LA (2017) Diabetes and wound angiogenesis. Int J Mol Sci 18(7). https://doi.org/10.3390/ijms18071419
3. Greaves NS, Ashcroft KJ, Baguneid M, Bayat A (2013) Current understanding of molecular and cellular mechanisms in fibroplasia and angiogenesis during acute wound healing. J Dermatol Sci 72(3):206–217. https://doi.org/10.1016/j.jdermsci.2013.07.008
4. Costa PZ, Soares R (2013) Neovascularization in diabetes and its complications. Unraveling the angiogenic paradox. Life Sci 92(22):1037–1045. https://doi.org/10.1016/j.lfs.2013.04.001
5. Lai L, Bohnsack BL, Niederreither K, Hirschi KK (2003) Retinoic acid regulates endothelial

cell proliferation during vasculogenesis. Development 130(26):6465–6474. https://doi.org/10.1242/dev.00887

6. Bohnsack BL, Lai L, Dolle P, Hirschi KK (2004) Signaling hierarchy downstream of retinoic acid that independently regulates vascular remodeling and endothelial cell proliferation. Genes Dev 18(11):1345–1358. https://doi.org/10.1101/gad.1184904
7. Gaetano C, Catalano A, Illi B, Felici A, Minucci S, Palumbo R, Facchiano F, Mangoni A, Mancarella S, Muhlhauser J, Capogrossi MC (2001) Retinoids induce fibroblast growth factor-2 production in endothelial cells via retinoic acid receptor alpha activation and stimulate angiogenesis in vitro and in vivo. Circ Res 88(4):E38–E47
8. Saito A, Sugawara A, Uruno A, Kudo M, Kagechika H, Sato Y, Owada Y, Kondo H, Sato M, Kurabayashi M, Imaizumi M, Tsuchiya S, Ito S (2007) All-trans retinoic acid induces in vitro angiogenesis via retinoic acid receptor: possible involvement of paracrine effects of endogenous vascular endothelial growth factor signaling. Endocrinology 148(3):1412–1423. https://doi.org/10.1210/en.2006-0900
9. Al Haj Zen A, Nawrot DA, Howarth A, Caporali A, Ebner D, Vernet A, Schneider JE, Bhattacharya S (2016) The retinoid agonist Tazarotene promotes angiogenesis and wound healing. Mol Ther 24 (10):1745–1759. https://doi.org/10.1038/mt.2016.153
10. Kubota Y, Kleinman HK, Martin GR, Lawley TJ (1988) Role of laminin and basement membrane in the morphological differentiation of human endothelial cells into capillary-like structures. J Cell Biol 107(4):1589–1598
11. Elkins JM, Fedele V, Szklarz M, Abdul Azeez KR, Salah E, Mikolajczyk J, Romanov S, Sepetov N, Huang XP, Roth BL, Al Haj Zen A, Fourches D, Muratov E, Tropsha A, Morris J, Teicher BA, Kunkel M, Polley E, Lackey KE, Atkinson FL, Overington JP, Bamborough P, Muller S, Price DJ, Willson TM, Drewry DH, Knapp S, Zuercher WJ (2016) Comprehensive characterization of the published kinase inhibitor set. Nat Biotechnol 34(1):95–103. https://doi.org/10.1038/nbt.3374
12. Bishop ET, Bell GT, Bloor S, Broom IJ, Hendry NF, Wheatley DN (1999) An in vitro model of angiogenesis: basic features. Angiogenesis 3(4):335–344
13. Mavria G, Vercoulen Y, Yeo M, Paterson H, Karasarides M, Marais R, Bird D, Marshall CJ (2006) ERK-MAPK signaling opposes rho-kinase to promote endothelial cell survival and sprouting during angiogenesis. Cancer Cell 9(1):33–44. https://doi.org/10.1016/j.ccr.2005.12.021
14. Sorrell JM, Baber MA, Caplan AI (2007) A self-assembled fibroblast-endothelial cell co-culture system that supports in vitro vasculogenesis by both human umbilical vein endothelial cells and human dermal microvascular endothelial cells. Cells Tissues Organs 186 (3):157–168. https://doi.org/10.1159/000106670
15. Duester G (2008) Retinoic acid synthesis and signaling during early organogenesis. Cell 134 (6):921–931. https://doi.org/10.1016/j.cell.2008.09.002

Chapter 6

Analysis of Retinoic Acid Receptor Signaling in Colorectal Cancer

Masamichi Imajo

Abstract

Retinoic acid receptor (RAR) signaling plays an important role in embryonic development and homeostasis of many tissues. At the cellular level, activation of RAR signaling often induces cell cycle arrest, differentiation, and apoptosis in many types of cells. Consequently, loss of normal RAR function in the presence of physiological levels of retinoic acid (RA) is often observed in cancers, and pharmacological reactivation of RAR signaling has been considered a promising strategy for cancer therapy and prevention. One of important mechanisms that regulate RAR activity in cancer cells is cross-talk with growth factor signaling, where activation of extracellular signal-regulated kinase (ERK) plays a major role in suppressing RAR transcriptional activity downstream of growth factor receptors. Conversely, strong activation of RAR can induce suppression of ERK activity by inducing expression of a phosphatase specific for ERK to exert tumor-suppressive activity in colorectal cancer. Here, we describe the basic methods to analyze interactions between RAR and ERK signaling in colorectal cancer cells.

Key words Retinoid, Retinoic acid receptor (RAR), Growth factor signaling, ERK MAP kinase, Colorectal cancer

1 Introduction

Retinoic acid (RA), an active derivative of vitamin A, plays divergent roles during embryonic development and homeostasis of many vertebrate tissues through regulating cell proliferation, differentiation, and apoptosis. Most of these effects are mediated by nuclear retinoid receptors, RA receptors (RARs), and retinoid X receptors (RsXRs) [1, 2]. In the absence of ligand, RAR recruits histone deacetylase (HDAC) complexes via binding to the nuclear receptor corepressor (NCoR) and the silencing mediator of retinoid and thyroid hormone receptors (SMRT). Ligand binding to RAR induces its conformational change, which causes replacement of corepressors by coactivators and transactivation of target genes. In addition to these classical corepressors and coactivators, several studies have also identified atypical coregulators including ligand-

Swapan K. Ray (ed.), *Retinoid and Rexinoid Signaling: Methods and Protocols*, Methods in Molecular Biology, vol. 2019, https://doi.org/10.1007/978-1-4939-9585-1_6,

dependent corepressors and ligand-independent regulators, adding another level of complexity to the regulation of RAR signaling [3–5].

Transactivation of RAR-target genes often induces cell cycle arrest, differentiation, and apoptosis in a wide variety of cell types. Thus, retinoids have tumor-suppressive activity, and loss of normal RAR function has been implicated in a diverse range of cancers [6–8]. Recently, we have shown that RAR and growth factor signaling mutually inhibit each other in colorectal cancer cells, where extracellular signal-regulated kinase (ERK), one of four groups of classical mitogen-activated protein kinase (MAPK), represses RAR transcriptional activity and is also suppressed by RAR-induced expression of MAP kinase phosphatase 4 (MKP4) [9]. Since constitutive activation of ERK is often caused by mutations in *RAS* or *RAF* genes during the progression of many cancers, ERK-dependent suppression of RAR likely represents a general mechanism through which cancer cells escape from the effects of retinoids. Moreover, ERK plays an essential role in cell proliferation and survival in many tissues, and RAR-mediated suppression of ERK might account for tumor-suppressive activity of retinoids. As RAR and ERK have been shown to regulate many physiological and pathological processes, interactions between the two signaling pathways would shed light on molecular bases for these processes. In this chapter, we describe basic methods that we used to analyze function and regulatory mechanisms of RAR signaling in colorectal cancer cells.

2 Materials

2.1 Cell Culture

1. Dulbecco's Modified Eagle Medium (DMEM) supplemented with 10% fetal bovine serum, 2 mM glutamine and antibiotics (penicillin/streptomycin).
2. Opti-MEM I Reduced Serum Medium.
3. Caco-2 human colon cancer cells (HTB-37, American Type Culture Collection (ATCC)).
4. HEK293T human embryonic kidney cells (CRL-3216, ATCC).

2.2 Reagent and Solutions

1. Phosphate buffered saline (PBS).
2. Dual-Luciferase Reporter Assay System (Promega).
3. pGL4.74[hRluc/TK] Vector (Promega).
4. $RARE_3$-Luc reporter plasmid (*see* **Note 1**).
5. FuGENE HD (Promega).
6. PEI MAX (Polysciences).

7. 5′-Biotinylated oligonucleotides. The sequence of oligonucleotides used is 5'-GGGTAGGGTTCACCGAAAGTTC ACTCG-3′ (sense strand).
8. 1× Annealing buffer: 125 mM Tris–HCl (pH 8.0), 40 mM EDTA, 100 mM NaCl.
9. Streptavidin agarose (Thermo Fisher Scientific).
10. Lysis buffer: 20 mM HEPES (pH 7.4), 150 mM NaCl, 2 mM EGTA, 1.5 mM $MgCl_2$, 1 mM phenylmethylsulfonyl fluoride (PMSF), 10 μg/mL aprotinin, 2 mM dithiothreitol, 10% glycerol, 0.5% Triton X-100.
11. SDS sample buffer: 62.5 mM Tris–HCl pH 6.8, 2% sodium dodecyl sulfate (SDS), 10% glycerol, 0.002% bromophenol blue, 5% 2-mercaptoethanol.
12. Retinoic acid.
13. Ch55.
14. 0.45 μm syringe filter (KURABO, Steradisc 25).
15. Dimethyl sulfoxide.
16. psPAX2 (Addgene, 12260).
17. pCMV-VSV-G-RSV-Rev (RIKEN, RDB04393).
18. CSII-EF-MCS (RIKEN, RDB04378).

2.3 Equipment

1. GloMax Discover System (Promega) or other luminometers.
2. CO_2 incubator.
3. Biosafety cabinet.
4. Microcentrifuge.

3 Methods

3.1 Analysis of the Effects of ERK Activation on RAR Transcriptional Activity in Colorectal Cancer Cells

Downstream of growth factor receptors, ERK MAP kinase is activated via the RAS/RAF/MEK kinase cascade (Fig. 1a). Since previous studies have identified many activating mutations in this pathway, these mutant proteins can be used to activate ERK. Among these mutant proteins, we used an oncogenic mutant of K-RAS (K-RAS^{V12}) and a constitutively active form of MEK1, which harbors phospho-mimetic mutations in two serine residues targeted by RAF (SDSE-MEK1) [10]. By transfecting expression plasmids for SDSE-MEK1 or K-RAS^{V12} together with a reporter plasmid for RAR, we analyzed the effects of ERK activation on RAR transcriptional activity. Among various colorectal cancer cell lines, we used Caco-2 human colon cancer cells in this study. This cell line still retains responsiveness to RA and growth factors, and therefore is suitable for analyzing interactions between RAR and ERK signaling.

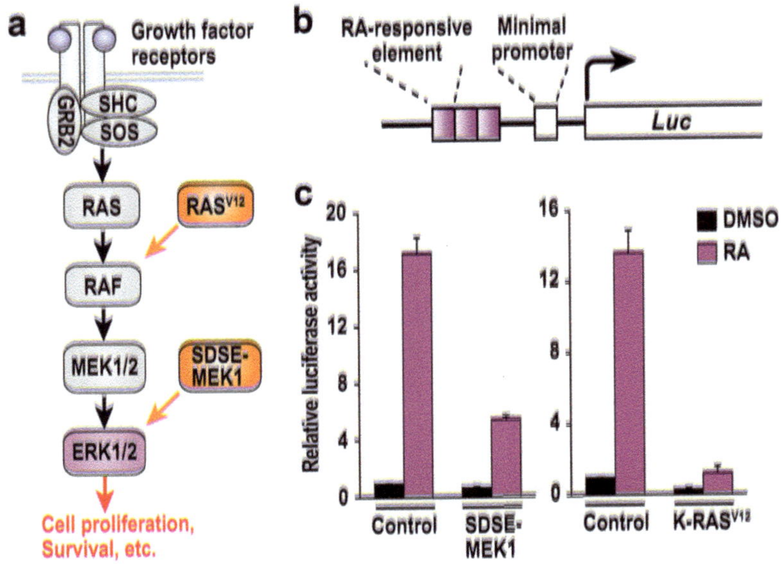

Fig. 1 Analysis of the effects of ERK activation on RAR transcriptional activity. (**a**) A schematic representation of the growth factor signaling pathway. (**b**) The structure of a reporter plasmid ($RARE_3$-Luc) to measure RAR transcriptional activity. (**c**) Examples of reporter assays with the $RARE_3$-Luc plasmid. The results show that SDSE-MEK1 or K-RAS^{V12}-induced ERK activation suppresses RAR-dependent transcription

1. Seed Caco-2 cells (0.25×10^5 cells per well) in the collagen type-I-coated 24 well microplate, and culture them for 24 h at 37 °C, 5% CO_2.
2. Transfection of DNA into Caco-2 cells by using a FuGENE HD transfection reagent can be performed according to the manufacture's protocol. In brief, first dilute DNA ($RARE_3$-Luc and pGL4.74[hRluc/TK] reporter plasmids: 0.1 μg, expression plasmids for SDSE-MEK or K-RAS^{V12}: 0.35 μg) with Opti-MEM I Reduced Serum Medium (26 μL). Add 1.65 μL of FuGENE HD to the solution, mix well, and incubate at room temperature for 15 min. After incubation, add the solution to Caco-2 cells and culture cells for 24 h (*see* **Note 2**). $RARE_3$-Luc expresses firefly luciferase under the control of RA-responsive elements (Fig. 1b), while pGL4.74[hRluc/TK] plasmid constitutively expresses *Renilla* luciferase and is used to normalize the transfection efficiency.
3. Add RA at a final concentration of 100 nM or vehicle (DMSO) to Caco-2 cells, and culture cells for 12 h.
4. Aspirate the culture medium, and wash cells with PBS once. (For long-term storage, store the plate at −30 °C until the measurement.)

5. Dilute 5× Passive Lysis Buffer with distilled water to make 1× working solution. Add 100 μL of 1× Passive Lysis Buffer to each well and incubate for 15 min. Then, transfer 20 μL of the cell lysates to 96 well microplate.
6. Measure firefly and *Renilla* Luciferase activity by luminometers. First, add 100 μL of Luciferase Assay Reagent II (LAR II) to the well, incubate for 2 s, and measure luminescence for 10 s to quantify the firefly luciferase activity. Next, add 100 μL of Stop & Glo Reagent to the same well, and measure the *Renilla* luciferase activity in the same manner.

Figure 1c shows an example of our analyses. The ratio of firefly luciferase activity over *Renilla* luciferase activity represents normalized RAR transcriptional activity under each condition. RA treatment increased RAR activity, whereas expression of SDSE-MEK or K-RAS^{V12} suppressed it.

3.2 In Vitro DNA Pull-Down Assay of RAR and RXR

Identification of RAR-target genes in cancer cells is critical to understand how RAR signaling exerts its tumor-suppressive activity. In this process, it is essential to determine the binding sites of RAR and RXR. In vitro DNA pull-down assay is an easy method to examine whether certain DNA sequences found in the promoter regions of genes of interest can be bound by RAR and RXR (Fig. 2a).

1. Seed HEK293T cells in 60 mm collagen type I-coated dishes, so that they will be 70–90% confluent next day, and culture for 24 h at 37 °C, 5% CO_2.

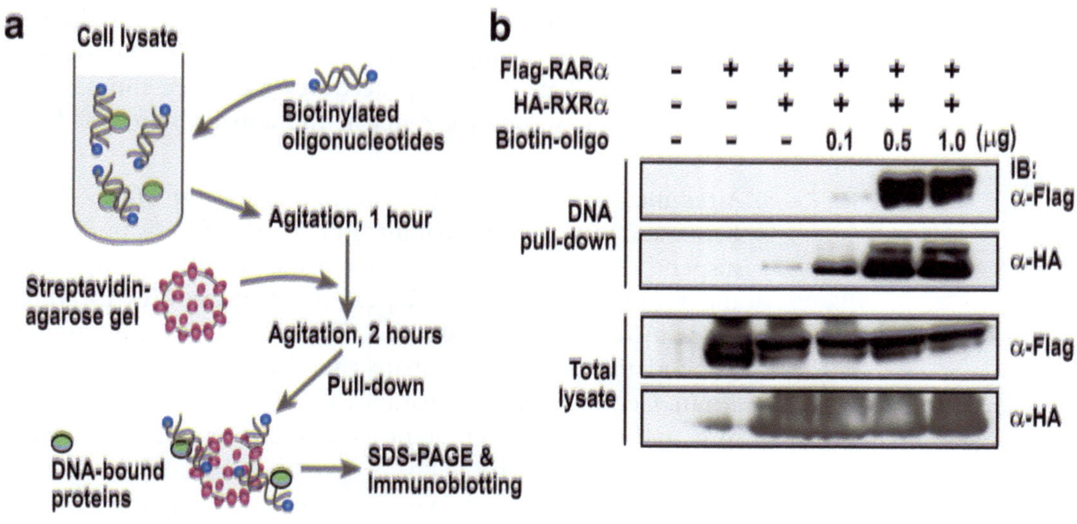

Fig. 2 In vitro DNA pull-down assay. (**a**) A schematic diagram of in vitro DNA pull-down assay. (**b**) An example of in vitro DNA pull-down assays. Flag-RARα and HA-RXRα expressed in HEK293T cells were precipitated with biotinylated oligonucleotides containing an RA-responsive element

2. Transfection of DNA into HEK293T cells can be performed by using a PEI MAX transfection reagent basically according to the manufacture's protocol. First, dilute expression plasmids for RARα and RXRα or control plasmids (4.5 μg each) with Opti-MEM I Reduced Serum Medium (570 μL). In another tube, dilute a PEI MAX transfection reagent (23 μL) with Opti-MEM I Reduced Serum Medium (570 μL), mix well, and incubate for 5 min. Then, mix the diluted DNA with the diluted PEI MAX solution, and incubate for 20 min. After incubation, add the transfection solution to the culture dishes, and culture cells for 3 h. Then, replace the culture medium with fresh one, and culture cells for 48 h.
3. Aspirate the culture medium, and wash cells with PBS once. From here, handle cells and cell lysates on ice or at 4 °C. Lyse cells by adding 300 μL of Lysis buffer. Transfer cell lysates to microtubes, vortex for 10 s, incubate on ice for 10 min, and then centrifuge at 10,000 × *g* for 10 min. Transfer supernatant to different microtubes as cell lysate samples.
4. Add 0.1–1.0 μg of pre-annealed 5′-biotinylated oligonucleotides containing a RA-responsive element to cell lysate samples, and incubate for 1 h with gentle agitation. For annealing of the oligonucleotides, dilute both sense and antisense oligonucleotides in 1× Annealing buffer to a concentration of 10 μM each, followed by sequential 5-min incubation at 95, 85, 75, 65, 55, 45, and 35 °C.
5. Add 20 μL of streptavidin agarose gel to the microtube, and incubate for 2 h with gentle agitation. Before adding, the gel should be washed more than 10 volumes of Lysis buffer at least three times as follows: add Lysis buffer to the gel, mix well by tapping, then centrifuge at 500 × *g* for 1 min and remove supernatant. Repeat these procedures for three times. After the last wash, suspend the gel in the equal volume of Lysis buffer.
6. After incubation, centrifuge the microtubes at 500 × *g* for 1 min and remove supernatant. Add 1 mL of Lysis buffer and suspend the gel. Then, centrifuge again at 500 × *g* for 1 min and remove supernatant. Repeat these procedures for three times. After the last wash, add 60 μL of SDS sample buffer, suspend the gel, and incubate the microtubes at 98 °C for 5 min. Centrifuge the microtubes at 500 × *g* at room temperature, and transfer supernatant to different microtubes as samples for immunoblotting.
7. Analyze the amount of RARα and RXRα precipitated with the oligonucleotides by performing SDS-PAGE and immunoblotting with a standard protocol. Figure 2b shows an example of our analyses. The amount of precipitated RARα and RXRα

proteins were increased with increasing concentration of the oligonucleotides. By changing the sequence of oligonucleotides, one can analyze whether certain DNA sequences of interest could be bound by RARα and RXRα.

3.3 Analysis of RAR-Dependent Transcription by Using a Dominant Negative Form of RAR

As described above, in vitro DNA pull-down assay is useful to identify putative RAR/RXR-binding motifs in the promoter regions of genes of interest. However, to identify novel RAR-target genes, it is essential to confirm that the genes are indeed expressed in an RAR-dependent manner. A dominant negative form of RARα (DN-RARα) can be used to address this issue. DN-RARα lacks a C-terminal domain essential for transactivation, but still harbors a DNA-binding domain (Fig. 3a). By competing with endogenous RAR for the binding to DNA, DN-RARα strongly suppresses RAR-dependent transcription (Fig. 3b). Here, we describe how we used DN-RARα to show that MAP kinase phosphatase 4 (MKP4) is a novel target gene of RAR. A schematic diagram of the whole procedure is shown in Fig. 3c. We first generated a lentivirus that expresses DN-RARα in HEK293T cells, and then used the virus for the transduction of DN-RARα to Caco-2 cells.

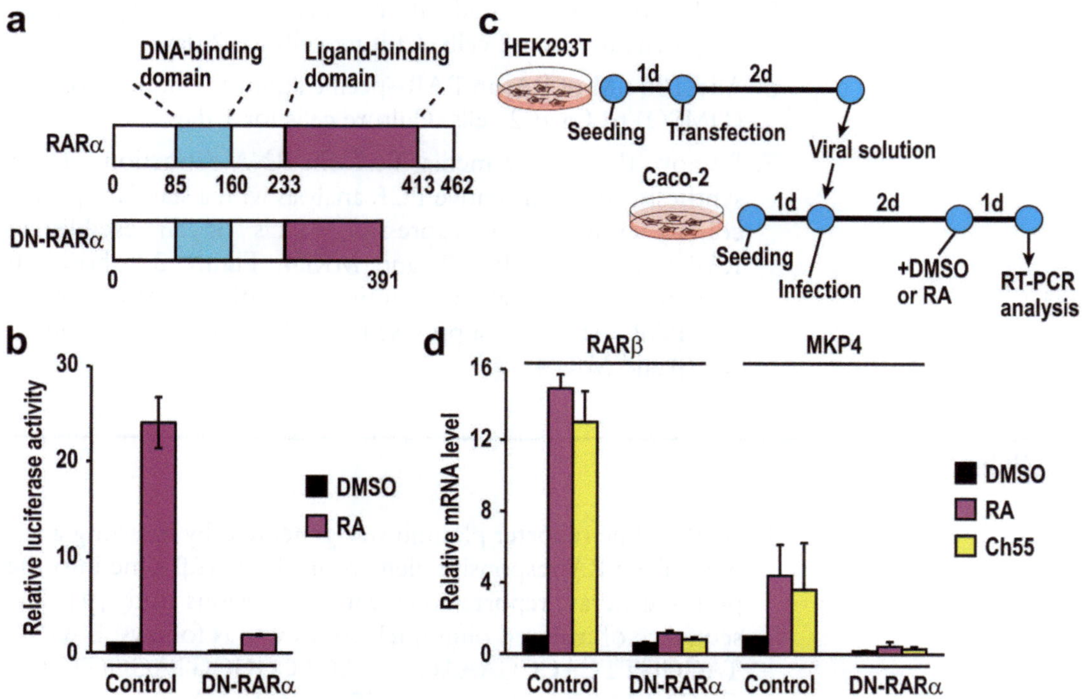

Fig. 3 Analysis of the effects of DN-RARα on RAR-target gene expression. (**a**) The structure of full-length and dominant-negative RARα. (**b**) A reporter assay with the $RARE_3$-Luc plasmid and an expression plasmid for DN-RARα. Expression of DN-RARα strongly suppressed RAR-dependent reporter gene expression. (**c**) Schedule of the lentivirus experiments. (**d**) An example of the results obtained from the experiments described in (**c**). Expression of DN-RARα suppressed the RA- and Ch55-induced expression of an RAR-target gene, RARβ, and MKP4

1. Seed HEK293T cells in 60 mm collagen type I-coated dishes, so that they will be 70–90% confluent next day, and culture for 24 h at 37 °C, 5% CO_2.
2. Perform transfection of DNA into HEK293T cells by using a PEI MAX transfection reagent. Dilute lentiviral packaging plasmids (psPAX2 and pCMV-VSV-G-RSV-Rev: 2.3 μg each) and transfer plasmids expressing DN-RARα (4.5 μg) (*see* **Note 3**) with Opti-MEM I Reduced Serum Medium (570 μL). In another tube, dilute PEI MAX (23 μL) with Opti-MEM I Reduced Serum Medium (570 μL), mix well, and incubate for 5 min. Then, mix the diluted DNA with the diluted PEI MAX solution and incubate for 20 min. After incubation, add the transfection solution to the culture dishes, and culture cells for 3 h. Then, replace the culture medium with fresh one, and culture cells for 48 h.
3. 1 day after transfection into HEK293T cells, seed Caco-2 cells in collagen type I-coated 12 well microplate (5.0×10^4 cells per well).
4. 2 days after transfection, collect the culture supernatant of HEK293T and filtrate with a 0.45 μm syringe filter to prevent potential contamination of HEK293T cells.
5. Add 1 mL of the filtrated culture supernatant and 10 μg/mL of Polybrene to Caco-2 cells. Culture cells for 2 days.
6. Add 100 nM of RA, an RAR-specific agonist, Ch55, or vehicle (DMSO) to Caco-2 cells. Culture cells for 1 day.
7. Remove the culture media. Perform RNA extraction, cDNA synthesis, and quantitative PCR analysis with a standard protocol to evaluate the expression levels of an established RAR-target gene, RARβ, and MKP4. Figure 3d shows an example of our analyses. Lentivirus-mediated expression of DN-RARα strongly suppressed RA-dependent induction of RARβ and MKP4.

4 Notes

1. $RARE_3$-Luc reporter plasmid was generated by inserting a trimer of the RA-responsive element in the RARβ gene into the pGL3 luciferase reporter vector in the previous study [9]. The sequence of inserted oligonucleotides was as follows: 5′-GGG TAGGGTTCACCGAAAGTTCACTCGGGGTAGGGTTCA CCGAAAGTTCACTCGGGGTAGGGTTCACCGAAAGTT CACTCG-3′
2. Usually, 24 h incubation is enough to achieve overexpression of transfected genes. If you want to examine the effects of knockdown of genes on RAR transcriptional activity, extend the

culture period up to 48–72 h until the target mRNA level is sufficiently decreased.

3. The lentiviral transfer vector expressing DN-RARα was generated by inserting a partial fragment of human RARα cDNA (encoding the 1–391 amino acid residues) (Fig. 3a) into the CSII-EF-MCS vector [11].

Acknowledgment

This work was supported by the Takeda Science Foundation and JSPS KAKENHI Grant Numbers 18H05100 and 18K06929.

References

1. Bastien J, Rochette-Egly C (2004) Nuclear retinoid receptors and the transcription of retinoid-target genes. Gene 328:1–16
2. Mark M, Ghyselinck NB, Chambon P (2006) Function of retinoid nuclear receptors. lessons from genetic and pharmacological dissections of the retinoic acid signaling pathway during mouse embryogenesis. Annu Rev Pharmacol Toxicol 46:451–480
3. Epping MT, Wang L, Edel MJ et al (2005) The human tumor antigen PRAME is a dominant repressor of retinoic acid receptor signaling. Cell 122:835–847
4. Fernandes I, Bastien Y, Wai T et al (2003) Ligand-dependent nuclear receptor corepressor LCoR functions by histone deacetylase-dependent and -independent mechanisms. Mol Cell 11:139–150
5. Imajo M, Nishida E (2010) Human Tribbles homolog 1 functions as a negative regulator of retinoic acid receptor. Genes Cells 15:1089–1097
6. Altucci L, Gronemeyer H (2001) The promise of retinoids to fight against cancer. Nat Rev Cancer 1:181–193
7. Freemantle SJ, Spinella MJ, Dmitrovsky E et al (2003) Retinoids in cancer therapy and chemoprevention: promise meets resistance. Oncogene 22:7305–7315
8. Ordóñez-Morán P, Dafflon C, Imajo M et al (2015) HOXA5 counteracts stem cell traits by inhibiting Wnt signaling in colorectal cancer. Cancer Cell 28:815–829
9. Imajo M, Kondoh K, Yamamoto T et al (2017) Antagonistic interactions between extracellular signal-regulated kinase mitogen-activated protein kinase and retinoic acid receptor signaling in colorectal cancer cells. Mol Cell Biol 37: e00012–e00017
10. Fukuda M, Gotoh I, Adachi M et al (1997) A novel regulatory mechanism in the mitogen-activated protein (MAP) kinase cascade. Role of nuclear export signal of MAP kinase kinase. J Biol Chem 272:32642–32648
11. Miyoshi H, Blömer U, Takahashi M et al (1998) Development of a self-inactivating lentivirus vector. J Virol 72:8150–8157

Chapter 7

Methods to Assess Activity and Potency of Rexinoids Using Rapid Luciferase-Based Assays: A Case Study with NEt-TMN

Peter W. Jurutka and Carl E. Wagner

Abstract

This chapter outlines the materials, methods, and procedures for the in vitro biological evaluation of retinoid-X-receptor (RXR) agonists including 6-(ethyl(5,5,8,8-tetramethyl-5,6,7,8-tetrahydronaphthalen-2-yl)amino)nicotinic acid (**NEt-TMN**), as well as several **NEt-TMN** analog compounds recently reported by our group. These methods have general applicability beyond this NEt-TMN case study, and can be employed to characterize and biologically evaluate other putative RXR agonists (rexinoids), and benchmarked against perhaps the most common rexinoid known as bexarotene (Bex), a drug awarded FDA approval for the treatment of cutaneous T-cell lymphoma in 1999 but that is also prescribed for non-small cell lung cancer and continues to be explored in multiple human cancer types. The side-effect profile of Bex treatment includes hypothyroidism and hypertriglyceridemia arising from the inhibition or activation of additional nuclear receptors that partner with RXR. Because rexinoids often exhibit selectivity for RXR activation, versus activating the retinoic-acid-receptor (RAR), rexinoid treatment avoids the cutaneous toxicity commonly associated as a side effect with retinoids. There are many examples of other potent rexinoids, where biological evaluation has contributed useful insight into qSAR studies on these compounds, often also benchmarked to Bex, as potential treatments for cancer. Because of differential pleiotropy in other pathways, even closely related rexinoids display unique side-effect and activity profiles. Notable examples of potent rexinoids in addition to Bex and NEt-TMN include CD3254, LGD100268, and 9-*cis*-UAB30. Indeed, the methods described herein to evaluate NEt-TMN and analogous rexinoids are generally applicable to a wider variety of potent, moderate, and even weak RXR ligands.

In terms of in vitro biological evaluation, methods for a rapid and preliminary assessment of rexinoid activity are described by employing a biologically relevant, RXR-responsive element (RXRE)-mediated transcription assay in mammalian cells. In addition, a second, more sensitive assay is also detailed that utilizes activation of RXR-RXR homodimers in the context of a mammalian two-hybrid (M2H) luciferase assay. Methods for applying the M2H assay at different rexinoid concentrations are further described for the determination of EC_{50} values for rexinoids from dose-response curves.

Key words Retinoid X receptor, Rexinoid, Nuclear receptor, Cutaneous T-cell lymphoma, Luciferase assay, Mammalian two hybrid assay

Swapan K. Ray (ed.), *Retinoid and Rexinoid Signaling: Methods and Protocols*, Methods in Molecular Biology, vol. 2019, https://doi.org/10.1007/978-1-4939-9585-1_7, © Springer Science+Business Media, LLC, part of Springer Nature 2019

1 Introduction

There are three known isoforms (α, β, and γ) [1, 2] for the retinoid X receptor (RXR), and at least one isoform is expressed in all human cell types. RXR-selective agonists, otherwise known as rexinoids, can be used as molecular signals that stimulate RXR to dimerize after binding to the RXR ligand binding domain (LBD) and act as a transcription factor for RXR responsive element (RXRE) controlled genes. RXR has been described as the "central" nuclear receptor, since it has the ability to partner with several other types of nuclear receptors, forming heterodimers that ultimately control those pathways. RXR partners the vitamin-D-receptor (VDR), the thyroid hormone receptor (TR), the peroxisome proliferator-activated receptor (PPAR), the liver-X-receptor (LXR), the retinoic acid receptor (RAR), the constitutive androstane receptor (CAR), the farnesoid-X-receptor (FXR), and the pregnane-X-receptor (PXR). The ability of a specific RXR-heterodimer to function as a transcription factor can depend on whether the RXR LBD is occupied or vacant, and two classifications for these cases have been described in literature (nonpermissive and permissive); however, additional subclassifications have been described under special circumstances. In permissive RXR-heterodimers, RXR may either possess a vacant or rexinoid-bound LBD. In nonpermissive RXR-heterodimers, the RXR-LBD must be vacant, and only the RXR-partner may bind to agonists selective for the activation of these heterodimers. TR, VDR, and RAR form generally nonpermissive RXR-heterodimers, though RXR-RAR has been reported to exhibit increased activity signaled by certain rexinoids and absent RAR-specific ligands [3]. Thus, RXR-RAR has been termed conditionally nonpermissive. In contradistinction to nonpermissive receptors, the FXR, LXR, and PPAR heterodimers formed with RXR have all been reported to be fully permissive.

At the time RXR was discovered and first reported [4], the naturally occurring (endogenous) ligand was not identified. Before a receptor ligand has been reported, the receptor is termed an "orphan" until it is "adopted" by a binding ligand. Whereas 9-*cis*-retinoic acid (9-*cis*-RA) has been reported to be an RXR-selective ligand, it is reasonable to expect that several additional RXR-selective endogenous ligands have since been identified, particularly since RXR participates in several permissive heterodimer pathways that regulate important functions such as metabolism and lipid homeostasis. Performing the structure-activity relationship (SAR) studies on the 9-*cis*-RA model compound has led to the reports of several potent rexinoids [5, 6]. The most important of these synthetic rexinoids is bexarotene (Bex) [7], a rexinoid that gained the Food and Drug Administration (FDA) approval in 1999 as a treatment for cutaneous T-cell lymphoma (CTCL). Bex

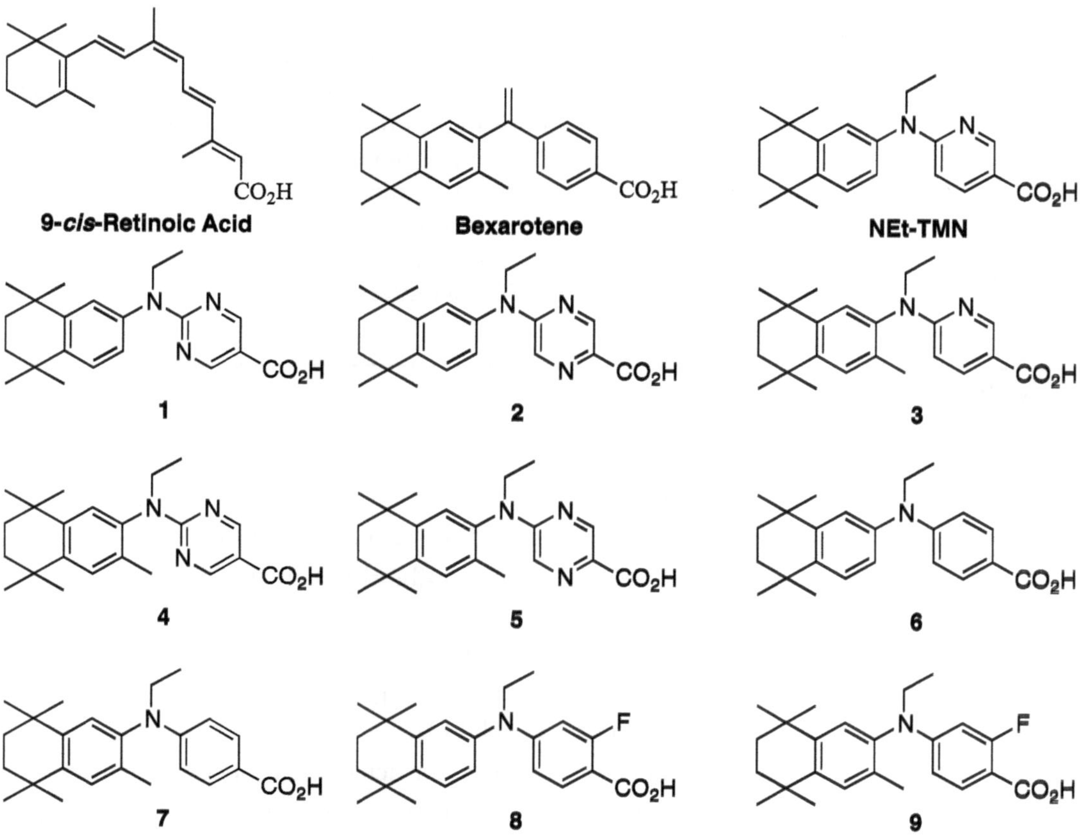

Fig. 1 Model retinoid-X-receptor agonists 9-*cis*-retinoic acid, bexarotene, and NEt-TMN, along with nine analogs of NEt-TMN

exhibits several features characteristic of 9-*cis*-RA (Fig. 1), and while other rexinoids such as LGD100268 [8] have displayed very favorable preclinical anticancer activity [9, 10], they haven't progressed in clinical trials for various reasons, such as lack of robust intellectual property (IP) due to length of time from initial disclosure and patent publication. Remarkably, Bex has been explored as a treatment for several cancers, including lung cancer [11, 12], glioblastoma multiforme [13], breast cancer [14], as well as numerous neurodegenerative conditions [15–18]. Two factors that contribute to the motivation to develop new rexinoids include the potential to reduce the side-effect profile associated with rexinoid treatment and to generate new compounds with stronger IP. Therefore, in addition to a straightforward strategy for creating new rexinoids from potent model compounds [19], a description of a standardized set of rapid methods for biologically evaluating such compounds would represent an important contribution to the field.

A potent rexinoid commonly referred to as **NEt-TMN** [20–24] recently served as a model compound for our group to synthesize and evaluate several closely related analogs, specifically compounds **1–9** (Fig. 1) [25]. Some of the results we have observed concerning two of these compounds in mouse models of breast cancer has motivated us to continue synthesizing analogs of **NEt-TMN** for evaluation of potency and therapeutic potential using in vitro assays. Thus, the purpose of this chapter will be to provide a codified set of methods that can be easily used to perform in vitro evaluations of potential RXR-selective molecules utilizing the set of **NEt-TMN** analogs as a case study.

2 Materials

2.1 Cell Culture Supplies and Reagents

1. High glucose Dulbecco's modified Eagle medium (DMEM) (Gibco, ThermoFisher).
2. Fetal Bovine Serum (FBS) (Atlanta Biologicals).
3. Penicillin-streptomycin antibiotics (Gibco, ThermoFisher).
4. Human embryonic kidney, 293 [HEK-293] (ATCC, CRL-1573) (*see* **Note 1**).
5. Cold-PBS (1 L): 8 g NaCl; 0.2 g KCl; 1.15 g $Na_2HPO_4 \cdot 7H_2O$; 0.2 g KH_2PO_4.
6. Trypsin–EDTA mixture, for detachment of cells in culture.
7. Trypan Blue solution (0.4%) and hemocytometer (for cell counting).
8. Inverted, phase-contrast microscope (Leica).
9. Cell culture incubator with 5% CO_2.
10. Cell culture plasticware (flasks, 24-well plates, pipets, etc.).

2.2 DNA Plasmids

1. RXRE-Luciferase plasmid (*see* **Note 2**).
2. Human RXRα mammalian expression vector, pSG5-hRXRα (*see* **Note 3**).
3. Human RXRα-AD (Stratagene M2H plasmid) (*see* **Note 4**).
4. Human RXRα-BD (Stratagene M2H plasmid) (*see* **Note 4**).
5. pFR-Luc Firefly M2H reporter plasmid (*see* **Note 5**).
6. Wild-type *Renilla* luciferase (*Rluc*) control reporter vector, pRL-Null (Promega).

2.3 Transient Transfection Assays

1. PolyJet in vitro DNA transfection reagent (SignaGen Laboratories) (*see* **Note 6**).
2. Transfection-grade (purified) plasmids (*see* subheading 2.2) (*see* **Note 7**).
3. Serum-free DMEM (no supplements).
4. Sterile, 15 mL conical tubes.
5. Vortexer.

2.4 Dosing

1. Dimethyl sulfoxide (DMSO) (Sigma), 100%, sterile-filtered, cell culture grade.
2. Ethanol (Sigma), cell culture grade.
3. Bexarotene (*see* **Note 8**).
4. Rexinoids to be assessed for activity (*see* **Note 9**).

2.5 Cell Lysis

1. Orbital shaker capable of rotation at 250–300 rpm.
2. 1× PBS wash (*see* Subheading 2.1).
3. Passive Lysis Buffer (PLB), 1×. Provided as a 5× concentrate (Promega) that is diluted to 1× with dH_2O.

2.6 Luciferase Assay

1. Luciferase Assay Buffer II (LARII) provided by manufacturer (Promega) must be prepared prior to use. Combine provided Luciferase Assay Buffer into the bottle containing Luciferase Assay Substrate. Mix thoroughly. Aliquot LARII into 1 mL aliquots and store at −80 °C. Thaw needed number of aliquots on day of assay (*see* **Note 10**).
2. Dilute 50× Stop & Glo Substrate to 1× with appropriate volume Stop & Glo buffer to make Stop & Glo Reagent (*see* **Note 11**).
3. 12 × 75 mm culture tubes (*see* **Note 12**).
4. FB12 Single Tube Luminometer (Titertek-Berthold) or similar luminometer (*see* **Note 13**).

2.7 Data Analysis

1. FB12 Titertek-Berthold Dual Assay PC analysis software (*see* **Note 14**).
2. Microsoft Excel.
3. GraphPad Prism (or equivalent) statistical software (*see* **Note 15**).

3 Methods

3.1 Cell Culture

1. Human 293 kidney cells are cultured and maintained in DMEM media supplemented with 10% FBS and antibiotics. The cell culture is incubated at 37 °C in a humidified atmosphere including 5% CO_2.
2. Cells should not exceed 90% confluency in cell culture flasks. When a sufficient number of cells are obtained, depending on size/scope of the experiment (*see* below), the cells are harvested from the flask via trypsinization, and an aliquot is counted in a hemocytometer to obtain total cell count. The cells are then gently pelleted at 300 × *g*.

3. The cell pellet is then resuspended at 75,000 cells/mL and the cells are plated in 24-well plates at 75,000 cells/well (1 mL per well) and maintained in DMEM/high glucose with 10% FBS and antibiotics (*see* Subheading 2.1). The total number of wells depends on the number of rexinoid compounds being tested.
4. In addition to the rexinoid compounds being evaluated, each experiment should also include a Bex positive control group, and vehicle (ethanol and/or DMSO) negative controls, and each "treatment group" should be performed in at least triplicate wells (*see* **Note 16**).
5. The cells are allowed to attach overnight and are observed under the microscope the following day to ensure proper adhesion, morphology and confluency (*see* **Note 17**).
6. All steps involving cell culture (above) and subsequent steps involving cell transfection and dosing are performed under aseptic conditions in a biological safety cabinet.

3.2 Plasmids

1. Depending on the type of assay being conducted, either the RXRE-mediated luciferase reporter protocol or the RXR-RXR M2H assay, different plasmids are employed as described below for each assay variant, but in either case high-quality plasmid DNA is purified via standard methods using a commercial kit such as the Qiagen Maxi-prep or equivalent.

3.2.1 RXRE Assay

1. The RXRE-based assay employs the following three plasmids:
   (a) RXRE-Luciferase plasmid, 250 ng/well (*see* **Note 2**)
   (b) Human RXRα mammalian expression vector, pSG5-hRXRα, 50 ng/well (*see* **Note 3**)
   (c) Wild-type *Renilla* luciferase (*Rluc*) control reporter vector, pRL-Null, 20 ng/well
2. The total amount of each plasmid required depends on the total number of wells in each experiment; thus, simply multiply the number of wells by the ng/well required (above) to calculate total plasmid amount (plus 10%).

3.2.2 RXR-RXR M2H Assay

1. The RXR M2H assay employs the following four plasmids:
   (a) Human RXR-AD, 50 ng/well (*see* **Note 4**)
   (b) Human RXR-BD, 50 ng/well (*see* **Note 4**)
   (c) pFR-Luc firefly M2H reporter plasmid (*see* **Note 5**)
   (d) Wild-type *Renilla* luciferase (*Rluc*) control reporter vector, pRL-Null, 20 ng/well
2. The total amount of each plasmid required depends on the total number of wells in each experiment; thus, simply multiply the number of wells by the ng/well required (above) to calculate total plasmid amount (plus 10%).

3.3 Transient Transfection Assays

1. Calculate the total volume of DMEM medium needed for transfection. Use only DMEM without supplements (no FBS, antibiotics). Multiply the number of wells in the experiment times 100 μL (plus 10%). Pipet this amount of DMEM into a microfuge or 15 mL conical tube (depending on the volume) and label DMEM-TR (transfection reagent).
2. Pipet an identical amount of DMEM into a second microfuge or 15 mL conical tube and label DMEM-DNA.
3. Add the appropriate amount of transfection reagent to the tube labeled DMEM-TR (*see* **Note 18**).
4. Add the appropriate amount of the DNA plasmids to the tube labeled DMEM-DNA. The plasmids used are those described in Subheading 3.2, **step 1** (RXRE assay) or Subheading 3.2, **step 2** (M2H assay) (*see* **Note 19**).
5. Vortex the tube DMEM-TR for 10 s. Add the tube labeled DMEM-DNA *into* the tube labeled DMEM-TR. After addition, pipet up and down gently to mix, but do not vortex. Incubate at room temperature for 20 min (*see* **Note 20**).
6. Remove the cell culture media from each well of the plate via aspiration and add 200 μL of the DNA-transfection reagent mixture from **step 5** (*see* **Note 21**).
7. Check cell morphology and record confluency after examination of multiple random wells under the microscope.
8. Incubate overnight at 37 °C in a humidified atmosphere including 5% CO_2.

3.4 Dosing

1. Prepare DMEM media supplemented with 10% FBS and antibiotics (*see* **Note 22**).
2. Supplement the appropriate volume of the DMEM from **step 1** with either: (1) 0.1% ethanol vehicle (for those rexinoids dissolved in ethanol), (2) 0.1% DMSO vehicle (for those rexinoids dissolved in DMSO), or (3) the desired concentration of rexinoid compounds (*see* below, depending on RXRE or M2H assay). Include bexarotene as a positive control in all assays.
3. Regardless of the rexinoid concentrations used, the stock concentration of rexinoid is usually 1000× and a 1:1000 dilution is accomplished into the DMEM media (*see* **Note 23**).
4. After preparation of the DMEM/rexinoid, add 1.0 mL of each mixture to the appropriate wells and incubate for 24 h (*see* **Note 24**).

3.4.1 RXRE Assay

1. For most rexinoids (including those of the NEt-TMN class, Fig. 1) that are predicted to bind with at least some affinity to the RXR, a useful starting concentration employed in the RXRE assay is 100 nM.

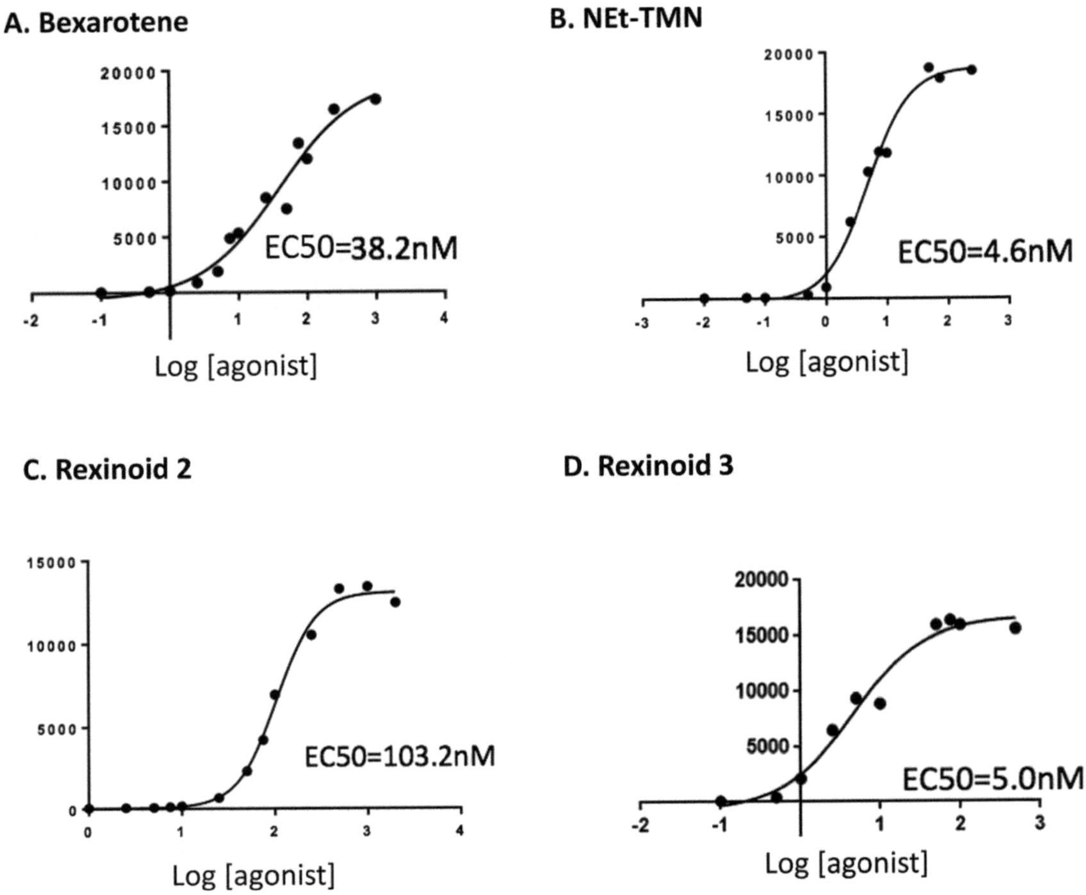

Fig. 2 Data from M2H assays across multiple rexinoid concentrations are plotted as the log of [rexinoid concentration] versus luciferase activity. The EC50 is estimated from these plots and is defined as the concentration of rexinoid that generates 50% of the maximal activity in the M2H assay for that particular compound

2. The assay can then be repeated at one log lower or higher concentration to obtain additional insights into RXR activity of selected rexinoids at different concentrations (*see* Reference [25]).

3.4.2 RXR-RXR M2H Assay

1. The M2H assay is conducted with a wide range of rexinoid concentrations that are then used to construct a dose-response curve (Fig. 2) and subsequent extrapolation of EC_{50} estimates for rexinoid binding to RXR (*see* Subheading 3.7).
2. For M2H assays, the range of rexinoid concentration that yield adequate dose-response curves includes 1.0, 2.5, 5.0, 7.5, 10, 25, 50, 75, 100, 250, 500 nM and 1, 2, 3 μM.

3.5 Cell Lysis

1. 1× Passive Lysis Buffer (PLB) is prepared from the manufacturer's concentrate (Subheading 2.5). Prepare 160 μL of 1× PLB/well to be analyzed (*see* **Note 25**).
2. Aspirate the media from each well, rinse once with PBS, aspirate to dryness, and add 150 μL of 1× PLB (*see* **Note 17**).
3. Incubate with shaking on an orbital shaker for 20 min at room temperature at 250–300 rpm (*see* **Note 26**).
4. After 20 min, transfer the lysates to labeled microfuge tubes and spin for 5 min at max speed. The cleared lysates are used in the luciferase assay (Subheading 3.6)

3.6 Luciferase Assay

1. All luciferase reagents must be thawed and allowed to equilibrate at room temperature.
2. The LARII reagent is prepared as described in Subheading 2.6, and the volume required is 25 μL per the number of wells being analyzed.
3. The Stop & Glo reagent is prepared as described in Subheading 2.6, and the volume required is 25 μL per the number of wells being analyzed.
4. Aliquot 25 μL each of LARII into 12 × 75 mm luminometer tubes, but only aliquot six tubes at a time (*see* **Note 12**).
5. Transfer 5 μL of the appropriate centrifuged cell lysate into 25 μL of LARII solution in each luminometer tube. Mix by pipetting up/down and place the tube directly in the luminometer (*see* **Note 27**).
6. After the firefly luciferase reading is recorded, remove the tube from the luminometer and add 25 μL Stop & Glo reagent, mix by pipetting up/down, and place the tube back in the luminometer to measure the Renilla luciferase (*see* **Note 28**).

3.7 Data Analysis

1. Obtain the data readout from the luminometer (via the FB12 PC software) and then transfer the data to a Microsoft Excel spreadsheet.
2. The luminescence generated by the rexinoid-inducible Firefly luciferase is divided by the luminescence from the constitutively active *Renilla* luciferase to obtain a normalized value that accounts for transfection efficiency, cell toxicity, and cell death due to ligand treatment.
3. The mean ratio of Firefly/*Renilla* of 3–6 well replicates is calculated for each experimental treatment group (rexinoid treated wells, Bex control, vehicle control(s), etc.), and standard deviation values are computed.
4. All RXRE-based data are reported as the average of all wells (3–6) within one treatment group and denote one representative experiment. In general, three or more independent

biological replicates are processed to generate a global mean RXRE activity. The mean activity (−/+ SD) of each rexinoid is expressed as a percentage of the Bex reference group, which is arbitrarily set to 100% (*see* Reference [25]).

5. Data for M2H assays across multiple rexinoid concentrations (*see* Subheading 3.4.2) are plotted as the log of [rexinoid concentration] versus percent maximal activity. The EC_{50} is estimated from these plots (Fig. 2) and is defined as the concentration of rexinoid that generates 50% of the maximal activity in the M2H assay for that particular compound (*see* Reference [25]). Each M2H assay is carried out with triplicate wells for every rexinoid concentration tested, and each M2H is subjected to at least 3 biological replicates in order to generate a mean (−/+SD) EC_{50} value.

4 Notes

1. We have also employed other cell lines besides HEK-293 but have found that the HEK-293 kidney cells represent an easy-to-grow and transfect cell line that is suitable for the in vitro assays described herein. However, other cells lines can also be utilized.
2. The RXRE (AAAATGAACTGTGACCTGTGACCTGTGACCTGTGAC; RXRE half-site underlined) from the naturally occurring responsive element in the rat cellular retinol binding protein II gene is cloned into pMCS-Luc to generate the RXRE-luciferase reporter vector that is employed in these protocols.
3. The human RXRα isoform is used throughout the screening and activity assays. However, there are three known isoforms (α, β, and γ) for RXR; thus, additional assays may be conducted by substituting RXRα with one of the other isoforms in the assays described to generate RXR-isoform-specific rexinoid activity profiles.
4. While we routinely use the original M2H system from Stratagene, other M2H-based assays should yield similar results (*see* **Note 5**).
5. The pFR-Luc has 5 tandem GAL4 DNA binding sites for the BD domain. This is important because it allows for synergistic activation of the luciferase reporter gene and leads to high sensitivity. If other M2H systems are utilized, they should also include multiple tandem DNA binding sites in the reporter vector.
6. Any liposome-mediated transfection reagent can be used, especially if HEK-293 cells are utilized since they are easy to

transfect. If other cell lines are used, then the transfection conditions (and reagent) may require further optimization.

7. We employ Qiagen plasmid preps for purification of transfection-grade (quality) DNA, but other suitable purification systems can be used as long as they generate transfection-quality plasmid DNA.
8. Bex is one of the most common and well-studied synthetic rexinoids, and it is used in our assay systems as the "positive" control to benchmark any putative tested rexinoids (Bex $EC_{50} = 28$, 25, and 20 nM for RXRα, β, and γ, respectively). It is commercially available from Cayman Chemical (Ann Arbor, Michigan) and other suppliers.
9. Bex and other rexinoids to be tested are routinely dissolved in the appropriate solvent (generally ethanol or DMSO) at a stock concentration of 10^{-4} M. This allows for a 1:1000 dilution into the cell culture media to generate a 10^{-7} M final concentration of rexinoid, which is an appropriate concentration for initial testing/screening. After preliminary testing, additional 1000× stock concentrations may be generated to allow for quick and simple dilution into cell culture media.
10. Thaw enough LARII for the entire assay. Generally, 25 μL of LARII per the number of wells being analyzed (plus 10% excess for any possible repeats). For a typical 2× 24-well plate experiment (48 wells total), 1325 μL of LARII are prepared.
11. The amount of Stop & Glo Reagent required is 25 μL per the number of wells being analyzed (plus small excess for any possible repeats). For a typical 2× 24-well plate experiment (48 wells total), add 25 μL of 50× Stop & Glo Substrate to 1225 μL of Stop & Glo buffer to make 1250 μL Stop & Glo Reagent (enough for 50 wells).
12. The 12 × 75 mm culture tubes are only required for single tube luminometers. Ensure that the tubes fit snugly (but not too tightly) into the tube holder of the instrument.
13. The assay can also be run directly in the plate if a luminescence plate reader is available, although such instruments are typically more expensive that a single tube luminometer.
14. The FB12 Titertek-Berthold Dual Assay PC analysis software allows for data collection in the dual assay format using both Firefly- and Renilla-generated luminescence.
15. Other comparable statistical software packages may also be utilized.
16. It is generally not recommended to manually perform more than 2–3 24-well plates at a time (72 wells total) because the assay becomes cumbersome and difficult to manage with greater than 72-wells.

17. It is important to observe the cells under the microscope each day and record cell adhesion, morphology, and confluency properties during the course of the entire experiment, not just on the first day.
18. The volume of transfection reagent (in our case PolyJet) required is 1.5 μL per well (plus 10%). For a typical 2-plate (48-well) experiment, we add 79.2 μL of PolyJet to 5.28 mL of DMEM in the tube labeled DMEM-TR. However, this volume may be different if using another brand of liposome-directed transfection reagent or if not using the 24-well/plate format.
19. The amount of DNA added (per well) is listed in section 3.2.1 (RXRE) and 3.2.2 (M2H). As a typical example, for an RXRE assay in a 2-plate (48-well total) experiment, add 13.2 μg RXRE-Luc plasmid, 2.64 μg pSG5-hRXRα, and 1.06 μg pRL-Null to 5.28 mL of DMEM in the tube labeled DMEM-DNA (all DNA amounts include 10% excess).
20. It is critical to add the tube labeled DMEM-DNA into the tube labeled DMEM-TR to facilitate maximal DNA-micelle formation. During this incubation period, do not disturb the contents of the mixture.
21. Add the 200 μL of DNA-micelle mix slowly and gently to each well. Since this may take some time, we generally only aspirate the media from one row (6-wells) at a time and then add the transfection mix. This also ensures that downstream wells (with cells) do not "dry out" during the pipetting of the transfection mix.
22. The volume of DMEM:FBS required is 1.0 mL per well. However, separate tubes must be prepared for each "treatment" group. In a typical 2-plate (48-well) experiment with triplicate wells per treatment, there are generally 16 treatment groups (*see* **Note 23**).
23. If each treatment group is performed in triplicate wells, then we usually prepare a 15 mL sterile conical tube that contains 3.5 mL of DMEM:FBS and we add 3.5 μL (1:1000 dilution) of each test compound (rexinoid) or controls (Bex, solvent, etc.).
24. The 1.0 mL of DMEM/rexinoid is added to each well after the transfection mix (from the previous 24-h transfection incubation) has been removed by aspiration. Alternatively, the 1.0 mL of DMEM/rexinoid can be added directly to the 200 μL of transfection mix already in the well (without aspiration), but the amount of ligand must be adjusted (increased) since the total volume in each well will instead be 1.2 mL.
25. In a typical 2-plate (48-well) experiment, we prepare a total of 7.68 mL of 1× PLB by mixing 1536 μL 5× PLB with 6144 μL dH_2O.

26. This step may be omitted if sensitivity is not an issue, although we have observed that the orbital shaking step enhances cellular lysis and leads to more consistent results.
27. The luminometer should be set to a delay time of 3 s and a counting time of 10 s for all samples.
28. When adding the LARII and Stop & Glo reagent, mixing, and placing the tubes in the luminometer, it is important to treat each sample in the same way. Thus, the operator should establish a "rhythm" such that the samples are consistently mixed, placed in the luminometer, and then removed in a uniform manner across all samples.

References

1. Mangelsdorf DJ, Evans RM (1994) The Retinoids. Academic Press, Orlando, FL
2. Leid M, Kastner P, Chambon P (1992) Multiplicity generates diversity in the retinoic acid signalling pathways. Trends Biochem Sci 17:427–433
3. Lala DS, Mukherjee R, Schulman IG, Koch SSC, Dardashti LJ, Nadzan AM et al (1996) Activation of specific RXR heterodimers by an antagonist of RXR homodimers. Nature 383:450
4. Mangelsdorf DJ, Ong ES, Dyck JA, Evans RM (1990) Nuclear receptor that identifies a novel retinoic acid response pathway. Nature 345:224
5. Jong L, Lehmann JM, Hobbs PD, Harlev E, Huffman JC, Pfahl M et al (1993) Conformational effects on retinoid receptor selectivity. 1. Effect of 9-double bond geometry on retinoid X receptor activity. J Med Chem 36:2605–2613
6. Dawson MI, Jong L, Hobbs PD, Cameron JF, Chao W-r, Pfahl M et al (1995) Conformational effects on retinoid receptor selectivity. 2. Effects of retinoid bridging group on retinoid X receptor activity and selectivity. J Med Chem 38:3368–3383
7. Boehm MF, Zhang L, Badea BA, White SK, Mais DE, Berger E et al (1994) Synthesis and structure-activity relationships of novel retinoid X receptor-selective retinoids. J Med Chem 37:2930–2941
8. Boehm MF, Zhang L, Zhi L, McClurg MR, Berger E, Wagoner M et al (1995) Design and synthesis of potent retinoid X receptor selective ligands that induce apoptosis in leukemia cells. J Med Chem 38:3146–3155
9. Liby K, Rendi M, Suh N, Royce DB, Risingsong R, Williams CR et al (2006) The combination of the rexinoid, LG100268, and a selective estrogen receptor modulator, either arzoxifene or acolbifene, synergizes in the prevention and treatment of mammary tumors in an estrogen receptor–negative model of breast cancer. Clin Cancer Res 12:5902
10. Cao M, Royce DB, Risingsong R, Williams CR, Sporn MB, Liby KT (2016) The rexinoids LG100268 and LG101506 inhibit inflammation and suppress lung carcinogenesis in A/J mice. Cancer Prev Res 9:105
11. Yen W-C, Corpuz MR, Prudente RY, Cooke TA, Bissonnette RP, Negro-Vilar A et al (2004) A selective retinoid x receptor agonist bexarotene (targretin) prevents and overcomes acquired paclitaxel (taxol) resistance in human non–small cell lung cancer. Clin Cancer Res 10:8656
12. Dragnev KH, Petty WJ, Shah SJ, Lewis LD, Black CC, Memoli V et al (2007) A proof-of-principle clinical trial of bexarotene in patients with non–small cell lung cancer. Clin Cancer Res 13:1794
13. Heo J-C, Jung T-H, Lee S, Kim HY, Choi G, Jung M et al (2016) Effect of bexarotene on differentiation of glioblastoma multiforme compared with ATRA. Clin Exp Metastasis 33:417–429
14. W-c Y, Prudente RY, Lamph WW (2004) Synergistic effect of a retinoid X receptor-selective ligand bexarotene (LGD1069, Targretin) and paclitaxel (Taxol) in mammary carcinoma. Breast Cancer Res Treat 88:141–148
15. Cramer PE, Cirrito JR, Wesson DW, Lee CYD, Karlo JC, Zinn AE et al (2012) ApoE-directed therapeutics rapidly clear β-amyloid and reverse deficits in AD mouse models. Science 335:1503
16. Tai LM, Bilousova T, Jungbauer L, Roeske SK, Youmans KL, Yu C et al (2013) Levels of

soluble apolipoprotein E/amyloid-β (Aβ) complex are reduced and oligomeric Aβ increased with APOE4 and alzheimer disease in a transgenic mouse model and human samples. J Biol Chem 288:5914–5926

17. McFarland K, Spalding TA, Hubbard D, Ma J-N, Olsson R, Burstein ES (2013) Low dose bexarotene treatment rescues dopamine neurons and restores behavioral function in models of Parkinson's disease. ACS Chem Nerosci 4:1430–1438
18. Cummings JL, Zhong K, Kinney JW, Heaney C, Moll-Tudla J, Joshi A et al (2016) Double-blind, placebo-controlled, proof-of-concept trial of bexarotene Xin moderate Alzheimer's disease. Alzheimer's Res Ther 8(4)
19. Wagner CEJ, Jurutka PW Methods to generate an array of novel rexinoids by SAR on a potent retinoid X receptor agonist: a case study with NEt-TMN. Methods Mol Biol. https://doi.org/10.2174/1568026616666160617091559
20. Fujii S, Ohsawa F, Yamada S, Shinozaki R, Fukai R, Makishima M et al (2010) Modification at the acidic domain of RXR agonists has little effect on permissive RXR-heterodimer activation. Bioorg Med Chem Lett 20:5139–5142
21. Ohsawa F, Morishita K-i, Yamada S, Makishima M, Kakuta H (2010) Modification at the lipophilic domain of RXR agonists differentially influences activation of RXR heterodimers. ACS Med Chem Lett 1:521–525
22. Kakuta H, Yakushiji N, Shinozaki R, Ohsawa F, Yamada S, Ohta Y et al (2012) RXR partial agonist CBt-PMN exerts therapeutic effects on type 2 diabetes without the side effects of RXR full agonists. ACS Med Chem Lett 3:427–432
23. Kakuta H, Ohsawa F, Yamada S, Makishima M, Tai A, Yasui H et al (2012) Feasibility of structural modification of retinoid X receptor agonists to separate blood glucose-lowering action from adverse effects: studies in KKAy type 2 diabetes model mice. Biol Pharm Bull 35:629–633
24. Ohsawa F, Yamada S, Yakushiji N, Shinozaki R, Nakayama M, Kawata K et al (2013) Mechanism of retinoid X receptor partial agonistic action of 1-(3,5,5,8,8-pentamethyl-5,6,7,8-tetrahydro-2-naphthyl)-1H-benzotriazole-5-carboxylic acid and structural development to increase potency. J Med Chem 56:1865–1877
25. Heck MC, Wagner CE, Shahani PH, MacNeill M, Grozic A, Darwaiz T et al (2016) Modeling, synthesis, and biological evaluation of potential retinoid X receptor (RXR)-selective agonists: analogues of 4-[1-(3,5,5,8,8-pentamethyl-5,6,7,8-tetrahydro-2-naphthyl)ethynyl]benzoic acid (Bexarotene) and 6-(ethyl(5,5,8,8-tetrahydronaphthalen-2-yl)amino)nicotinic acid (NEt-TMN). J Med Chem 59:8924–8940

Chapter 8

Methods to Generate an Array of Novel Rexinoids by SAR on a Potent Retinoid X Receptor Agonist: A Case Study with NEt-TMN

Carl E. Wagner and Peter W. Jurutka

Abstract

The methods described in this chapter concern procedures for the design, synthesis, and in vitro biological evaluation of an array of potent retinoid-X-receptor (RXR) agonists employing 6-(ethyl(5,5,8,8-tetramethyl-5,6,7,8-tetrahydronaphthalen-2-yl)amino)nicotinic acid (**NEt-TMN**), and recently reported **NEt-TMN** analogs, as a case study. These methods have been extensively applied beyond the present case study to generate several analogs of other potent RXR agonists (rexinoids), particularly the RXR agonist known as bexarotene (Bex), a Food and Drug Administration (FDA) approved drug for cutaneous T-cell lymphoma that is also often prescribed, off-label, for breast, lung, and other human cancers. Common side effects with Bex treatment include hypertriglyceridemia and hypothyroidism, because of off-target activation or inhibition of other nuclear receptor pathways impacted by RXR. Because rexinoids are often selective for RXR, versus the retinoic-acid-receptor (RAR), cutaneous toxicity is often avoided as a side effect for rexinoid treatment. Several other potent RXR agonists, and their analogs, have been reported in the literature and rigorously evaluated (often in comparison to Bex) as potential cancer therapeutics with unique activity and side-effect profiles. Some of the more prominent examples include LGD100268, CD3254, and 9-*cis*-UAB30, to name only a few. Hence, the methods described herein are more widely applicable to a diverse array of RXR agonists.

In terms of design, the structure-activity relationship (SAR) study is usually performed by modifying three distinct areas of the rexinoid base structure, either of the nonpolar or polar sides of the rexinoid and/or the linkage that joins them. For the synthesis of the modified base-structure analogs, often identical synthetic strategies used to access the base-structure are applied; however, reasonable alternative synthetic routes may need to be explored if the modified analog intermediates encounter bottlenecks where yields are negligible for a given step in the base-structure route. In fact, this particular problem was encountered and successfully resolved in our case study for generating an array of **NEt-TMN** analogs.

Key words Retinoid X receptor, Rexinoid, Nuclear receptor, Cutaneous T-cell lymphoma, Synthetic methods, Luciferase assay, Mammalian two hybrid assay

Swapan K. Ray (ed.), *Retinoid and Rexinoid Signaling: Methods and Protocols*, Methods in Molecular Biology, vol. 2019, https://doi.org/10.1007/978-1-4939-9585-1_8, © Springer Science+Business Media, LLC, part of Springer Nature 2019

1 Introduction

The retinoid-X-receptor (RXR) is a nuclear receptor expressed as one of three identified isoforms (α, β, and γ) [1, 2] in every human tissue type. In the presence of an RXR-selective agonist—a rexinoid—that binds to RXR's ligand binding domain (LBD), RXR can dimerize and function as a transcription factor for genes controlled by the RXR responsive element. RXR is a rather unique nuclear receptor because it also partners with other nuclear receptors to form heterodimers that may ultimately regulate those receptor pathways. The receptors with which RXR partners include the thyroid hormone receptor (TR), the vitamin-D-receptor (VDR), the liver-X-receptor (LXR), the peroxisome proliferator-activated receptor (PPAR), the retinoic acid receptor (RAR), the farnesoid X receptor (FXR), the constitutive androstane receptor (CAR), and the pregnane X receptor (PXR). There are two primary classifications for the LBD occupancy status for RXR-heterodimers that function as transcriptional factors, permissive and nonpermissive, though nuanced activity has led to additional subclassifications. For permissive RXR-heterodimers, RXR may possess a rexinoid in the LBD. For nonpermissive RXR-heterodimers, the RXR-LBD cannot be occupied and only agonists selective for the RXR-partner can activate these heterodimers. VDR, TR, and RAR generally form nonpermissive heterodimers with RXR, though the RXR-RAR heterodimer has been observed to function in the presence of certain rexinoids and absence of RAR-specific ligands [3]. Hence, the RXR-RAR heterodimer has been designated conditionally nonpermissive in this respect. In contrast to VDR, TR, and RAR, the PPAR, LXR, and FXR heterodimers with RXR are all fully permissive.

When RXR was first identified [4], the endogenous ligand that signals its activation was not yet known. While 9-*cis*-retinoic acid (9-*cis*-RA) was later reported to bind and activate RXR, and 9-*cis*-13,14-dihydroretinoic acid has been identified as an endogenous RXR ligand in mice, there are other endogenous RXR ligands that signal RXR or RXR-heterodimer pathways. Several structure-activity relationship studies for rexinoids developed using 9-*cis*-RA as a base model have been reported [5, 6]. Perhaps the most prominent synthetic rexinoid described in these studies is bexarotene (Bex) [7], a rexinoid that was approved to treat cutaneous T-cell lymphoma (CTCL) in 1999. Bex possesses several structural features in common with 9-*cis*-RA as comparison of their structures, side-by-side, clearly shows (Fig. 1). In addition to treating CTCL, Bex has also been explored in treating lung cancer [8, 9], breast cancer [10], glioblastoma multiforme [11], and it has recently been examined in neurodegenerative diseases [12–15], as well. Other structurally similar rexinoids have appeared, and some—such as LGD100268 [16]—have shown greater efficacy in

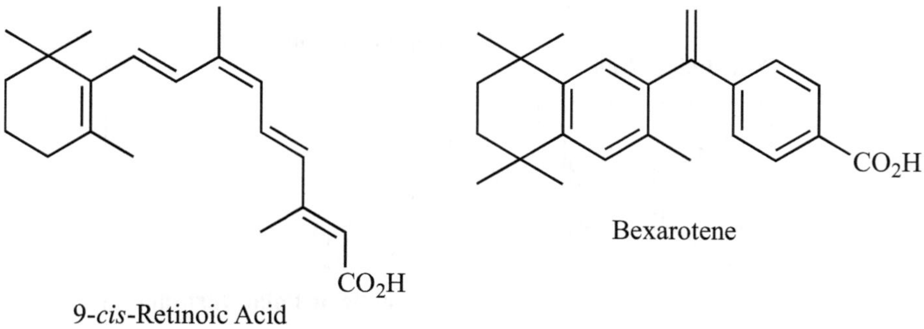

Fig. 1 Model retinoid-X-receptor agonists: 9-*cis*-retinoic acid and bexarotene

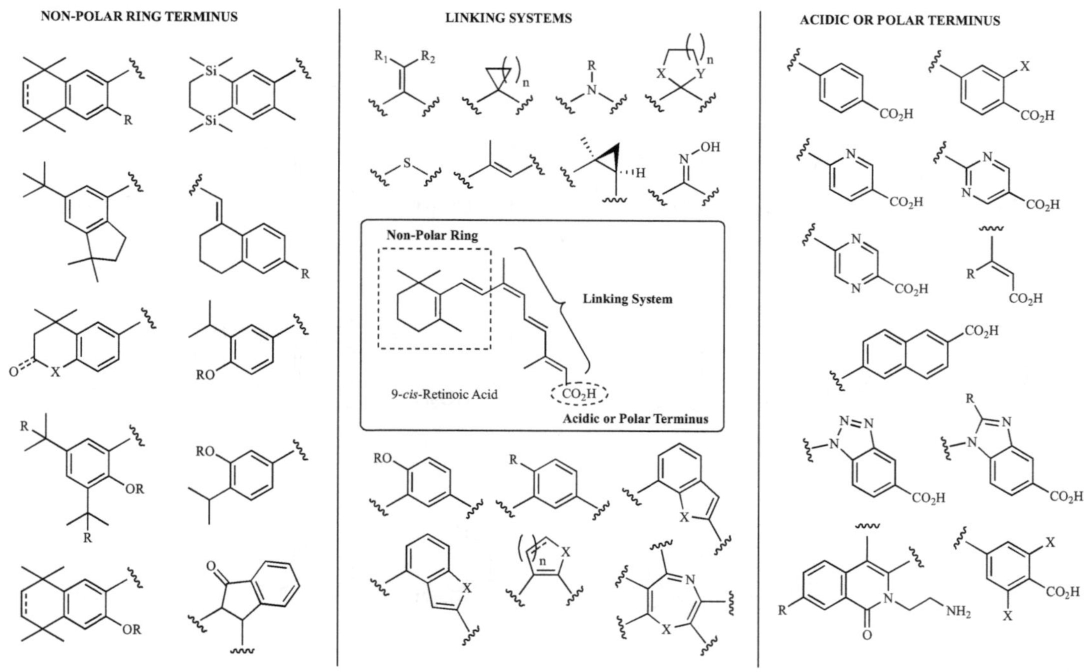

Fig. 2 Retrosynthetic analysis of 9-*cis*-retinoic acid as a model compound for the design of synthetic rexinoids

treating in vivo cancer models [17, 18] though, none have progressed through clinical trials to gain FDA approval. In addition to concerns for the side-effect profiles of potent rexinoids, there is a great need to develop new rexinoids with robust intellectual property protection to compel investment for further clinical development. Hence, a straightforward strategy to create new compounds from base-model compounds is highly desirable.

Kakuta and coworkers reported **NEt-TMN** [19–23] as a particularly potent agonist of RXR, and our group recently synthesized and evaluated a number of analogs [24] of **NEt-TMN**. Similar to the way 9-*cis*-RA was de-constructed as a base model for the construction of synthetic rexinoids (Fig. 2), a similar approach

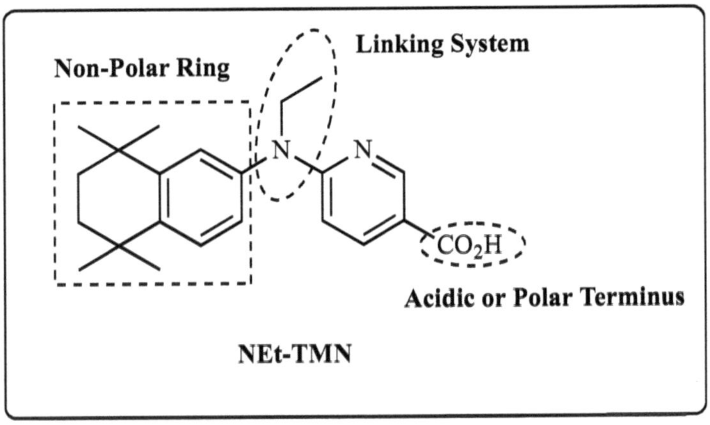

Fig. 3 Retrosynthetic analysis of NEt-TMN

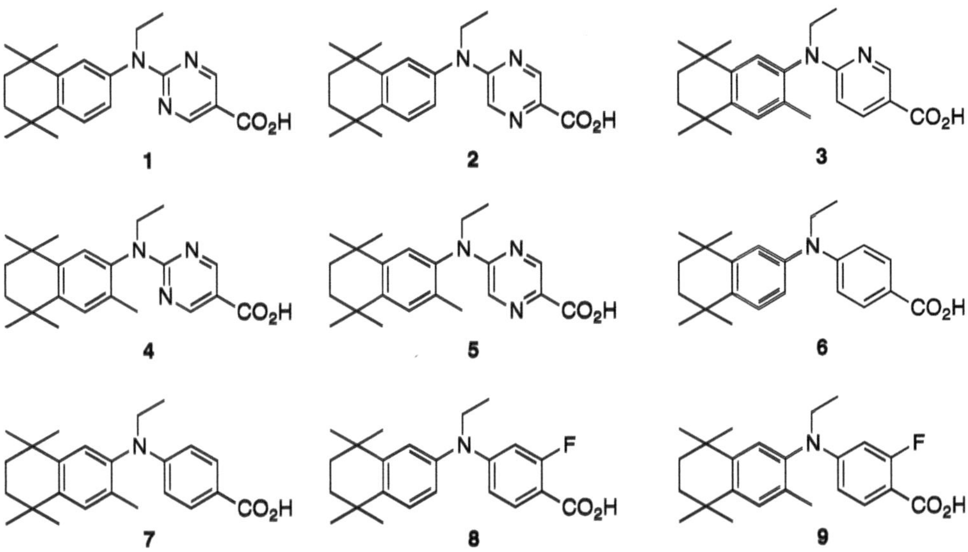

Fig. 4 Nine rexinoid analogs of NEt-TMN

can be used to deconstruct **NEt-TMN** and assemble new structural analogs by making changes in the three critical areas of **NEt-TMN**—the nonpolar ring system, the linking system, or the acidic terminus (Fig. 3). Analogs that exhibit modified acidic terminus and nonpolar ring systems include **1–5**, the reported **6** [25, 26] and **7** [25] and the fluorinated **8** and **9** (Fig. 4).

To construct analogs of **NEt-TMN** that possess novel acidic terminus ring systems, a similar route to the synthetic route for **NEt-TMN** can be applied (Fig. 5). In the route to NEt-TMN, commercially available nitro-aromatic **10** is reduced to **11** with hydrogen gas over 10% Pd/C. Compound **11** can then be combined with commercially available **12**, *para*-toluene sulfonic acid monohydrate, and refluxed in dioxane to provide **13**. Diaryl amine

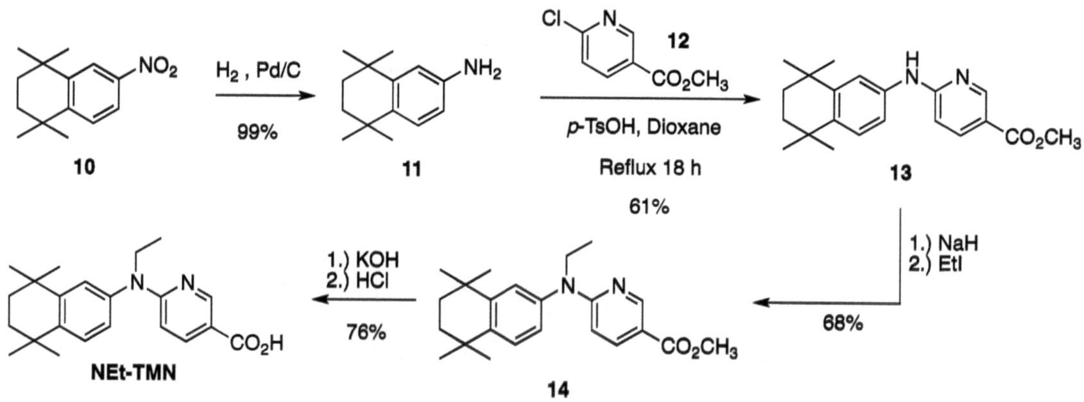

Fig. 5 Synthetic route to produce NEt-TMN

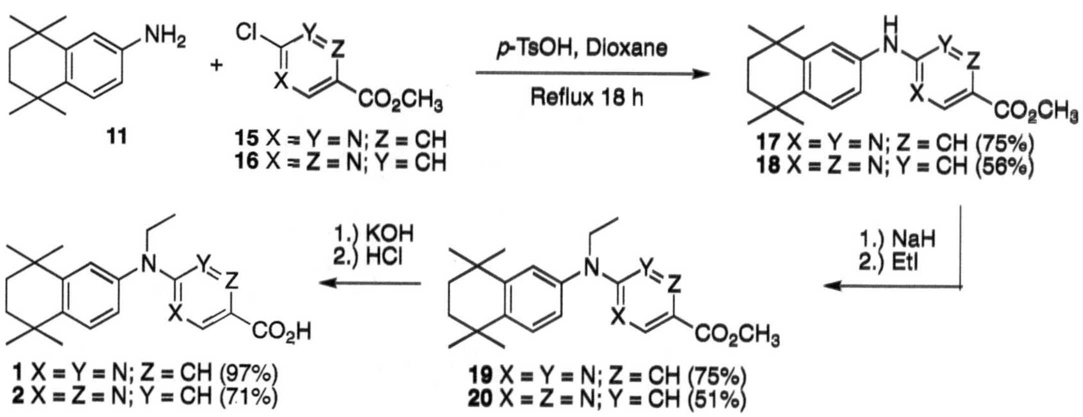

Fig. 6 Synthetic route to analogs 1 and 2

13 is then deprotonated and treated with ethyl iodide to give methyl ester **14** which can then be saponified with potassium hydroxide and worked up with hydrochloric acid to give **NEt-TMN**. Changing the pyridine ring in **NEt-TMN** to a pyrimidine or pyrazine ring follows a similar synthetic route with similar methods (Fig. 6).

To construct analogs of **NEt-TMN** possessing a nonpolar ring system incorporating a methyl group, a different synthetic methodology was needed, since the nucleophilic aromatic substitution step for the synthesis of **NEt-TMN** is sterically prohibited. Hence, a new route relying on a critical palladium catalyzed Buchwald-Hartwig amination reaction [27] was employed (Fig. 7). The bromo-aromatic **21** is coupled with commercially available amines **22–24** to give diaryl amines **25–27** in reasonable yield that are then alkylated to give **28–30** that are finally saponified to provide **3–5** in good yields. The known **6** and **7**, and their fluorinated analogs **8** and **9**, were prepared using a similar Buchwald-Hartwig

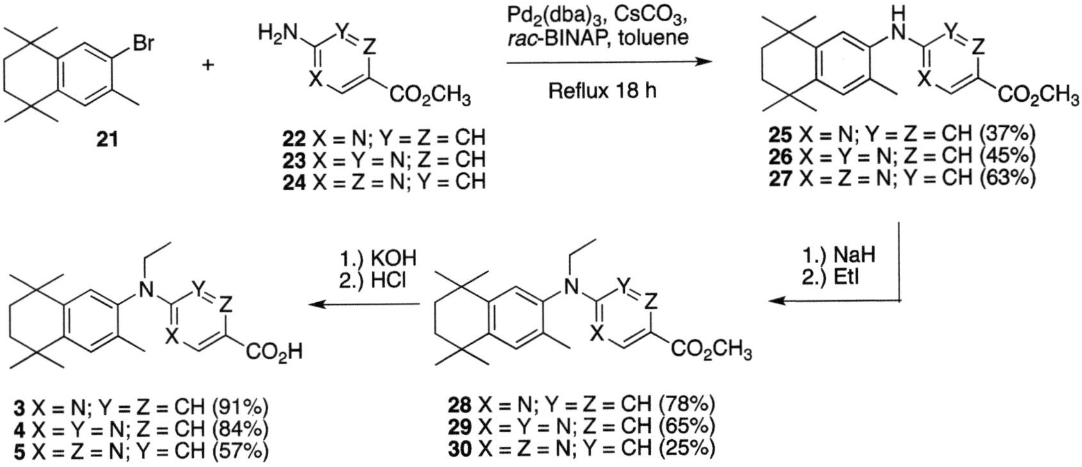

Fig. 7 Synthetic route to analogs 3–5

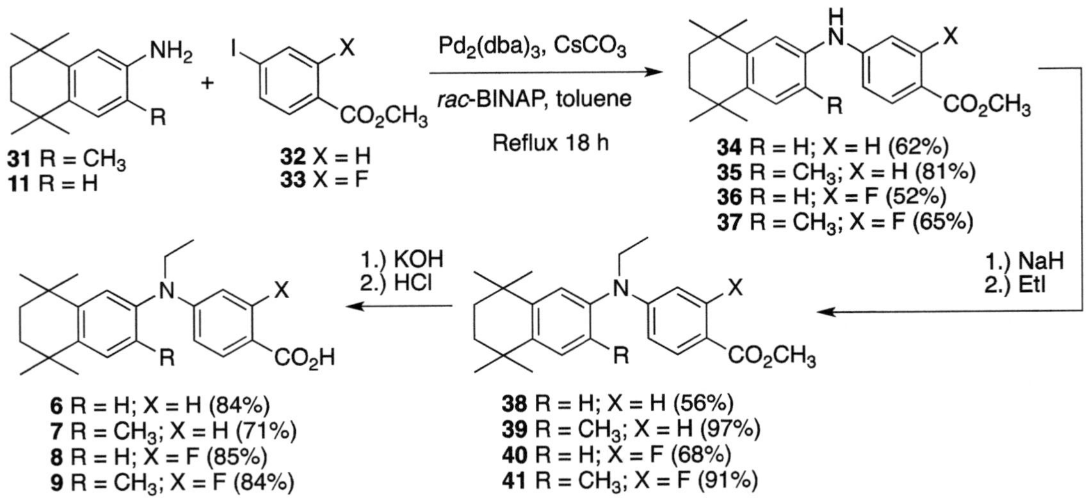

Fig. 8 Synthetic route to analogs 6–9

amination coupling either amine **31** or **11** with commercially available **32** and **33** to give diarylamines **34–37** that were alkylated to give methyl esters **38–41** that could finally be saponified to give compounds **6–9** (Fig. 8).

2 Materials

1. Hexanes: mixture of isomers ≥98.5%.
2. Ethyl Acetate: ≥99.5%.
3. Silica Gel: 32–63 particle size.
4. 1,1,4,4-Tetramethyl-6-nitro-1,2,3,4-tetrahydronaphthalene (**10**): ≥95%.

5. 3,5,5,8,8,-Pentamethyl-5,6,7,8-tetrahydronaphthalen-2-amine (**31**): 95+%.
6. Hydrogen gas: Grade 5.0, Ultra High Purity, 99.999%.
7. Palladium on activated carbon, 10% Pd (10% Pd/C).
8. Methyl 6-chloronicotinate (**12**): 98%.
9. *para*-Toluenesulfonic acid Monohydrate: ≥98.5%.
10. 1,4-Dioxane: 99.8%, anhydrous.
11. Sodium Hydride: 60% dispersion in mineral oil.
12. Iodoethane: 99%.
13. Potassium Hydroxide: 85–90%, Technical Flakes.
14. Methanol: 99.9%.
15. Hydrochloric Acid (1 N): 0.9995–1.005 N.
16. Methyl 2-Chloropyrimidine 5-carboxylate (**15**): 98%.
17. Methyl 5-Chloropyrazine 5-carboxylate (**16**): 98%.
18. 6-Bromo-1,1,4,4,7-pentamethyl-1,2,3,4-tetrahydronaphthalene (**21**): 95+%.
19. Methyl 6-aminonicotinate (**22**): 97%.
20. Methyl 2-aminopyrimidine 5-carboxylate (**23**): 97%.
21. Methyl 5-aminopyrazine-2-carboxylate (**24**): 97%.
22. Methyl 4-iodobenzoate (**32**): 97%.
23. Methyl 2-fluoro-4-iodobenzoate (**33**): 95–98%.
24. Tris(dibenzylideneacetone)dipalladium(0): 97%.
25. Cesium Carbonate: 99%.
26. *Rac*-BINAP: 97%.
27. Toluene: 99.8% dry, anhydrous.
28. Sodium sulfate: ≥99.0%, anhydrous powder.

3 Methods

3.1 Hydrogenation

1. Method for hydrogenation of 1,1,4,4-tetramethyl-6-nitro-1,2,3,4-tetrahydronaphthalene (**10**): to a solution of **10** (2.5 g, 10.7 mmol) in ethyl acetate (25 mL) in a 24/40 neck 300 mL round bottom flask was added 10% Pd/C (0.52 g) and a magnetic stir bar.
2. A T-joint glass stop cock was fitted to the 300 mL round bottom flask and to one of the two external attachments to the T-joint glass stop cock was fitted a tygon tube to vacuum and to the other attachment was fitted a tygon tube with a large balloon filled with hydrogen gas (~4 L) attached to the tygon tube by electrical tape.

3. The atmosphere over the solution was evacuated through the vacuum attachment until the solution appeared to boil at room temperature, and then back-filled with hydrogen from the balloon. This evacuation and back-fill procedure was performed three times, and then the reaction was allowed to stir at room temperature overnight.
4. The solution was filtered through cotton, and the ethyl acetate was removed in vacuo to give **11** in quantitative yield that was used without further purification.

3.2 Nucleophilic Aromatic Substitution Reaction

1. To a 100 mL, 14/20 one-neck round bottom flask equipped with a magnetic stir bar and charged with **11** (~4.0 mmol), chloro-aromatic (**12, 15,** or **16**, ~4.0 mmol), and *para*-toluene sulfonic acid monohydrate (~4.0 mmol) was added 1,4-dioxane (15 mL).
2. The flask was fitted with a reflux condenser, evacuated and back-filled with nitrogen, and then heated to reflux with stirring in an oil bath at 111 °C for 14 h.
3. After cooling the reaction to room temperature, the reaction mixture was poured into water (50 mL) and extracted with ethyl acetate.
4. The organic phase was washed with brine, dried over sodium sulfate, filtered, and concentrated in vacuo to yield crude aryl amines that were purified by column chromatography (150 mL silica gel, 10% ethyl acetate: hexanes) to give pure aryl amines (**13**, **17**, or **18**).

3.3 Alkylation of 13, 17, or 18

1. To a flame-dried, 100 mL 14/20 round bottom flask equipped with a magnetic stir bar was added a 60% dispersion of sodium hydride in mineral oil (~5.25 mmol).
2. The dispersion of sodium hydride was washed with hexanes (3 mL, twice) and dried under vacuum (*see* **Note 1**) followed by suspension in DMF (~3.1 mL) under a nitrogen atmosphere.
3. To this solution of sodium hydride in DMF was added a solution of aryl amines (**13**, **17**, or **18**, ~2.5 mmol) in DMF (~8.3 mL), and the reaction was stirred for 15 min at room temperature, and then ethyl iodide (~4.25 mmol) was added, and the reaction was stirred for 1 h.
4. The reaction was poured into water and extracted with ethyl acetate. The organic phase was washed with brine, dried over sodium sulfate, filtered, and concentrated in vacuo to yield a crude product that was purified by column chromatrography (150 mL silica gel, 6% ethyl acetate:hexanes) to give ethyl aryl amines (**14**, **19**, and **20**).

3.4 Saponification of 14, 19, or 20

1. To a 100 mL 14/20 one-neck round bottom flask equipped with a stir bar and charged with ethyl aryl amine (**14**, **19**, or **20**, ~1.0 mmol) suspended in methanol (~3.5 mL) was added a solution of potassium hydroxide (0.18 g, ~3.2 mmol) in water (0.22 mL).
2. This reaction solution was stirred at reflux in an oil bath at 87 °C for 1 h.
3. The reaction was then cooled to room temperature and acidified with hydrochloric acid (~80 mL, 1 N) to give a crude precipitate.
4. The crude precipitate was filtered and washed with cold water to give crude rexinoids that were purified by column chromatography (25 mL silica gel, 30% ethyl acetate:hexanes to pure ethyl acetate to 2% methanol:ethyl acetate) to give rexinoids (**NEt-TMN**, **1**, and **2**).

3.5 Buchwald-Hartwig Amination of 21

1. To a solution of **21** (~4.0 mmol) and aryl amine (**22**, **23**, or **24**, ~3.9 mmol), cesium carbonate (~9.6 mmol), *rac*-BINAP (~0.3 mmol) in toluene (4.5 mL) in a 14/20 one-neck 100 mL round bottom flask with a magnetic stir bar was added tris (dibenzylideneacetone)dipalladium(0) (~0.2 mmol).
2. Nitrogen gas was bubbled through the reaction solution for 3 min, a water-cooled reflux condenser was attached to the flask, and the apparatus was evacuated under vacuum and backfilled with nitrogen three times.
3. The round bottom flask was placed in an oil bath and heated to reflux (125–130 °C) with stirring for 18 h (*see* **Note 2**).
4. After cooling to room temperature, cesium carbonate salts were removed by filtering the solution through a medium-porosity filter paper in a Buchner filter funnel, and the salts were washed with ethyl acetate.
5. The filtrate was concentrated in vacuo, and the crude product was purified by column chromatography (150 mL silica gel, 6.5% to 10% ethyl acetate:hexanes) to give pure aryl amines (**25**, **26**, or **27**).

3.6 Alkylation of 25, 26, or 27

1. To a flame-dried, 100 mL 14/20 round bottom flask equipped with a magnetic stir bar was added a 60% dispersion of sodium hydride in mineral oil (~6.6 mmol).
2. The dispersion of sodium hydride was washed with hexanes (3.7 mL, twice) and dried under vacuum followed by suspension in DMF (~3.8 mL) under a nitrogen atmosphere.
3. To this solution of sodium hydride in DMF was added a solution of aryl amines (**25**, **26**, or **27**, ~3.0 mmol) in DMF (~11.4 mL), and the reaction was stirred for 15 min at room

temperature, and then ethyl iodide (~4.5 mmol) was added, and the reaction was stirred for 1 h.

4. The reaction was poured into water and extracted with ethyl acetate. The organic phase was washed with brine, dried over sodium sulfate, filtered, and concentrated in vacuo to yield a crude product that was purified by column chromatrography (150 mL silica gel, 6% ethyl acetate:hexanes) to give ethyl aryl amines (**28**, **29**, and **30**).

3.7 Saponification of 28, 29, or 30

1. To a 100 mL 14/20 one-neck round bottom flask equipped with a stir bar and charged with ethyl aryl amine (**28**, **29**, or **30**, ~1.3 mmol) suspended in methanol (~3.7 mL) was added a solution of potassium hydroxide (0.24 g, ~4.9 mmol) in water (0.29 mL).
2. This reaction solution was stirred at reflux in an oil bath at 87 °C for 1 h.
3. The reaction was then cooled to room temperature and acidified with hydrochloric acid (~80 mL, 1 N) to give a precipitate.
4. The crude precipitate was filtered and washed with cold water to give crude rexinoids that were purified by column chromatography (25 mL silica gel, 30% ethyl acetate:hexanes to pure ethyl acetate to 2% methanol:ethyl acetate) to give rexinoids (**3**, **4**, and **5**). In the case of **3**, solid precipitate that could be filtered was not observed after adding acid, so the solution was extracted with ethyl acetate, dried over sodium sulfate, and concentrated in vacuo to give a crude product that was then purified by column chromatography (*see* **Note 3**).

3.8 Buchwald-Hartwig Amination of 11 or 31

1. To a solution of **11** or **31** (~4.1 mmol) and iodo-aryl-methyl-ester (**32**, or **33**, ~4.2 mmol), cesium carbonate (~9.6 mmol), *rac*-BINAP (~0.3 mmol) in toluene (4.5 mL) in a 14/20 one-neck 100 mL round bottom flask with a magnetic stir bar was added tris(dibenzylideneacetone)dipalladium(0) (~0.2 mmol).
2. Nitrogen gas was bubbled through the reaction solution for 3 min, a water-cooled reflux condenser was attached to the flask, and the apparatus was evacuated under vacuum and back-filled with nitrogen three times.
3. The round bottom flask was placed in an oil bath and heated to reflux (125–130 °C) with stirring for 18 h.
4. After cooling to room temperature, cesium carbonate salts were removed by filtering the solution through a medium-porosity filter paper in a Buchner filter funnel, and the salts were washed with ethyl acetate.

5. The filtrate was concentrated in vacuo, and the crude product was purified by column chromatography (150 mL silica gel, 6.5% to 10% ethyl acetate:hexanes) to give pure aryl amines (**34**, **35**, **36**, or **37**).

3.9 Alkylation of 34, 35, 36, or 37

1. To a flame-dried, 100 mL 14/20 round bottom flask equipped with a magnetic stir bar was added a 60% dispersion of sodium hydride in mineral oil (~4.8 mmol).
2. The dispersion of sodium hydride was washed with hexanes (2.6 mL, twice) and dried under vacuum followed by suspension in DMF (~2.7 mL) under a nitrogen atmosphere.
3. To this solution of sodium hydride in DMF was added a solution of aryl amines (**34**, **35**, **36**, or **37**, ~2.1 mmol) in DMF (~8.0 mL), and the reaction was stirred for 15 min at room temperature, and then ethyl iodide (~3.3 mmol) was added, and the reaction was stirred for 1 h.
4. The reaction was poured into water and extracted with ethyl acetate. The organic phase was washed with brine, dried over sodium sulfate, filtered, and concentrated in vacuo to yield a crude product that was purified by column chromatrography (150 mL silica gel, 6% ethyl acetate:hexanes) to give ethyl aryl amines (**38**, **39**, **40**, and **41**).

3.10 Saponification of 38, 39, 40, and 41

1. To a 100 mL 14/20 one-neck round bottom flask equipped with a stir bar and charged with ethyl aryl amine (**38**, **39**, **40** or **41**, ~1.0 mmol) suspended in methanol (~3.6 mL) was added a solution of potassium hydroxide (0.20 g, ~3.6 mmol) in water (0.24 mL).
2. This reaction solution was stirred at reflux in an oil bath at 87 °C for 1 h.
3. The reaction was then cooled to room temperature and acidified with hydrochloric acid (~80 mL, 1 N) to give a precipitate.
4. The crude precipitate was filtered and washed with cold water to give crude rexinoids that were purified by column chromatography (25 mL silica gel, 20% ethyl acetate:hexanes to 50% ethyl acetate:hexanes) to give rexinoids (**6**, **7**, **8**, and **9**).

4 Notes

1. To wash the 60% dispersion of sodium hydride with hexanes, add the hexanes to the dispersion and stir the suspension under nitrogen in the 100 mL round bottom flask. Next, halt the stirring, and allow the sodium hydride to settle to the bottom of the flask. Carefully remove the hexanes with a pipette until

the hexane level is just above the sodium hydride. After washing twice, evaporate remaining hexanes under vacuum.

2. Better yields were observed for the Buchwald-Hartwig amination procedures when the reaction was run for 18 h or more, and lower yields were observed when the reaction was run for only 12 h.
3. Whenever a solid precipitate for any final product rexinoid cannot be isolated by filtration, the solution is extracted with ethyl acetate, the organic layers dried over sodium sulfate, and concentrated in vacuo to give a crude product that is then purified by column chromatography.

References

1. Mangelsdorf DJ, Evans RM (1994) The retinoids. Academic Press, Orlando, FL
2. Leid M, Kastner P, Chambon P (1992) Multiplicity generates diversity in the retinoic acid signalling pathways. Trends Biochem Sci 17:427–433
3. Lala DS, Mukherjee R, Schulman IG, Koch SSC, Dardashti LJ, Nadzan AM et al (1996) Activation of specific RXR heterodimers by an antagonist of RXR homodimers. Nature 383:450
4. Mangelsdorf DJ, Ong ES, Dyck JA, Evans RM (1990) Nuclear receptor that identifies a novel retinoic acid response pathway. Nature 345:224
5. Jong L, Lehmann JM, Hobbs PD, Harlev E, Huffman JC, Pfahl M et al (1993) Conformational effects on retinoid receptor selectivity. 1. Effect of 9-double bond geometry on retinoid X receptor activity. J Med Chem 36:2605–2613
6. Dawson MI, Jong L, Hobbs PD, Cameron JF, Chao W-r, Pfahl M et al (1995) Conformational Effects on retinoid receptor selectivity. 2. Effects of retinoid bridging group on retinoid X receptor activity and selectivity. J Med Chem 38:3368–3383
7. Boehm MF, Zhang L, Badea BA, White SK, Mais DE, Berger E et al (1994) Synthesis and structure-activity relationships of novel retinoid X receptor-selective retinoids. J Med Chem 37:2930–2941
8. Yen W-C, Corpuz MR, Prudente RY, Cooke TA, Bissonnette RP, Negro-Vilar A et al (2004) A selective retinoid X receptor agonist bexarotene (targretin) prevents and overcomes acquired paclitaxel (taxol) resistance in human non–small cell lung cancer. Clin Cancer Res 10:8656
9. Dragnev KH, Petty WJ, Shah SJ, Lewis LD, Black CC, Memoli V et al (2007) A proof-of-principle clinical trial of bexarotene in patients with non–small cell lung cancer. Clin Cancer Res 13:1794
10. W-c Y, Prudente RY, Lamph WW (2004) Synergistic effect of a retinoid X receptor-selective ligand bexarotene (LGD1069, Targretin) and paclitaxel (Taxol) in mammary carcinoma. Breast Cancer Res Treat 88:141–148
11. Heo J-C, Jung T-H, Lee S, Kim HY, Choi G, Jung M et al (2016) Effect of bexarotene on differentiation of glioblastoma multiforme compared with ATRA. Clin Exp Metastasis 33:417–429
12. Cramer PE, Cirrito JR, Wesson DW, Lee CYD, Karlo JC, Zinn AE et al (2012) ApoE-directed therapeutics rapidly clear β-amyloid and reverse deficits in AD mouse models. Science 335:1503
13. Tai LM, Bilousova T, Jungbauer L, Roeske SK, Youmans KL, Yu C et al (2013) Levels of soluble apolipoprotein E/amyloid-β (Aβ) complex are reduced and oligomeric Aβ increased with APOE4 and Alzheimer disease in a transgenic mouse model and human samples. J Biol Chem 288:5914–5926
14. McFarland K, Spalding TA, Hubbard D, Ma J-N, Olsson R, Burstein ES (2013) Low dose bexarotene treatment rescues dopamine neurons and restores behavioral function in models of parkinson's disease. ACS Chem Nerosci 4:1430–1438
15. Cummings JL, Zhong K, Kinney JW, Heaney C, Moll-Tudla J, Joshi A et al (2016) Double-blind, placebo-controlled, proof-of-concept trial of bexarotene Xin moderate Alzheimer's disease. Alzheimer's Res Ther 8(4)
16. Boehm MF, Zhang L, Zhi L, McClurg MR, Berger E, Wagoner M et al (1995) Design and

synthesis of potent retinoid X receptor selective ligands that induce apoptosis in leukemia cells. J Med Chem 38:3146–3155

17. Liby K, Rendi M, Suh N, Royce DB, Risingsong R, Williams CR et al (2006) The combination of the rexinoid, LG100268, and a selective estrogen receptor modulator, either arzoxifene or acolbifene, synergizes in the prevention and treatment of mammary tumors in an estrogen receptor–negative model of breast cancer. Clin Cancer Res 12:5902
18. Cao M, Royce DB, Risingsong R, Williams CR, Sporn MB, Liby KT (2016) The rexinoids LG100268 and LG101506 inhibit inflammation and suppress lung carcinogenesis in A/J mice. Cancer Prev Res 9:105
19. Fujii S, Ohsawa F, Yamada S, Shinozaki R, Fukai R, Makishima M et al (2010) Modification at the acidic domain of RXR agonists has little effect on permissive RXR-heterodimer activation. Bioorg Med Chem Lett 20:5139–5142
20. Ohsawa F, Morishita K-i, Yamada S, Makishima M, Kakuta H (2010) Modification at the lipophilic domain of RXR agonists differentially influences activation of RXR heterodimers. ACS Med Chem Lett 1:521–525
21. Kakuta H, Yakushiji N, Shinozaki R, Ohsawa F, Yamada S, Ohta Y et al (2012) RXR partial agonist CBt-PMN exerts therapeutic effects on type 2 diabetes without the side effects of RXR full agonists. ACS Med Chem Lett 3:427–432
22. Kakuta H, Ohsawa F, Yamada S, Makishima M, Tai A, Yasui H et al (2012) Feasibility of structural modification of retinoid X receptor agonists to separate blood glucose-lowering action from adverse effects: studies in KKAy type 2 diabetes model mice. Biol Pharm Bull 35:629–633
23. Ohsawa F, Yamada S, Yakushiji N, Shinozaki R, Nakayama M, Kawata K et al (2013) Mechanism of retinoid X receptor partial agonistic action of 1-(3,5,5,8,8-pentamethyl-5,6,7,8-tetrahydro-2-naphthyl)-1H-benzotriazole-5-carboxylic acid and structural development to increase potency. J Med Chem 56:1865–1877
24. Heck MC, Wagner CE, Shahani PH, MacNeill M, Grozic A, Darwaiz T et al (2016) Modeling, synthesis, and biological evaluation of potential retinoid X receptor (RXR)-selective agonists: analogues of 4-[1-(3,5,5,8,8-pentamethyl-5,6,7,8-tetrahydro-2-naphthyl)ethynyl]benzoic acid (bexarotene) and 6-(ethyl(5,5,8,8-tetrahydronaphthalen-2-yl)amino)nicotinic acid (NEt-TMN). J Med Chem 59:8924–8940
25. Ohta K, Tsuji M, Kawachi E, Fukasawa H, Hashimoto Y, Shudo K et al (1998) Potent retinoid synergists with a diphenylamine skeleton. Biol Pharm Bull 21:544–546
26. Ohta K, Kawachi E, Fukasawa H, Shudo K, Kagechika H (2011) Diphenylamine-based retinoid antagonists: regulation of RAR and RXR function depending on the N-substituent. Bioorg Med Chem 19:2501–2507
27. Wolfe JP, Wagaw S, Buchwald SL (1996) An improved catalyst system for aromatic carbon–nitrogen bond formation: the possible involvement of bis(phosphine) palladium complexes as key intermediates. J Am Chem Soc 118:7215–7216

Chapter 9

Promoting Primary Myoblast Differentiation Through Retinoid X Receptor Signaling

Jihong Chen and Qiao Li

Abstract

The differentiation and fusion of primary myoblasts into myotubes is tightly regulated through muscle-specific transcription networks and can be enhanced by small molecular inducers, which allow us to identify novel genetic targets and molecular interactions. As the pressing issue is to develop pharmacotherapy to prevent and treat muscle-related diseases, we describe how to efficiently direct the differentiation of primary myoblasts by using a nuclear receptor agonist for the development of muscle therapeutics.

Key words Gene regulation, Myogenic differentiation, Nuclear receptor, Primary myoblast, Skeletal myocytes

1 Introduction

Many diseases and underlying conditions, including cancer, AIDS, and aging, develop muscle atrophy, which can be extremely debilitating affecting both patients and care givers. However, existing treatment option is limited and palliative in nature. Understanding the molecular mechanisms underlying the differentiation and fusion of myoblasts is a critical step in developing the best strategy to direct muscle regeneration in the clinic. In an effort to develop these strategies, we recently found that bexarotene, a clinically approved retinoid X receptor (RXR) agonist [1], enhances the differentiation and fusion of myoblasts through the function of RXR as a transcription factor to directly regulate the expression of muscle master regulator MyoD [2, 3]. In addition, bexarotene is able to sustain myoblast differentiation following pro-atrophic insult [2]. Postnatal muscle regeneration is an important physiological process that maintains the homeostasis of the muscle tissue and repairs muscle following injury or pro-atrophic insult [4]. Satellite cells (SCs), a population of stem cells residing beneath the muscle basal lamina, provide the tissue with a remarkable ability to regenerate myofibers [5]. In the absence of stimuli, the SCs are

Swapan K. Ray (ed.), *Retinoid and Rexinoid Signaling: Methods and Protocols*, Methods in Molecular Biology, vol. 2019, https://doi.org/10.1007/978-1-4939-9585-1_9, © Springer Science+Business Media, LLC, part of Springer Nature 2019

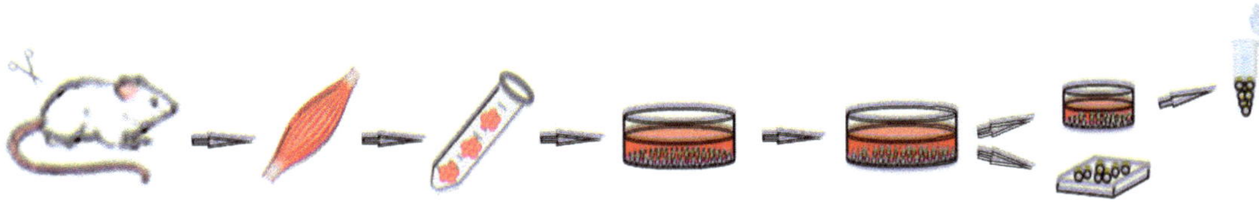

Fig. 1 Steps of primary myoblasts isolation. Muscle removed from young female mice were fragmented and subjected to enzymatic digestion. Following removal of fibroblasts through preplating, primary myoblasts are maintained with growth media in a coated culture dish. Efficacies of myoblast differentiation and muscle-specific gene expression can be determined by harvesting the cells at different time points for quantitative immunofluorescence microscopy, RT-qPCR, and Western blot analysis

quiescent and marked with Pax7 that is expressed in all SCs, but lack the expression of any muscle regulatory factors [6, 7]. In the presence of cellular stimulus, quiescent SCs become activated and begin to express MyoD, and as a result to differentiate that is marked by myogenin expression [5, 8–10]. Therefore, primary myoblast is an excellent experimental system for providing molecular insights into the signaling pathways of muscle regeneration. Here, we describe a protocol for isolating primary myoblasts from young mice (Fig. 1), to study the potential of small molecules including bexarotene in promoting myoblast differentiation and in overcoming muscle atrophy pathology. Ultimately, these types of studies will provide fundamental insights and enable the rational development of the best strategy to exploit multi-signaling pathways for promoting muscle regeneration.

2 Materials

1. Young C57BL/6 mice (*see* **Note 1**).
2. Horse serum (HS), store at −70 °C.
3. Fetal bovine serum (FBS), store at −70 °C.
4. Plating media: Dulbecco's Modified Eagle Medium (DMEM), supplemented with 10% HS, 5 ng/mL bFGF, and 1× penicillin/streptomycin (pen/strep), store at 4 °C.
5. Growth media: DMEM supplemented with 20% FBS, 10% HS, 10 ng/mL bFGF, 2 ng/ml HGF, and 1× pen/strep, store at 4 °C.
6. Differentiation media: DMEM supplemented with 10% HS and 1× pen/strep, store at 4 °C.
7. Phosphate-buffered saline (PBS) pH 7.4: 137 mM NaCl, 2.7 mM KCl, 10 mM Na_2HPO_4, 2 mM KH_2PO_4 (PBS may be supplemented with 1 mM $CaCl_2$ and 0.5 mM $MgCl_2$ for divalent cations), autoclave and store at room temperature.

8. Trypsin 10×, store at −20 °C.
9. Syringe and syringe filter (0.22 μm).
10. Scissors, forceps, hemostats, and spray bottle of 70% ethanol.
11. Nitrex 74 μm mesh filter.
12. Shaker, temperature control for 37 °C.
13. Cell culture incubator, 5% CO_2 at 37 °C.
14. Bexarotene: dissolve in ethanol, aliquot into light-safe tubes, and store at −20 °C.

3 Methods

3.1 Reagent Reconstitution and Plate Coating

3.1.1 Dispase Reconstitution

1. Dissolve 500 mg dispase (Roche) in 5 mL sterile PBS to make a 100 mg/mL solution.
2. Dilute 5 Ml of 100 mg/mL dispase solution with 20 mL autoclaved ddH_2O for a final concentration of 20 mg/mL.
3. Sterilize the diluted solution with a 0.22 μm filter.
4. Aliquot and store at −20 °C.

3.1.2 Growth Factor Reconstitution

1. Dissolve 15 mg of BSA in 15 mL autoclaved ddH_2O to make a 1 mg/mL BSA solution.
2. Dissolve each growth factor with 5 mL of 1 mg/mL BSA solution.
3. Sterilize the growth factor solutions with a 0.22 μm filter.
4. Aliquot and store at −80 °C.

3.1.3 Matrigel Coating

1. Dilute 5 mL Matrigel in 45 mL of ice-cold DMEM.
2. Aliquot and store at −20 °C (*see* **Note 2**).
3. Briefly rinse the wells of a six-well plate with the diluted Matrigel on ice (*see* **Note 3**).
4. Leave the plate to set in a cell culture incubator for at least 15 min.

3.2 Mouse Muscle Dissection

1. Prepare a fresh solution of 0.2% collagenase in DMEM (8 mL per mouse), mix with 500 μL of dispase, warm to 37 °C, and sterilize by syringe filtration (0.22 μm filter) into a 50-mL Falcon tube.
2. Prepare a 15-mL Falcon tube with ~10 mL of PBS with 1× pen/strep.
3. Euthanize the mouse with isoflurane in a bell jar, followed by cervical dislocation to ensure decease. Spray with 70% ethanol to keep the fur down.

4. Remove the right and left tibialis anterior, gastrocnemius, and quadriceps muscles (or any muscles desired) with scissors. Peel away the fascia.
5. Cut the muscle into small fragments (*see* **Note 4**).

3.3 Primary Myoblast Isolation

1. Place the muscle fragments into the 50-mL Falcon tube containing collagenase/dispase solution.
2. Incubate the Falcon tube in shaker at 75 rpm, 37 °C for 2 h.
3. Triturate the muscle fragments with a 10-mL pipette until muscle fibers are dissociated as determined under microscope, and incubate the mixture for another 15 min.
4. Triturate the mixture again to homogenize into a single cell suspension.
5. Add DMEM to a total of 20 mL and pass the supernatant through a Nitrex filter (74 μm) into a 50-mL Falcon tube to remove undigested connective tissue (*see* **Note 5**).
6. Centrifuge at 500 × *g* for 5 min.
7. Discard the supernatant carefully, leaving ~1 mL above the cell pellet.
8. Resuspend the cell pellet with 10 mL of fresh DMEM.
9. Centrifuge again at 500 × *g* for 5 min and discard the supernatant.
10. Resuspend the cell pellet in plating media (*see* **Note 6**).
11. Transfer the supernatant to a regular 10 cm tissue culture dish and preplate by leaving the cell suspension in the CO_2 incubator for 1 h.
12. While cell preplating, coat a six-well plate with Matrigel on ice.
13. Transfer the entire cell suspension to a single well of the six-well plate coated with Matrigel. Rinse the 10 cm dish carefully to ensure that all cell suspensions are transferred.
14. Maintain the cell culture for 2 days. Do not change the media during this period.
15. After the 2 days, if the cells reaching 80% confluence, passage the cells. Otherwise, simply switch to growth media (*see* **Note** 7).
16. Change the growth media daily afterward (*see* **Note 8**).
17. Preplate the cells once every two passages, or as necessary to remove fibroblasts (*see* **Note 9**).
18. After 5–6 days, 20–25 million cells can be harvested for freezing down (*see* **Note 10**).

3.4 Primary Myoblast Maintenance and Differentiation

1. Primary myoblasts should be maintained in growth media on Matrigel-coated plates, and passage at 70–80% confluency.
2. Preplate the cells periodically to remove fibroblasts.
3. For differentiation, seed the cells at 70% confluency in growth media and switch to differentiation media with or without 30–50 nM of bexarotene the following day.
4. Change the differentiation media and bexarotene every other day.
5. Fix and stain the cells at different time point of differentiation with antibodies for specific markers of skeletal myocytes, and use immunofluorescence microscopy to determine the efficacies of myoblast differentiation (*see* **Note 11**).
6. Harvest the cells at different time point for RT-qPCR or Western blotting analysis to examine myogenic expression-upregulated by bexarotene (*see* **Note 12**).

3.5 RT-qPCR and Western Analysis

1. The cell pellets can be used to prepare total RNA for RT-qPCR analysis.
2. Alternatively, the cell pellets can be used to prepare whole cell extracts or nuclear extracts for Western blotting analysis.
3. The percentage of SDS gel required for the Western blotting is depending on the molecular weight of protein of interest.

4 Notes

1. Preferably female mice (3–6 weeks old).
2. Diluted Matrigel solution should be thawed at 4 °C overnight when needed.
3. Matrigel must be kept cold at all times to prevent permanently set.
4. Large muscle fragments would not be readily for enzymatic break down.
5. There should be no large muscle fragments at this point.
6. About 2×10^6 cells can be harvested from muscles of one mouse.
7. Cells should never be allowed to grow over 80% confluency. At this point, cells grow rapidly and daily passaging may be necessary, but they should not be seeded below 30% confluency.
8. Growth factors only last for 24 h in culture.
9. Although fibroblast contamination is usually <0.1%, the numbers may increase along with passage. To eliminate the contamination, seed freshly trypsinized cells in growth media in

uncoated tissue culture plates for 1 h, and then transfer the cells into a coated plate. Fibroblasts adhere to the uncoated plate, while myoblasts in suspension.

10. Freeze 1 million cells in 1 mL of growth media and 10% sterile DMSO.
11. Cells can be stained simultaneously with myosin heavy chain antibody and MyoD antibody, followed with corresponding secondary antibodies for different filters, as well with Hoechst for DNA [11].
12. Cells can be used for RT-qPCR and Western blot analysis of MyoD and myogenin gene expression at different time points [2].

Acknowledgments

This work was sponsored by operating grant from the Natural Sciences and Engineering Research Council of Canada.

References

1. Gniadecki R, Assaf C, Bagot M et al (2007) The optimal use of bexarotene in cutaneous T-cell lymphoma. Br J Dermatol 157:433–440
2. AlSudais H, Aabed K, Nicola W, Dixon K, Chen J, Li Q (2016) Retinoid X receptor-selective signaling in the regulation of Akt/-protein kinase B isoform-specific expression. J Biol Chem 291:3090–3099
3. Hamed M, Khilji S, Dixon K, Blais A, Ioshikhes I, Chen J, Li Q (2017) Insights into interplay between rexinoid signaling and myogenic regulatory factor-associated chromatin state in myogenic differentiation. Nucleic Acids Res 45:11236–11248
4. Sincennes MC, Brun CE, Rudnicki MA (2016) Concise review: epigenetic regulation of myogenesis in health and disease. Stem Cells Transl Med 5:282–290
5. Dumont NA, Bentzinger CF, Sincennes MC, Rudnicki MA (2015) Satellite cells and skeletal muscle regeneration. Compr Physiol 5:1027–1059
6. Sambasivan R, Yao R, Kissenpfennig A, Van Wittenberghe L, Paldi A, Gayraud-Morel B, Guenou H, Malissen B, Tajbakhsh S, Galy A (2011) Pax7-expressing satellite cells are indispensable for adult skeletal muscle regeneration. Development 138:3647–3656
7. Lepper C, Partridge TA, Fan CM (2011) An absolute requirement for Pax7-positive satellite cells in acute injury-induced skeletal muscle regeneration. Development 138:3639–3646
8. Chakroun I, Yang D, Girgis J, Gunasekharan A, Phenix H, Kærn M, Blais A (2015) Genome-wide association between Six4, MyoD, and the histone demethylase Utx during myogenesis. FASEB J 29:4738–4755
9. Musarò A (2014) The basis of muscle regeneration. Adv Biol 2014:1–16
10. Dilworth FJ, Blais A (2011) Epigenetic regulation of satellite cell activation during muscle regeneration. Stem Cell Res Ther 2:18
11. Le May M, Mach H, Lacroix N, Hou C, Chen J, Li Q (2011) Contribution of retinoid X receptor signaling to the specification of skeletal muscle lineage. J Biol Chem 286:26,806–26,812

Chapter 10

Mapping Retinoic Acid-Dependant 5mC Derivatives in Mouse Embryonic Fibroblasts

Haider M. Hassan, T. Michael Underhill, and Joseph Torchia

Abstract

Methylase-assisted bisulfite sequencing (MAB-seq) is a derivatization technique to evaluate the presence of 5-formylcytosine (5-fC) and 5-carboxylcytosine (5-caC) at base-pair resolution. Although MAB-seq was originally designed to study these metabolites under steady-state conditions, we have developed an alternative protocol to evaluate the dynamics of 5-fC/5-caC accumulation in response to agonists, such as all-trans retinoic acid (ATRA). In addition, this protocol utilizes a lower quantity of the *M.SssI* enzyme without compromising methylation efficiency and requires less bench time. Herein, we describe the use of MAB-seq assay to evaluate the generation of 5-fC/5-caC in response to ATRA in mouse embryonic fibroblasts, using the hypermethylated in cancer 1 (*Hic1*) locus as a model system.

Key words *M.SssI* methyltransferase, TDG, Active DNA demethylation, MAB-seq, TET, DNMT, SAM, Retinoic acid, RAR, HIC1

1 Introduction

The methylation of cytosine at the 5th carbon (5mC) is an epigenetic modification that occurs predominantly in the context of CpG dinucleotides and is important for a variety of biological processes including genomic imprinting, X-chromosome inactivation, and the suppression of mobile genetic elements [1, 2]. During cell proliferation, the 5mC mark is faithfully reproduced by the DNA methyltransferase 1 protein (DNMT1) protein, which catalyzes the transfer of a methyl group from S-adenosylmethionine (SAM) to the cytosine nucleotide [3]. In contrast, the de novo methyltransferases (DNMT3a/b) can deposit the 5mC mark in a cell-cycle-independent manner. Although 5mC has long been considered a relatively stable epigenetic mark, recent studies have demonstrated that DNA methylation is highly dynamic [4]. In mammalian systems, there are two corroborated mechanisms that have been proposed for 5mC removal: passive dilution and active DNA demethylation. In passive dilution, the 5mC mark is diluted over

Swapan K. Ray (ed.), *Retinoid and Rexinoid Signaling: Methods and Protocols*, Methods in Molecular Biology, vol. 2019, https://doi.org/10.1007/978-1-4939-9585-1_10,

successive replication cycles by the deactivation or nuclear exclusion of maintenance DNMT1 or its associated cofactor, UHRF1 [5]. In contrast, active demethylation of DNA is an energy-consuming process that involves the oxidation of 5mC to 5-hydroxymethylcytosine (5-hmC) by the Ten eleven translocase (TET) family of dioxygenases [1, 4, 6, 7]. The 5-hmC is further oxidized by TETs to 5-formylcytosine (5-fC) and 5-carboxylcytosine (5-caC). The 5-fC and 5-caC metabolites are removed by the enzyme Thymine DNA glycosylase (TDG). This generates an abasic residue, which is converted to a single-strand break by APE I. POLβ/XRCC1 and DNA ligase 3 then ligate the nick to restore the unmethylated cytosine.

TDG is a pivotal enzyme in active demethylation and also serves as a coactivator for the retinoic acid receptor (RAR), as well as other transcription factors [8–11]. Retinoid signaling is mediated by the RAR nuclear receptor, which heterodimerizes with the retinoid X receptor (RXR) and undergoes a conformational change in response to agonists, such as all-*trans* retinoic acid (ATRA). The ligand-bound RAR:RXR complex facilitates recruitment of TDG and other co-activators to activate transcription of target genes. Several RAR-dependent target genes, such as Hypermethylated in Cancer I (*Hic1*) and *Rarβ*, function as tumor suppressors by causing cell cycle arrest and apoptosis [8, 12]. Hence, RA signaling is often compromised in cancer resulting in DNA hypermethylation of its target genes [8, 13]. Our labs have shown that the activation of *Hic1* by the RAR/RXR involves active demethylation and that the disruption of this process targets *Hic1* for silencing in vivo [10, 11, 14]. Importantly, defective active DNA demethylation is a characteristic feature of cancer cells that may predispose a subset of tumor suppressor genes to silencing via promoter DNA hypermethylation. Therefore, the tracing of 5mC and, more importantly, the active DNA demethylation metabolites 5-fC and 5-caC could serve as important diagnostic markers of cancer progression, and manipulation of these processes may represent suitable chemotherapeutic strategies.

Herein, we demonstrate that the generation of 5-fC and 5-caC in response to RA can be traced using *M.SssI*-assisted bisulfite sequencing (MAB-seq) [1, 15]. Although conventional bisulfite sequencing provides base-pair resolution of 5mC localization, it is unable to distinguish between the unmethylated cytosine, 5-fC or 5-caC. MAB-seq is a derivatization method which consists of enzymatic pretreatment of genomic DNA with the *M.SssI* methyltransferase prior to bisulfite conversion. *M.SssI* is a bacterial enzyme that methylates only unmodified cytosines within a double-stranded CpG context. This is followed by bisulfite treatment and sequencing of the *M.SssI*-treated DNA, which results in unmodified C, 5mC, and 5-hmc to be read as cytosine, whereas 5-fC and 5-caC cannot be methylated and are converted to thymine. Therefore, in

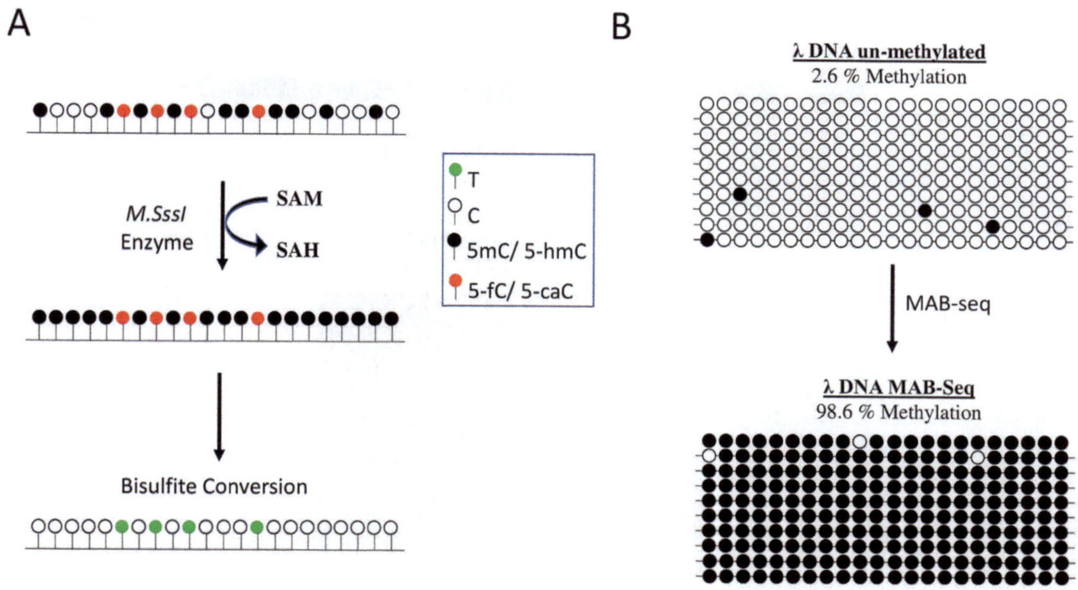

Fig. 1 *M.SssI*-assisted bisulfite sequencing of the lambda phage DNA. (**a**) In the MAB-seq protocol, the genomic DNA is treated with the *M.SssI* enzyme which converts unmethylated cytosines to 5mC. Upon bisulfite conversion of the methylated DNA, the unmethylated cytosine/5mC/5hmC are all read as cytosine, whereas the 5-fC and 5-caC cannot be methylated and are sequenced as thymine. (**b**) Methylation of 100 ng lambda phage DNA was performed using the *M.SssI* enzyme, followed by bisulfite conversion and sequencing. A total of 10 clones were sequenced which depicted a methylation efficiency of 98.6%. T: Thymine, C: cytosine, 5mC: 5-methylcytosine, 5-hmC: 5-hydroxymethylcytosine, 5-fC: 5-formylcytosine, 5-caC: 5-carboxylcytosine

the MAB-seq protocol, any nucleotide that is read as thymine within a CpG context must be considered 5-fC or 5-caC (Fig. 1a). For a successful MAB-seq assay, the efficient methylation of DNA is essential. Hence, we include a spike-in lambda (λ) DNA control, which lacks DNA methylation, with our experimental DNA and routinely observed over 98% methylation efficiency in the MAB-seq assay (Fig. 1b). The MAB-seq protocol we describe is a modification of a previously published protocol [16]. We present a novel methodology that requires significantly less *M.SssI* enzyme and has been thoroughly tested for tracing 5-fC/5-caC at genomic locations in response to RAR agonists. We have also incorporated reagents that simplify sample preparation and enhance DNA precipitation. Lastly, our protocol is considerably less time consuming and it has been thoroughly tested to deliver a more robust assay with increased reproducibility and efficiency of methylation.

The methods described here have been used to assess 5-fC/5-caC accumulation at the *Hic1* locus in response to ATRA, but has also been adapted to examine 5-mC derivatives from ES cells as well as whole tissues [1, 15]. *Hic1* is a transcriptional repressor and a candidate tumor suppressor that is silenced in many cancers as a result of promoter DNA hypermethylation [12, 17–19]. There are

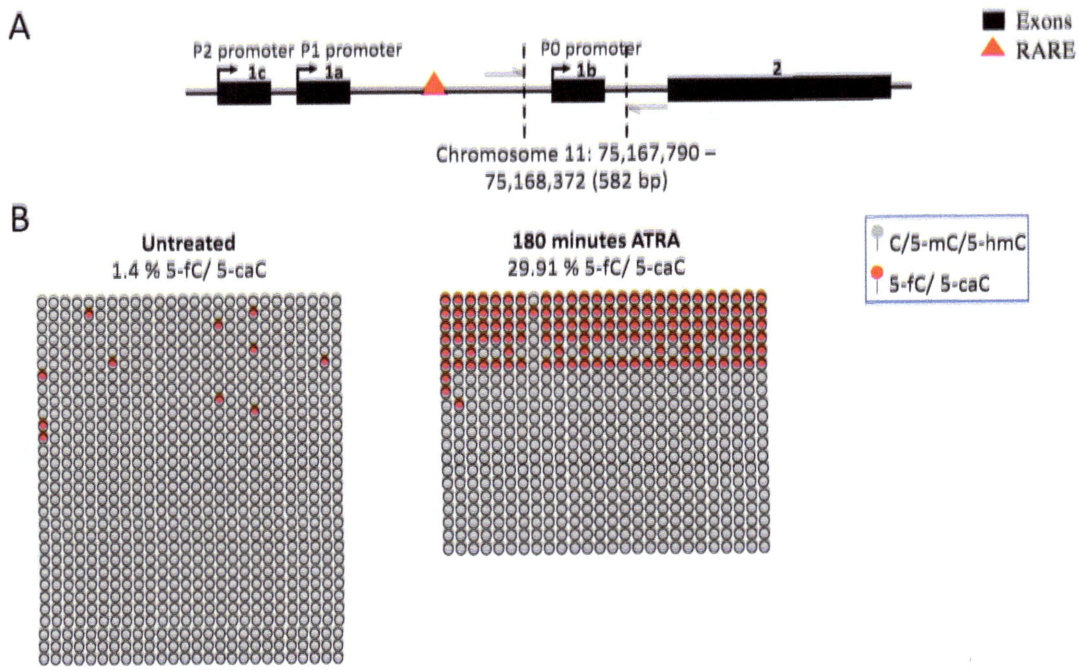

Fig. 2 MAB-seq analysis of the *Hic1* locus in response to 3-hr ATRA treatment. (**a**) Schematic of the mouse *Hic1* locus. The *Hic1* gene comprises three distinct promoters that give rise to alternate exons (1a, 1b, and 1c) and a common exon 2. Transcript containing large exon 2 and 1b, which is associated with the P0 promoter, is the most prevalent variant. The bisulfite sequencing primers listed in this protocol amplify the region encompassing the *Hic1* P0 promoter. (**b**) MAB-seq analysis of the *Hic1* locus demonstrates a significant accumulation of 5-fC/5-caC metabolites in response to a 3-h ATRA treatment. A total of 26 CpG sites were sequenced and the percentage of 5-fC/5-caC were calculated by dividing the total number of thymine nucleotides in CpG sites and total number of CpGs sequenced. A greater than 20-fold increase in 5-fC/5-caC metabolites was found after 3 h ATRA treatment in MEFs

three separate promoters within *Hic1* locus (Fig. 2a) and studies have shown that the methylation status of the P0 promoter strongly correlates with *Hic1* gene transcription. We have shown that this region is targeted for active DNA demethylation in response to ATRA (Fig. 2b) [10]. In general, the levels of 5-fC/5-caC in DNA are quite low in the untreated state [1, 9, 20], so a large number of clones need to be sequenced. However, upon ATRA treatment the 5-fC/5-caC metabolites accumulate to a significant level and can be easily detected with a lower sequencing depth.

Other derivatization methods have been developed to analyze 5-fC and 5-caC at single nucleotide resolution [21, 22]. However, MAB-seq requires a single step involving DNA pretreatment before bisulfite conversion, and importantly the calling of modified cytosines does not require subtractive analysis between two sequences which can reduce false discovery rate. Therefore, it is well suited for examining 5-fC/5-caC genome-wide using next generation sequencing [1, 15]. A major limitation of MAB-seq is its inability to distinguish between the 5-fC and 5-caC metabolites, although this can be overcome by performing additional derivatization assays such as treatment with sodium borohydride [16, 21]. Overall, the MAB-seq technique provides a simple and highly reproducible assay for base-pair resolution mapping of 5-fC and 5-caC metabolites and serves as a powerful tool for the study of dynamic DNA methylation in response to ATRA or other DNA-methylation modifying processes.

2 Materials

Prepare all solutions using ultra-pure milli Q water and store at −20 °C. ATRA is used as an agonist to stimulate retinoid signaling. The SAM reagent stock must be aliquoted and thawed only once per in vitro methylation cycle.

2.1 Retinoic Acid Treatment

1. ATRA solution: Dissolve the ATRA powder in 100% DMSO to a stock concentration of 300 mM. Then aliquot in 10 μL fractions into 1.5 mL tubes and store at −80 °C.
2. Prepare adherent cells for ATRA treatment: add penicillin/streptomycin solution, glutamine, and 50 mL of charcoal-stripped FBS to 500 mL of DMEM in a laminar flow-hood. Mix and store the media at 4 °C.
3. Trypsin (0.25%).
4. Sterile glass pipettes (2, 5, and 10 mL).
5. Dulbecco's phosphate buffer (DPBS).

2.2 The Methylation of DNA by M.SssI

1. *M.SssI* enzyme: the enzyme was purchased from NEB labs at 4 U/μL. Store the enzyme at −20 °C and store on ice when using at the bench.
2. SAM: the SAM stock is provided with the *M.SssI* enzyme at a concentration of 32 mM. Thaw out the SAM stock at 4 °C and aliquot in 10 μL fractions. Then store the SAM stock at −20 °C and avoid repeated freeze thaws.
3. Phenol:chloroform:isoamyl (PCI): The PCI solution was purchased from Invitrogen and stored at 4 °C. Warm to room temperature (23 °C) for 30 min immediately prior usage.
4. Ethanol precipitation: 95% ethanol solution (v/v) was prepared by adding 95 mL of anhydrous ethanol in 5 mL of milli Q water. This was stored at −20 °C. For the ethanol precipitation, a 3.0 M ammonium acetate, pH 5.5 solution is used. For the washes, a 75% ethanol solution needs to be made using milli Q water and stored at 4 °C.
5. Glycogen: glycogen was purchased from SIGMA. It was prepared to a 20 mg/mL stock solution using milli Q water, aliquoted in 10 μL fractions and stored at −20 °C.

2.3 Bisulfite Conversion and TA Cloning

1. DNA Extraction: The DNA was extracted using the SIGMA GenElute™ Mammalian Genomic DNA Miniprep kit.
2. Bisulfite conversion kit: the bisulfite kit was purchased from Qiagen EpiTect. The bisulfite powder was dissolved in 800 μL milli Q water, placed in −20 °C fridge, and stored for a maximum of 1 month.
3. Polymerase chain amplification (PCR): The PCR was performed using the Invitrogen Platinum *Taq* DNA polymerase kit as per manufacturer guidelines.
4. For the PCR amplification of lambda bisulfite-converted DNA, the following primers were used:
   (a) λ forward primer: 5′-TTTGGGTTATGTAAGTTGATTTTATG-3′.
   (b) λ reverse primer: 5′-CACCCTACTTACTAAAATTTACACC-3′.
5. For PCR amplification of the *Hic1* P0 promoter, the following bisulfite primers were used:
   (a) *Hic1* promoter forward primer: 5′-GAGTTGGGGAAGAGATGTGGAG-3′.
   (b) *Hic1* promoter reverse primer: 5′-CCAACCCCAAATACTCCTAAACAC-3′.
6. Amplicon cloning: The TA cloning kit was purchased from Invitrogen and stored at −20 °C prior to usage. The TOP10 bacteria were stored in −80 °C and thawed on ice prior to usage.
7. Qiagen quick lyse miniprep kit.

3 Methods

3.1 Treatment of Cells with ATRA

1. Incubate the charcoal-stripped DMEM cell culture media and DPBS in 37 °C water bath for 30 min. Also, thaw the ATRA stock solution (300 mM) at room temperature for 10 min.
2. Make a working stock of 1 μM ATRA in 25 mL of charcoal-stripped DMEM at room temperature (*see* **Note 1**).
3. Wash the plated adherent cells (*see* **Note 2**) with 5 mL of DPBS solution.
4. Aspirate the DPBS, and then add the appropriate amount of DMEM + ATRA stock media. Add the media in order to liberally cover the layer of cells insuring that no areas of the plate are without media.
5. Treat the cells with ATRA for 180 min in a 37 °C, 5% CO_2 incubator.

3.2 Isolation of the Genomic DNA for M. SssI Treatment

1. Aspirate the media from the cells and wash once with DPBS solution at room temperature. Repeat the wash with DPBS.
2. Scrape the cells in DPBS solution and pipette it into an appropriate 1.5 mL microcentrifuge tube.
3. Pellet the cells by centrifugation at 2000 × *g* for 5 min at 4 °C.
4. Remove the supernatant and resuspend the cells in DPBS solution. Then re-pellet the cells by centrifugation at 2000 × *g* for 5 min at 4 °C.
5. Remove the supernatant and resuspend the pellet in SIGMA resuspension buffer to prepare cells for DNA extraction.
6. DNA extraction was performed using the SIGMA GenElute kit as per manufacturer guidelines. For proteinase K digestion, the 1.5 mL microcentrifuge tube was incubated in a temperature-controlled thermal block with shaking at 600 × *g* for 10 min. At the end of the extraction, the DNA was eluted in 200 μL of TE buffer, pH 8.0.
7. The quality and concentration of DNA was assessed using NanoDrop. The DNA was then used immediately for in vitro methylation, or flash-frozen and kept at – 80 °C.

3.3 Methylation of the DNA In Vitro

1. Aliquot 1 μg of the genomic DNA into a 1.5 mL microcentrifuge tube and top up the volume to 200 μL. Then add 100 ng of λ-phage DNA for quality control to assess the methylation efficiency.
2. Add 1 μL of 20 mg/mL glycogen and mix by vortexing (*see* **Note 3**).
3. For ethanol precipitation of the DNA add 50 μL of 3.0 M ammonium acetate (pH 5.5) solution and mix by vortexing for 15 s. Then pipette 700 μL of ice-cold 95% ethanol (−20 °C) and mix by vortexing.
4. Incubate at −80 °C overnight to allow the DNA to precipitate.
5. After the overnight incubation, spin the microcentrifuge tube at 13,000 × *g* for 20 min at 4 °C to pellet the DNA.
6. Wash the pellet by adding 1 mL of cold 75% ethanol (4 °C) followed by centrifugation at 13,000 × *g* for 5 min at 4 °C and carefully remove the supernatant without dislodging the pellet.
7. Repeat **step 6** in Subheading 3.3 two more times. After removing the supernatant, spin the tube down one more time and carefully remove the remaining supernatant without disturbing the DNA pellet.
8. Let the DNA pellet air-dry for 5–10 min at room temperature. When the DNA pellet is completely dry, it will become

somewhat translucent. Do not over-dry which will hamper the subsequent aqueous reconstitution of the DNA.

9. Resuspend the DNA in 14 μL of milli Q water.
10. Following the NEB protocol, add 2 μL of NEBuffer 2, 1 μL of 4 U/μL *M*.SssI, and 1 μL SAM (working concentration of 160 μM). Mix the volume by pipetting up and down five times, then briefly centrifuge the microcentrifuge tube.
11. Incubate the microcentrifuge tube for 2 h in a 37 °C water bath.
12. Add an additional 4 U of *M.SssI* enzyme and 160 μM SAM (Fresh) to the sample, gently mix, centrifuge, and incubate for an additional 4 hours in a 37 °C water bath.
13. After the *M.SssI* incubation is complete, increase the volume to 100 μL using milli Q water. Then perform PCI extraction as follows: Add 100 μL of PCI solution (*see* **Note 4**) and vortex for 30 s. Then centrifuge the mixture at 13,000 × *g* for 5 min at room temperature and transfer upper phase to a clean 1.5 mL microcentrifuge tube. Please be careful to avoid pipetting the bottom phenol layer as it will inhibit the *M.SssI* methylase activity and any downstream analysis. Add an additional 100 μL of milli Q water to the phenol layer and vortex for 30 s at high speed. Spin the microcentrifuge tube at 13,000 × *g* for 5 min at room temperature and combine the two upper phase collections.
14. The **steps** outlined from **3** to **13** in Subheading 3.3 get repeated an additional three times. The entire protocol turns out to be a three-day procedure with overnight precipitations.
15. Then perform the final ethanol precipitation of the methylated DNA as outlined from **steps 3** to **8** in Subheading 3.3.
16. Resuspend the DNA in 20 μL of milli Q water and immediately proceed to bisulfite conversion.

3.4 Bisulfite Conversion of the DNA

1. The bisulfite conversion of the DNA was performed using the Qiagen Epitect bisulfite kit. Briefly, 20 μL of the *M.SssI*-treated DNA was transferred to a 200 μL PCR tube. Then 85 μL of the bisulfite mix and 35 μL of the DNA protection buffer were added to the PCR tube. The color of the solution at this point should be dark blue.
2. The PCR tube was capped and vortexed followed by a brief centrifugation.
3. The subsequent cleanup of the bisulfite-converted DNA was performed as per the manufacturers protocol. DNA was eluted from the provided mini columns using two volumes of 20 μL each (total volume of 40 μL) and immediately used for analysis. The remaining DNA was flash-frozen and kept at −80 °C.

3.5 Quality Control of the MAB-seq Assay

1. PCR was performed using the Platinum *Taq* DNA polymerase kit. 5 μL of the bisulfite-converted DNA and 1 μL of lambda phage primers were combined in a 50 μL reaction. The cycle conditions for PCR amplification were as follows: 94 °C, 60 s; 30 × [94 °C, 30 s; 60 °C, 30 s, 72 °C, 30 s]. 72 °C, 7 min; 4 °C, indefinitely.
2. After amplification, 2 μL of the PCR product was used for the TA cloning as per the manufacturer's instruction (*see* **Note 5**). For these purposes, the ligation reaction was performed for 30 min at room temperature.
3. The ligations were transformed into TOP10 bacteria. The transformed bacteria were streak plated on agar media supplemented with 50 μg/mL ampicillin and grown in a 37 °C incubator overnight (*see* **Note 6**).
4. Following emergence of colonies, 10–15 individual white colonies from blue white selection were inoculated into 5 mL liquid broth (LB) media supplemented with 50 μg/mL ampicillin and grown overnight in a 37 °C incubator, with shaking at 225 rpm.
5. Plasmids were isolated from the LB media using the QIAGEN plasmid mini-purification kit and sequenced using the M13-reverse primer.
6. The efficiency of DNA methylation was assessed using the following formula:

$$\left(\frac{\text{(Total Number of methylated CpG)}}{\text{(Total CpG in the region being sequenced)} \times \text{(Number of clones sequenced)}}\right) \times 100\%.$$

A lambda phage DNA methylation efficiency of greater than 95% is required in the MAB-seq assay prior to analysis of 5-fC/5-caC accumulation at the genomic target of interest

3.6 Analysis of 5-fC/5-caC at the Hic1 P0 Promoter in Response to RA

1. PCR amplification was performed using the Platinum *Taq* DNA polymerase kit. 5 μL of the bisulfite-converted DNA and 0.5 μL of each of the *Hic1* primers were combined in a 50 μL reaction (*see* **Note 7**). The cycle conditions were as follows: 94 °C, 60 s; 30 × [94 °C, 30 s; 61 °C, 30 s, 72 °C, 40 s], 72 °C, 7 min; 4 °C, indefinitely.
2. Following amplification, 2 μL of the PCR product was cloned and transformed into TOP10 bacteria as outlined in **steps 2** and **3** in Subheading 3.5.
3. After incubation, 20–30 single white colonies from blue/white selection were inoculated into LB media supplemented with 50 μg/mL ampicillin and grown overnight in a 37 °C incubator, with shaking at 225 rpm and the plasmids were

isolated as outlined in **steps 4** and **5** in Subheading 3.5 (*see* **Note 8**).

4. The plasmids were then sequenced by standard sanger-sequencing using the M13-reverse primer.
5. To evaluate the accumulation of 5-fC/5-caC metabolites, the total number of thymines in a CpG island were pooled from all individual clones and then divided by the total number of CpGs sequenced (*see* **Note 9**). This can be represented using the following formula:

$$\%\frac{5-\mathrm{fC}}{5-\mathrm{caC}}=\frac{(\text{Total number } of\ T \text{ in target CpG sites [from all clones]})}{(\text{Total CpG's in target region})\times(\text{Number of clones sequenced})}\times 100\%.$$

This analysis is to be performed for each treatment in three independent experiments. We find that the *Hic1* P0 promoter has negligible 5-fC/5-caC in the untreated state; however, these metabolites were found to accumulate upon RA treatment. This would be indicative of the genomic region undergoing active DNA demethylation.

Although we were able to consistently observe an increase in 5-fC/5-caC accumulation at the *Hic1* locus in response to ATRA, there was significant variability observed between different experiments. This can be attributed to inefficient methylation of the DNA, or variable oxidation of 5mC in different cells. This can be overcome by serum-starvation of the cells for cell cycle synchronization prior to treatment and including an extra cycle of DNA methylation. In addition, it is important to repeat the experiment a minimum of three times and include a lambda DNA spike in control with each MAB-seq experiment as a measure of methylation efficiency.

4 Notes

1. The ATRA solution is light sensitive. Wrap the tube in aluminum foil when carrying it to the laminar flow-hood and use reduced light in the hood. Extra precautions should be taken when at the bench, such as prepping the ATRA in DMEM stock under dimmed conditions. Fresh ATRA solution should be prepared at least once a month.
2. The plate of adherent cells must be grown for a minimum of 48 h in charcoal-stripped FBS supplemented media prior to the ATRA treatment. This is important because normal FBS can partially stimulate the retinoid receptors and modulate transcription of target genes. The charcoal-stripped serum is deficient for many lipophilic molecules. In addition, the cells must

be 90% confluent at the time of treatment since the levels of proteins involved in the active DNA demethylation process tend to fluctuate with cell confluency. Lastly, it is recommended that the MEFs must not be passaged more than three times prior to treatment, as that can induce an onset of replicative senescence.

3. Glycogen assists with the visualization of DNA during the PCI purification steps and also enhances precipitation. This is crucial when performing ethanol precipitation of DNA and prepping for subsequent methylation assays. We have determined through extensive analysis that glycogen at the amounts being used does not interfere with the MAB-seq assay.
4. Make sure that the PCI has been equilibrated to room temperature and properly mixed prior to usage. The incubation should take roughly 30 min.
5. Ensure that no more than 2 μL of the PCR reaction is used for the TA cloning. More DNA can induce artifacts such as self-ligation of the PCR fragments or the plasmids and confound downstream sequencing. Also, it is important to ensure that the PCR reaction generates a single product, as the presence of unwanted amplification can complicate the colony selection process. The PCR product should be run on a 1% agarose gel to ensure the presence of a single band.
6. Do not incubate the agar plate for more than 16 h at 37 °C as it will accelerate the degradation of ampicillin leading to the growth of non-transformed or satellite colonies. Excessive incubation of the plate can also make it hard to distinguish colonies, thereby preventing successful colony picking prior to sequencing. Once the plate is removed from the incubator, they can be stored in an appropriate 4 °C fridge for up to 1 month.
7. We previously identified a 155 base-pair (bp) core region within the *Hic1* P0 promoter (containing 15 CpG sites) that undergoes active DNA demethylation in response to RA [10]. The primers outlined in the materials section 1.3.5 amplify a 582 bp region (Chromosome 11: 75,167,790–75,168,372) within the mouse genome that encompasses *Hic1* P0 promoter and the flanking DNA sequences. These primers were utilized for conventional bisulfite and MAB-seq analysis of the *Hic1* P0 promoter in response to RA treatment in MEFs.
8. The 5-fC/5-caC metabolites are present at a significantly lower frequency compared to 5-mC and 5-hmC. Therefore, a much higher sequencing depth is required for a meaningful estimate of RA-mediated 5-fC and 5-caC accumulation at a genomic locus. We recommend sequencing between 20 and 30 individual colonies for MAB-seq analysis of genomic target of interest.

9. In addition to efficient methylation by the *M.SssI* enzyme, it is important that the bisulfite conversion rate is ≥98%. A defective bisulfite conversion will lead to a high incidence of "N" in the clones sequenced, which is indicative of an indistinguishable nucleotide sequence. These clones often possess a low bisulfite conversion efficiency and must be disregarded from the analysis.

References

1. Wu H, Zhang Y (2014) Reversing DNA methylation: mechanisms, genomics, and biological functions. Cell 156:45–68
2. Zhu J-K (2009) Active DNA demethylation mediated by DNA glycosylases. Annu Rev Genet 43:143–166
3. Kohli RM, Zhang Y (2013) TET enzymes, TDG and the dynamics of DNA demethylation. Nature 502:472–479
4. Wu X, Zhang Y (2017) TET-mediated active DNA demethylation: mechanism, function and beyond. Nat Rev Genet 18:517–534
5. Piccolo FM, Fisher AG (2014) Getting rid of DNA methylation. Trends Cell Biol 24:136–143
6. He Y-F, Li B-Z, Li Z, Liu P, Wang Y, Tang Q, Ding J, Jia Y, Chen Z, Li L et al (2010) Tet-mediated formation of 5-carboxylcytosine and its excision by TDG in mammalian DNA. Science 333:742–745
7. Hashimoto H, Zhang X, Cheng X (2013) Selective excision of 5-carboxylcytosine by a thymine DNA glycosylase mutant. J Mol Biol 425:971–976
8. Tang X-H, Gudas LJ (2011) Retinoids, retinoic acid receptors, and cancer. Annu Rev Pathol Mech Dis 6:345–364
9. Léger H, Smet-Nocca C, Attmane-Elakeb A, Morley-Fletcher S, Benecke AG, Eilebrecht S (2014) A TDG/CBP/RARα ternary complex mediates the retinoic acid-dependent expression of DNA methylation-sensitive genes. Genomics Proteomics Bioinformatics 12:8–18
10. Hassan HM, Kolendowski B, Isovic M, Bose K, Dranse HJ, Sampaio AV, Underhill TM, Torchia J (2017) Regulation of active DNA demethylation through RAR-mediated recruitment of a TET/TDG complex. Cell Rep 19:1685–1697
11. Le May N, Mota-Fernandes D, Vélez-Cruz R, Iltis I, Biard D, Egly JM (2010) NER factors are recruited to active promoters and facilitate chromatin modification for transcription in the absence of exogenous genotoxic attack. Mol Cell 38:54–66
12. Chen WY, Zeng X, Carter MG, Morrell CN, Chiu Yen R-W, Esteller M, Watkins DN, Herman JG, Mankowski JL, Baylin SB (2003) Heterozygous disruption of Hic1 predisposes mice to a gender-dependent spectrum of malignant tumors. Nat Genet 33:197–202
13. Um S, Harbers M, Benecke A, Pierrat B, Losson R, Chambon P (1998) Retinoic acid receptors interact physically and functionally with the T:G mismatch-specific thymine-DNA glycosylase. J Biol Chem 273:20728–20736
14. Thillainadesan G, Chitilian JM, Isovic M, Ablack JNG, Mymryk JS, Tini M, Torchia J (2012) TGF-β-dependent active demethylation and expression of the p15ink4b tumor suppressor are impaired by the ZNF217/CoREST complex. Mol Cell 46:636–649
15. Neri F, Incarnato D, Krepelova A, Rapelli S, Anselmi F, Parlato C, Medana C, Dal Bello F, Oliviero S (2015) Single-base resolution analysis of 5-formyl and 5-carboxyl cytosine reveals promoter DNA methylation dynamics. Cell Rep 10:674–683
16. Wu H, Wu X, Zhang Y (2016) Base-resolution profiling of active DNA demethylation using MAB-seq and caMAB-seq. Nat Protoc 11:1081–1100
17. Zheng J, Xiong D, Sun X, Wang J, Hao M, Ding T, Xiao G, Wang X, Mao Y, Fu Y et al (2012) Signification of hypermethylated in cancer 1 (HIC1) as tumor suppressor gene in tumor progression. Cancer Microenviron 5:285–293
18. Chen WY, Wang DH, Yen RC, Luo J, Gu W, Baylin SB (2005) Tumor suppressor HIC1 directly regulates SIRT1 to modulate p53-dependent DNA-damage responses. Cell 123:437–448
19. Jenal M, Trinh E, Britschgi C, Britschgi A, Roh V, Vorburger S a, Tobler A, Leprince D, Fey MF, Helin K et al (2009) The tumor suppressor gene hypermethylated in cancer 1 is transcriptionally regulated by E2F1. Mol Cancer Res 7:916–922

20. Shen L, Wu H, Diep D, Yamaguchi S, D'Alessio AC, Fung H-L, Zhang K, Zhang Y (2013) Genome-wide analysis reveals TET- and TDG-dependent 5-methylcytosine oxidation dynamics. Cell 153:692–706
21. Lu X, Song C-X, Szulwach K, Wang Z, Weidenbacher P, Jin P, He C (2013) Chemical modification-assisted bisulfite sequencing (CAB-Seq) for 5-carboxylcytosine detection in DNA. J Am Chem Soc 135:9315–9317
22. Song CX, Szulwach KE, Dai Q, Fu Y, Mao SQ, Lin L, Street C, Li Y, Poidevin M, Wu H et al (2013) Genome-wide profiling of 5-formylcytosine reveals its roles in epigenetic priming. Cell 153:678–691

Chapter 11

Hemodynamics-Based Strategy of Using Retinoic Acid Receptor and Retinoid X Receptor Agonists to Induce MicroRNA-10a and Inhibit Atherosclerotic Lesion

Ding-Yu Lee and Jeng-Jiann Chiu

Abstract

The protocols in this chapter describe methods for identifying the functional roles of retinoic acid receptor (RAR) and retinoid X receptor (RXR) signaling in atherosclerosis and developing RARα/RXRα-specific agonists as hemodynamics-based therapeutic components for atherosclerosis treatment. In vitro cell culture flow system is used to elucidate the effects of different flow patterns and shear stresses, i.e., atherogenic oscillatory shear stress (OS) vs. atheroprotective pulsatile shear stress (PS), on RAR/RXR signaling and inflammatory responses in vascular endothelial cells (ECs). Western blotting, nuclear and cytoplasmic protein extraction, immunoprecipitation, and in situ proximity ligation assay are used to examine the expression, location, and association of RARs (i.e., RARα, RARβ, and RARγ) and RXRs (i.e., RXRα, RXRβ, and RXRγ) in ECs in response to OS vs. PS. Chromatin immunoprecipitation is used to examine the binding activity of RARα/RA-responsive elements (RARE). RT-microRNA (miR) quantitative real-time PCR and RT-PCR are used to detect the expressions of miR-10a and pro-inflammatory molecules, respectively. Specific siRNAs of RARα and RXRα, precursor miR-10a (PreR-10a), and antagomiR-10a (AMR-10a) are used to elucidate the regulatory roles of RARα, RXRα, and miR-10a in pro-inflammatory signaling in ECs. RARα/RXRα-specific agonists are used to induce miR-10a expression and inhibit OS-induced pro-inflammatory signaling in ECs in vitro. Apolipoprotein E-deficient (ApoE$^{-/-}$) mice are used as an atherosclerotic animal model. Administration of ApoE$^{-/-}$ mice with RARα/RXRα-specific agonists results in inhibitions in atherosclerotic lesion formation. Co-administration of ApoE$^{-/-}$ mice with RARα/RXRα agonists and AMR-10a is performed to identify the role of miR-10a in RARα/RXRα agonists-mediated inhibition in atherosclerotic lesions. Oil Red O staining and H&E staining are used to examine the levels of atherosclerotic lesions in the vessel wall. In situ miR hybridization and immunohistochemical staining are used to detect the expression of miR-10a and pro-inflammatory molecules and the infiltration of inflammatory cells in the vessel wall. RARα/RXRα-specific agonists are used to mimic the atheroprotective effects of PS to induce endothelial miR-10a and hence repress OS-induced pro-inflammatory signaling and atherosclerotic lesion formation in vivo. The results indicate that RAR/RXR-specific agonists have great potential to be developed as hemodynamics-based therapeutic components for atherosclerosis treatment.

Key words Atherosclerosis, Endothelial cell, MicroRNA, Retinoic acid receptor, Retinoid X receptor, Shear stress

Swapan K. Ray (ed.), *Retinoid and Rexinoid Signaling: Methods and Protocols*, Methods in Molecular Biology, vol. 2019, https://doi.org/10.1007/978-1-4939-9585-1_11, © Springer Science+Business Media, LLC, part of Springer Nature 2019

1 Introduction

In the circulation system, hemodynamic forces have been defined as vital factors to modulate vascular endothelial cell (EC) function and dysfunction [1]. Pulsatile (PS) and oscillatory shear stresses (OS) are well-recognized hemodynamic forces to elicit atheroprotective and atherogenic signaling in ECs, respectively [1, 2]. OS develops in branches and curvatures of arterial tree to exert atherogenic effects, which induce EC dysfunction and hence atherosclerosis [1–4]. In contrast, PS occurs in the straight segments of arterial tree to play atheroprotective roles in regulating EC function to prevent atherosclerosis [1–4].

Retinoic acid receptors (RARs, i.e., RARα, RARβ, and RARγ) and retinoid X receptors (RXRs, i.e., RXRα, RXRβ, and RXRγ), which are members of nuclear hormone receptor superfamily, are involved in the modulation of EC functions and vascular biology [5–8]. RARs serve as transcriptional factors, which can bind to specific DNA sequences, i.e., RA-responsive elements (RAREs), in the enhancer regions of target genes to induce their transcriptional activation. RARs can also recruit various co-regulators to form hetero-complexes to modulate their transcriptional activity [5]. Among these co-regulators, RXRs are identified as "enhancers" to associate with RARs to enhance their transcriptional activity [5]. In contrast, histone deacetylases (HDACs), which can remove acetyl groups from critical transcriptional factors or signaling molecules to suppress their functions [9], are defined as "repressors" to suppress their transcriptional activity [10]. RAR-α-specific agonist AM580 has been found to activate RARα to release "repressors" and recruit "enhancers" through repositioning of the helical motif H12 to make structure changes of RARα [11, 12]. In addition, co-addition of RAR- and RXR-specific ligands has synergistic effects on enhancing the transcriptional activity of RARs [13].

MicroRNA (miR)-10a, which is the miR binding to 3'UTR of target genes to repress their expression [14], has been shown to be transcriptionally regulated by RARs through binding to RARE in the enhancer region of miR-10a [15]. RAR antagonist can repress miR-10a expression and inhibit metastatic behavior of tumor cells [15]. In vascular ECs, miR microarray has identified miR-10a as the miR with the lowest expression in athero-susceptible region in normal adult swine in vivo [16]. Moreover, miR-10a can inhibit EC expressions of pro-inflammatory molecules involved in the NF-κB-mediated signaling in vitro [16].

In this article, we describe the experimental protocols for studying the functional roles of RAR and RXR signaling in hemodynamics-associated miR-10a expression and inflammatory responses in ECs and atherosclerotic lesion formation. We also

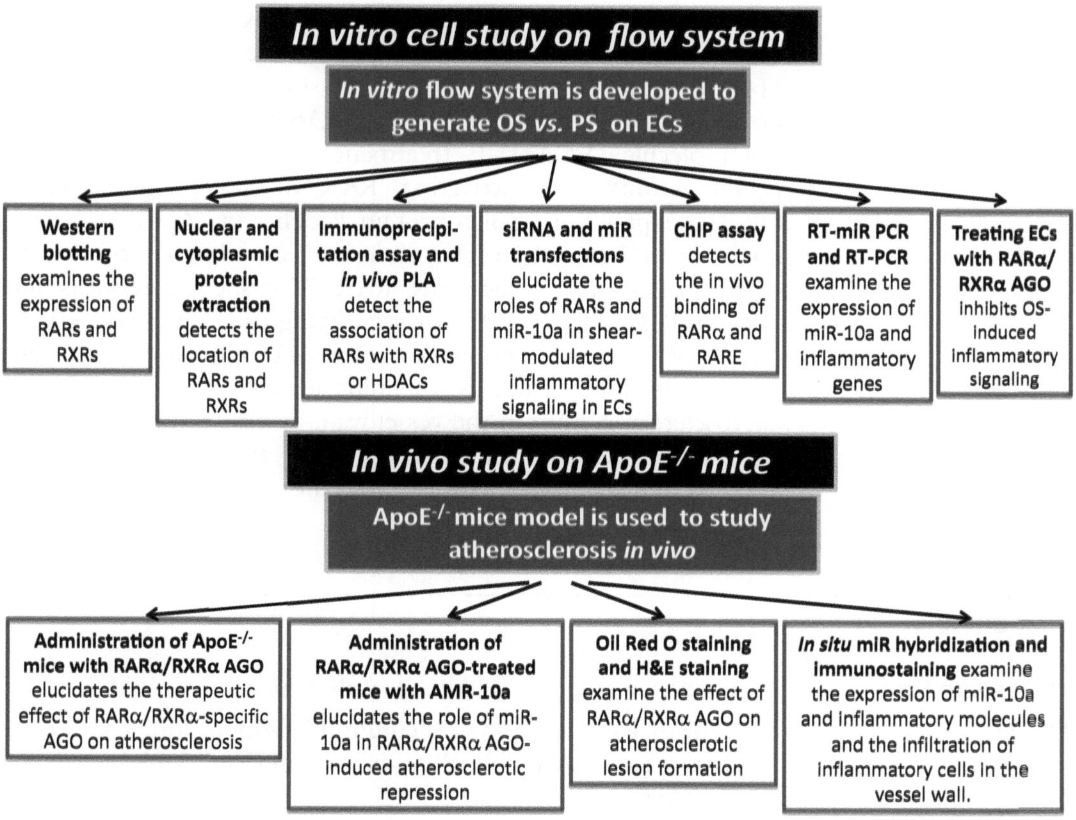

Fig. 1 Schematic diagram of comprehensive approaches for elucidating the roles of RAR and RXR signaling in ECs and atherosclerosis and for developing RAR/RXR-specific agonists as therapeutic components for atherosclerosis treatment. *PS* pulsatile shear stress, *OS* oscillatory shear stress, *ECs* endothelial cells, *RARs* retinoic acid receptors, *RXRs* retinoid X receptors, *PLA* proximity ligation assay, *miR* microRNA, *ChIP* chromatin immunoprecipitation, *ApoE*$^{-/-}$ *mice* apolipoprotein E-deficient mice, *AGO* agonist, *AMR-10a* antagomiR-10a

describe the procedures of developing RARα- and RXRα-specific agonists as hemodynamics-based therapeutic components for atherosclerosis treatment [17, 18]. These experimental procedures include the following steps (Fig. 1): in vitro cell studies on flow system are performed to identify the effects of hemodynamic forces on RAR signaling and miR-10a expression in ECs, in vitro flow system for generating OS vs. PS on ECs to simulate their pathophysiological and physiological hemodynamic conditions, Western blotting and nuclear and cytoplasmic protein extraction for detecting the expressions and locations of RARs and RXRs, immunoprecipitation assay for detecting the associations of RAR with their co-regulator proteins, i.e., RXRs and HDACs, in situ proximity ligation assay (PLA) for visualizing the association of RAR and RXR in the cell nucleus in vivo, small interfering RNA (siRNA) and miR transfections for elucidating the roles of RARs, RXRs, and miR-10a in the pro-inflammatory signaling, chromatin immunoprecipitation

(ChIP) assay for detecting the in vivo binding of RARα with RARE in the enhancer region, RT-miR quantitative real-time PCR and RT-PCR for detecting the expressions of miR-10a and pro-inflammatory molecules, i.e., GATA6 and vascular cell adhesion molecule (VCAM)-1, treatment of ECs with RAR/RXR-specific agonists for activating RAR or RXR and inducing miR-10a expression, apolipoprotein E-deficient ($ApoE^{-/-}$) mice model for studying atherosclerotic lesion formation in vivo, administration of $ApoE^{-/-}$ mice with RAR/RXR-specific agonists and antagomiR-10a (AMR-10a) for identifying the therapeutic effects of RAR/RXR-specific agonists on atherosclerotic lesion formation, Oil Red O staining and H&E staining for examining the levels of atherosclerotic lesions in the vessel wall, in situ miR hybridization and immunohistochemical staining for detecting the expressions of miR-10a and pro-inflammatory molecules and the infiltration of inflammatory cells in the vessel wall.

2 Materials

2.1 In Vitro Flow Apparatus

1. Human aortic ECs (HAECs) (Cell Applications, CA).
2. Endothelial Cell Growth Medium (Cell Applications, CA) supplemented with 10% fetal bovine serum (FBS) and 1% penicillin/streptomycin (P/S).
3. Glass slides (75 × 38 mm, Corning, CA).
4. Fibronectin (30 μg/mL).
5. Parallel-plate flow chamber: silicon gasket with dimensions of 2.5 cm in width (w), 5.0 cm in length, and 0.025 cm in height (h), polycarbonate base plate, glass slides cultured with ECs, and a stainless plate are used to create flow chamber (Fig. 2) [1, 2].
6. Piston pump.
7. 5% CO_2 in air.
8. Temperature controller (Tm controller).
9. Reservoirs.
10. In vitro flow apparatus containing parallel-plate flow chamber, piston pump, 5% CO_2 in air, and Tm controller (*see* Fig. 3) [1, 2].

2.2 Western Blotting

1. Lysis buffer: a buffer contains 1% Nonidet P-40, 0.5% sodium deoxycholate, 0.1% SDS, and a protease inhibitor mixture, including PMSF, aprotinin, and sodium orthovanadate.
2. 2× loading dye: add 5 mL 0.5 M Tris–HCl (pH 6.5), 5 mL glycerol, 10 mL 10% SDS, 5 mL 1% bromophenol blue, and ddH_2O to 50 mL. Before use, add β-mercaptoethanol to make 2%.

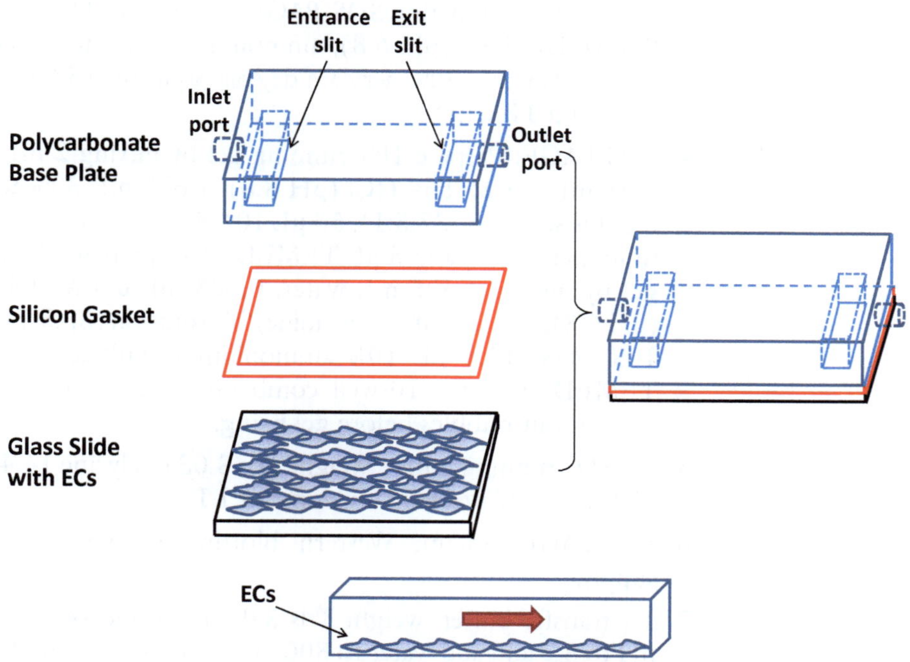

Fig. 2 Schematic diagram shows the composition of the parallel-plate flow chamber. The polycarbonate base plate (Top), silicon gaskets (middle), and the glass slide with EC monolayer (Bottom) are used to create the parallel-plate flow chamber. *ECs* endothelial cells

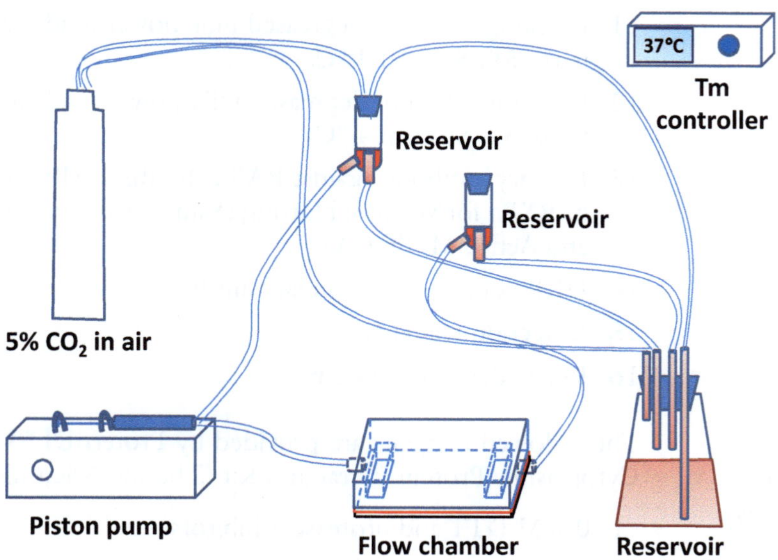

Fig. 3 The composition of in vitro system for generating different flow pattern, i.e., OS vs. PS, on ECs. *Tm controller* temperature controller

3. Materials for preparing SDS-PAGE: 1.5 M Tris–HCl (pH 8.8), 0.5 M Tris–HCl (pH 6.8), ammonium persulfate (10% solution in water), 30% acrylamide/bis solution (37.5:1), 10% SDS, and TEMED.
4. SDS-PAGE: Prepare 10% running gel by mixing 2 mL water, 1.25 mL 1.5 M Tris–HCl (pH 8.8), 1.675 mL 30% acrylamide/bis solution (37.5:1), 50 μL 10% SDS, 25 μL 10% ammonium persulfate, and 5 μL TEMED. Also prepare 4% stacking gel by mixing 1.525 mL water, 0.625 mL 0.5 M Tris–HCL (pH 6.8), 325 μL 30% acrylamide/bis solution (37.5:1), 25 μL 10% SDS, 12.5 μL 10% ammonium persulfate, and 5 μL TEMED. Insert a 10-well comb immediately into stacking gel without bubbles before gel fixing.
5. 1× SDS running buffer: weight Tris 3.03 g, glycine 14.4 g, and SDS 1 g. Add water to a volume of 1 L.
6. SDS-PAGE system, Western blotting system, and power supply.
7. 1× transfer buffer: weight Tris 3.03 g, glycine 14.4 g. Adjust pH to 8.3 and add water to 800 mL. Add 200 mL methanol to a volume of 1 L.
8. Nitrocellulose membranes and filter papers.
9. 10× Tris-buffered saline (10× TBS): weight 12.114 g Tris base and 87.66 g NaCl, adjust PH to 7.4, and add ddH_2O to 1 L.
10. TBST: Tween 20 is added to 1× TBS buffer to make 0.05%.
11. Blocking solution: degreased milk power is added to TBST to make 5%. Store at 4 °C.
12. Dilution solution: degreased milk power is added to TBST to make 5%. Store at 4 °C.
13. Primary antibody against RARα, RARβ, RARγ, RXRα, RXRα, or RXRγ for Western blotting (Santa Cruz Biotechnology, CA; and Active Motif, CA).
14. HRP-conjugated secondary antibody.
15. Detection reagents.
16. Image detection system.

2.3 Nuclear and Cytoplasmic Protein Extraction

The followed reagents are provided by ProteoJET™ Nuclear and Cytoplasmic Protein Extraction Kit (Thermo Scientific, MA):

1. 0.1 M DTT and protease inhibitors.
2. Cell lysis buffer with protease inhibitors and DTT.
3. Nuclei washing buffer with protease inhibitors and DTT.
4. Nuclei storage buffer with protease inhibitors and DTT.
5. Nuclei lysis reagent.

2.4 Immunoprecipitation

1. Lysis buffer: 25 mM HEPES, pH 7.4, 1% Triton X-100, 1% deoxycholate, 0.1% SDS, 0.125 M NaCl, 5 mM EDTA, 50 mM NaF, 1 mM Na_3VO_4, 1 mM PMSF, 10 mg/mL leupeptin, and 2 mM β-glycerophosphate (BGP).
2. Protein A/G plus agarose.
3. Antibody against RARα for immunoprecipitation.
4. Boiling sample buffer: 62 mM Tris–HCl, pH 6.7, 1.25% SDS, 10% (vol/vol) glycerol, 3.75% (vol/vol) mercaptoethanol, and 0.05% bromophenol blue.

2.5 In Situ PLA

1. 4% (wt/vol) paraformaldehyde.
2. PBS.
3. PBS containing 0.1% Triton X-100.
4. Primary antibodies: rabbit anti-human RARα antibody and mouse anti-human RXRα antibody (Santa Cruz Biotechnology, CA).
5. PBS containing 10% or 1% Goat serum.
6. The followed reagents are provided by in situ PLA assay kit (Duolink II, Olink Bioscience, Sweden), except goat serum and primary antibody:
   (a) PLA probe solution: 8 μL PLA PLUS stock and 8 μL PLA MINUS stock are mixed rotationally with 24 μL PBS containing 1% Goat serum for 20 min at 4 °C.
   (b) Wash buffer A.
   (c) Ligase (1 U/μL) and ligation buffer stock (5×).
   (d) Ligation solution containing ligase: dilute ligation buffer stock in water (1:4) to make 40 μL and add 1 μL ligase (*see* **Note 1**).
   (e) Polymerase (10 U/μL) and amplification buffer stock (5×).
   (f) Amplification solution containing polymerase: dilute the amplification buffer stock in water (1:4), and then add the polymerase (i.e., 8 μL 5× amplification buffer stock, 31 μL water, and 1 μL polymerase) (*see* **Note 2**).
   (g) 1x wash buffer B and 0.01× wash buffer B.
   (h) Duolink II mounting medium with DAPI.
   (i) Fluorescence microscopy.

2.6 siRNA or miR Transfection

1. Opti-MEM I Reduced Serum Medium (Thermo Scientific, MA).
2. Specific siRNA for RARα (siRARα), RXRα (siRXRα), GATA6 (siGATA6) and control siRNA (siCL).

3. MiRs, i.e., control miR (CL-miR), precursor miR (PreR-10a), and AMR-10a.
4. Lipofectamine RNAiMAX (Thermo Scientific, MA).
5. CO_2 incubator.

2.7 Treatment of ECs with RARα- and RXRα-Specific Agonists in Response to OS or PS

1. RARα-specific agonist stock solution (AM580, 10 mM): weight 3.5144 mg AM580. Add DMSO to a volume of 1 mL. Votex, and filter by 0.45 μm filter.
2. RXRα-specific agonist stock solution (CD3254, 10 mM): weight 3.7799 mg CD3254. Add DMSO to a volume of 1 mL. Votex, and filter by 0.45 μm filter.
3. DMSO.

Add RARα- or RXRα-specific agonist stock solution to growth medium with 1000× dilution to make the final concentration of 10 μM.

2.8 ChIP Assay

Most of the following reagents are provided by EZ ChIP TM assay kit (Upstate, NY).

1. Growth medium containing 1% formaldehyde: add 550 μL 37% formaldehyde to 20 mL growth medium.
2. Cold PBS containing proteinase inhibitor: 5 μL proteinase inhibitor cocktail is added to 1 mL PBS.
3. SDS lysis buffer containing proteinase inhibitor: 5 μL proteinase inhibitor cocktail is added to 1 mL 1× SDS lysis buffer.
4. 37% formaldehyde, proteinase inhibitor cocktail, and 10× glycine.
5. Sonicator.
6. Dilution buffer containing proteinase inhibitor: 5 μL proteinase inhibitor cocktail is added to 1 mL dilution buffer.
7. Protein G agarose.
8. Anti-RARα antibody for ChIP.
9. Low salt wash buffer, high salt wash buffer, LiCl wash buffer, and TE buffer.
10. 1 M $NaHCO_3$.
11. Water bath.
12. Elution buffer: 10 μL 20% SDS, 20 μL 1 M $NaHCO_3$, and 170 μL ddH_2O are mixed to make 200 μL elution buffer.
13. 5 M NaCl.
14. RNase A.
15. 0.5 M EDTA, 1 M Tris–HCl, and proteinase K.

16. Spin filter/collection tube: take the spin filter and put into the collection tube.
17. Binding reagent A.
18. Wash reagent B.
19. Elution buffer C.
20. PCR primers are used for detecting human DR5 type 10 RARE: 5′-ACAGACAAGGTGGACTGATGCAG-3′ and 5′-CATTCAGGTCGGCTGCAGAGCCC-3′ [15].
21. PCR machine (GeneAmp® System 9700 (PE Biosystems, CA)).

2.9 RNA Isolation

1. Trizol reagents (Thermo Scientific, MA).
2. Cold PBS.
3. Chloroform.
4. Isopropanol.
5. 75% ethanol.
6. RNase-free water.

2.10 RT-MiR Quantitative Real-Time PCR

1. 2 μM miR-specific RT primer (miR-10a RT-stem loop) (Table 1) or oligo dT and 10 mM dNTP Mix.
2. RNase-free water.
3. 5× FS buffer, 0.1 M DTT, and RNase out.
4. Superscript II reverse transcriptase.
5. Water bath.
6. 10 μM miR-10a forward primer and 10 μM universal primer (Table 1).
7. iQ SYBR Green PCR master mix (Bio-Rad, CA).

2.11 RT-PCR

1. Oligo dT and 10 mM dNTP Mix.
2. RNase-free water.
3. 5× FS buffer, 0.1 M DTT, and RNase out.
4. Superscript II reverse transcriptase.
5. Water bath.

Table 1
Primer sequences of RT-miR quantitative real-time PCR for detecting the level of miR-10a

MiR name		Sequence (5′–3′)
miR-10a	RT-stem loop	GTTGGCTCTGGTGCAGGGTCCGAGGTATTCGCACCAGAGCCAACCACAAA
	Forward	GACTTACCCTGTAGATCCGAA
Universal	Reverse	GTGCAGGGTCCGAGGT

Table 2
Primer sequences of RT-PCR for detecting pro-inflammatory gene expression in ECs

Gene name	Sequence (5′–3′)	Annealing Tm (°C)
GATA6 Forward	TTCCCATGACTCCAACTTCC	60
GATA6 Reverse	CGCCTATGTAGAGCCCATCT	60
VCAM-1 Forward	CATTCAGCGTCACCTTGG	60
VCAM-1 Reverse	CGCATCCTTCAACTGGCCTT	60
GAPDH Forward	CAACTACATGGTTTACATGTTCC	60
GAPDH Reverse	GGACTGTGGTCATGAGTCCT	60

6. 10 μM forward primers and 10 μM reverse primers (Table 2).
7. 10× Standard Taq Reaction Buffer, 2.5 mM dNTPs, and Taq DNA polymerase (5 U/μL) (Promega, WI).
8. PCR machine (GeneAmp® System 9700 (PE Biosystems, CA).
9. 1% agarose gel and ethidium bromide.

2.12 Administration of ApoE$^{-/-}$ Mice with RARα/RXRα Agonists and AMR-10a [18]

1. ApoE$^{-/-}$ mice and western diet (WD) of mice.
2. RARα- and RXRα-specific agonists stock solution.
   (a) RARα-specific agonist AM580.
   (b) RXRα-specific agonist CD3254.
   (c) Dimethyl sulfoxide (DMSO).
   (d) First cold-press olive oil.
   (e) AM580 and CD3254 are dissolved in first cold-press olive oil containing 5% DMSO to make the stock solution (1.5 mg/mL); The stock solution is stored at −20 °C and protected from light.
3. miR-Invivofectamine mixture of CL-miR or AMR-10a.
   (a) CL-miR and AMR-10a.
   (b) Invivofectamine reagent (Life Technologies, CA).
   (c) Invivofectamine (Life Technologies, CA) is used according to the manufacturer's protocol.
   (d) miR-Invivofectamine mixture is made by miR solution (0.75 mg/mL) and Invivofectamine reagent (500 μL).
   (e) miR-Invivofectamine mixture is further dialyzed in 2 L of PBS (PH7.4) with gentle agitation for at least 6 h.
   (f) Add PBS to miR-Invivofectamine mixture to 3 mL.
   (g) For ApoE$^{-/-}$ mouse study, the concentration of miR/-Invivofectamine complex is constantly 125 μg/mL (12.5 μg/mouse body, and 100 μL injection volume/mouse).

2.13 ApoE$^{-/-}$ Mice Tissue Preparation

1. Lukewarm saline and 10% neutral-buffered zinc-formalin (Thermo Scientific, MA).
2. Fixative solution.
3. PBS and 70% ethanol.
4. Microtome.

2.14 Oil Red O Staining

1. 60% isopropanol.
2. Oil Red O stock solution: weight 0.5 g Oil Red O into 100 mL isopropanol.
3. Oil Red O solution: dilute 30 mL Oil Red O stock solution to 50 mL with sterile ddH_2O.
4. Digital camera.

2.15 miR-10a In Situ Hybridization

1. Oven.
2. Xylene and 100%, 95%, 75%, 50% ethanol.
3. 0.01 M sodium citrate (pH 6.0): weight 2.94 g Tri-sodium citrate. Adjust pH to 6.0 and add water to a volume of 1 L.
4. 0.5% Triton X-100 in PBS containing 25 μL RNase out.
5. Pre-hybridization buffer: weight 0.3 g BSA and add 4× SSC to 10 mL.
6. Hybridization buffer: 10% dextran in 4× SSC.
7. 5′ DIG-labeled locked-nucleic acid probes (LNA miRCURY probe, Exiqon)
8. Wash buffer I (4× SSC with Tween): Add 50 μL Tween 20 into 50 mL 4× SSC.
9. Wash buffer II (2× SSC): Dilute 5 mL 20× SSC to 50 mL with ddH_2O.
10. Wash buffer III (1× SSC): Dilute 2.5 mL 20× SSC to 50 mL with ddH_2O.
11. 3% H_2O_2 in PBS: dilute 1 mL 30% H_2O_2 to 10 mL with PBS.
12. TN buffer: weight 121.14 g Tris base and 87.7 g NaCl. Adjust PH to 7.5 and add ddH_2O to 1 L.
13. TNB blocking buffer (Thermo Scientific, MA): dissolve 0.5 g blocking reagent in 100 mL TN buffer.
14. HRP-conjugated anti-DIG antibody (Roche Applied Science, Germany).
15. TNT buffer: 0.2% Triton X-100 in TN buffer.
16. TSA solution: dilute TSA in 1× Plus amplification diluents (1:50) (Perkin Elmer, MA).
17. DAPI.
18. Mounting medium.
19. Epifluorescence microscope.

2.16 Immunohistochemical Staining

1. Oven.
2. Prepare xylene and 100%, 95%, 75%, 50% ethanol.
3. 0.01 M sodium citrate containing Tween 20 (pH 6.0): weight 2.94 g Tri-sodium citrate (dehydrate). Adjust pH to 6.0 and add water to a volume of 1 L. Add Tween 20 to make 0.05%.
4. Blocking solution: mix blocking reagent (Merck Millipore, CA) with antibody dilute with background (Thermo Scientific, MA) (1:1).
5. Working solution: dilute blocking solution in PBS (1:10).
6. Primary antibody: rabbit or mouse antibody against pro-inflammatory molecules (GATA6 or VCAM-1), EC marker (vWF), or CD45 for immunohistochemical staining.
7. Secondary antibody (i.e., rhodamine- or FITC-conjugated secondary antibody).
8. DAPI.
9. Mounting medium.
10. Epifluorescence microscope.

3 Methods

3.1 In Vitro Flow Apparatus

1. Glass slide is pre-coated with 1 mL fibronectin (30 μg/mL) for 2 h.
2. Confluence ECs are trypsinized and cultured onto glass slides for 24 h before the flow experiment.
3. Cultured ECs are subjected to OS vs. PS in a parallel-plate flow chamber by in vitro flow apparatus.
   (a) Cell-seeded glass slide and the gasket are fastened between a polycarbonate base plate and a stainless plate to create the flow chamber (Fig. 2) [1, 2].
   (b) Flow chamber is connected to a perfusion loop system supplied with 5% CO_2 in air to maintain pH 7.4 and Tm controller to keep 37 °C (Fig. 3) [1, 2].
   (c) OS (0.5 ± 4 dynes/cm^2) (*see* **Note 3**) and PS (12 ± 4 dynes/cm^2) are generated by in vitro flow apparatus to apply to ECs (*see* **Note 4**).
   (d) The fluid shear stress subjected to cells can be estimated by $6\,Q\mu/wh^2$.
   (Q is flow rate and μ is the dynamic viscosity of the perfusate).
   (e) The mean shear stresses at 0.5 dynes/cm^2 and 12 dynes/cm^2 are generated by in vitro flow apparatus for OS and PS, respectively.

(f) The piston pump with a frequency of 1 Hz is used to generate a peak-to-peak amplitude of 4 dynes/cm^2 for fluid flow with oscillation.

(g) The osmolality of perfusate is 285–295 mOsm/kg H_2O.

4. In parallel experiments, ECs are kept under static condition for the same shear time.

3.2 Western Blotting

1. ECs kept in static condition or subjected to OS vs. PS are lysed with lysis buffer and resuspended with syringe 5–10 times.
2. Centrifuge with a microcentrifuge at 16,500 × *g* for 5 min at 4 °C to collect the supernatant containing cell lysate proteins and discard the pellet.
3. The same mounts of cell lysate proteins (100 μg proteins) from ECs are mixed with equal volume of 2× loading dye.
4. The mixture is boiled at 100 °C for 5 min, put on ice for 1 min, and loaded into each well of SDS-PAGE. Add standard marker into the first lane.
5. The SDS-PAGE is set into SDS-PAGE system containing 1× SDS running buffer to separate different proteins depending on their molecular weight.
6. Electrophoresis at 60 volts for 30 min and then continue at 100 volts for around 1 h until the dye reaches the bottom of the gel.
7. After electrophoresis, using a spatula to get the running gel containing cell lysate proteins and remove the stacking gel.
8. Rinse the running gel with 1× transfer buffer and put in the container containing the 1× transfer buffer for 3 min with slightly shaking.
9. Take the nitrocellulose membrane with the same size of running gel and put in the container containing 1× transfer buffer.
10. Get two filter papers with the same size of running gel.
11. Lift filter paper-gel-nitrocellulose membrane-filter paper as sandwich without bubbles.
12. Secure filter paper-gel-nitrocellulose membrane-filter paper sandwich with the clamp of Western blotting system.
13. Put the clamp containing filter paper-gel-nitrocellulose membrane-filter paper sandwich into container of Western blotting system containing 1× transfer buffer.
14. Connect the Western blotting system to power supply with 300 mM for 2 h to transfer protein from gel to nitrocellulose membrane.
15. Wash the transferred membrane with TBST for 10 min 3 times.

16. Block the transferred membrane with blocking solution for 30 min under shaking (40–50 rpm/min) in the room temperature.
17. Wash the transferred membrane with TBST for 10 min under shaking (60–90 rpm/min) 3 times in the room temperature.
18. Incubate the membrane with primary antibody diluted in the dilution solution (1:1000) under shaking (40–50 rpm/min) at 4 °C for overnight.
19. Wash the transferred membrane with TBST under shaking (60–90 rpm/min) for 10 min 3 times in the room temperature.
20. Incubate the membrane with HRP-conjugated secondary antibody diluted in the dilution solution (1:5000) under shaking (40–50 rpm/min) for 1 h at room temperature.
21. Wash the transferred membrane with TBST under shaking (60–90 rpm/min) for 10 min 3 times.
22. Incubate with the detection reagents for 3 min.
23. Acquire the image by image detection system.

3.3 Nuclear and Cytoplasmic Protein Extraction

Nuclear and cytoplasmic protein extraction is performed by using the ProteoJET™ Nuclear and Cytoplasmic Protein Extraction Kit in accordance with manufacturer's protocols. In brief,

1. The cell pellets of ECs are collected by microcentrifuge at 250 × *g* for 5 min, and discard the supernatant.
2. Add 100 μL cell lysis buffer with protease inhibitors and DTT to cell pellet (*see* **Note 5**).
3. Votex for 10 s and keep on the ice for 10 min twice.
4. Centrifuge at 500 × *g* for 7 min at 4 °C.
5. Separate supernatant (cytoplasmic protein fraction (followed by **step 6**)) and pellet (nuclear protein fraction (followed by **steps 7** and **8**)) to two different tubes and put on ice.
6. Cytoplasmic protein extraction: take the tube containing supernatant from **step 5**, and centrifuge at 20,000 × *g* for 15 min at 4 °C. The supernatant is transferred to the new tube to get cytoplasmic protein extracts, and stored at −70 °C.
7. Washing nuclear protein: add 500 μL nuclei washing buffer with protease inhibitors and DTT to the pellet from **step 5**, put on ice for 2 min, and centrifuge at 500 × *g* for 7 min at 4 °C to remove supernatant. The pellet is mixed with 150 μL cold nuclei storage buffer with protease inhibitors and DTT. Pipette the pellet with cold nuclei storage buffer with protease inhibitors and DTT 5–10 times to prevent clumps.
8. Nuclear protein extraction: Add nuclei lysis reagent with 1/10 volume to the nuclei suspension from **step 7**. Vortex and shake

at 4 °C for 15 min (1,500–2,000 × *g*). Nuclear protein extracts are collected by centrifugation at 20,000 × *g* for 5 min at 4 °C. The supernatant containing nuclear protein extracts is transferred to the new tube and stored at −70 °C.

9. The nuclear and cytoplasmic protein extracts are analyzed by Western blotting to detect the distributions and locations of RARs and RXRs in ECs.

3.4 Immunoprecipitation

1. ECs are lysed with 130 μL lysis buffer and resuspended with syringe 5–10 times.
2. Centrifuge with a microcentrifuge at 16,500 × *g* for 5 min at 4 °C, collect the supernatant containing cell lysate proteins, and discard the pellet.
3. Pre-clean: the same amounts of cell lysate proteins from ECs subjected to shear or static condition are incubated rotationally with 40 μL protein A/G plus agarose for 1 h at 4 °C.
4. Centrifuge with a microcentrifuge at 16,500 × *g* for 20 s at 4 °C, collect the supernatant containing cell lysate proteins, and discard the pellet.
5. The supernatant containing cell lysate proteins is incubated rotationally with 3 μL antibody against RARα for overnight at 4 °C.
6. The supernatant-antibody mixture is incubated rotationally with 40 μL protein A/G plus agarose at 4 °C for 2 h.
7. Centrifuge with microcentrifuge at 16,500 × *g* for 1 min at 4 °C, collect the agarose-bound immunoprecipitates, and discard the supernatant.
8. Wash: add 1 mL lysis buffer without SDS to agarose-bound immunoprecipitates and incubate rotationally for 10 min at 4 °C.
9. Centrifuge with microcentrifuge at 16,500 × *g* for 20 s at 4 °C, collect the agarose-bound immunoprecipitates, and discard the supernatant.
10. Repeat **steps 8** and **9** three times.
11. The agarose-bound immunoprecipitates are mixed with a boiling sample buffer and boiled at 100 °C for 5 min.
12. The supernatant containing immunoprecipitates is collected by centrifugation at 16,500 × *g* for 5 min at 4 °C, followed by subjecting to Western blotting with primary antibody against RAR's co-regulator, i.e., enhancer (i.e., RXRα, RXRβ, or RXRγ) or repressor (i.e., HDAC1, HDAC2, HDAC3, HDAC5, or HDAC7).

3.5 In Situ PLA

The protocol is performed in accordance with the manufacturer's instruction of in situ PLA kit. In brief,

1. ECs are fixed with 4% (wt/vol) paraformaldehyde for 20 min, and permeabilized with PBS containing 0.1% Triton X-100 (*see* **Note 6**).
2. The samples are blocked with PBS containing 10% goat serum for 30 min.
3. The samples are washed with PBS three times.
4. The samples are incubated with PBS containing 1% goat serum, rabbit anti-human RARα antibody, and mouse anti-human RXRα antibody (1:100) at 37 °C for 2 h (*see* **Note 7**).
5. The samples are washed with PBS three times.
6. The samples are incubated with PLA probe solution at 37 °C for 1 h.
7. Wash the samples with wash buffer A for 5 min twice.
8. Incubate the samples with ligation solution containing ligase at 37 °C for 30 min.
9. Wash the samples with wash buffer A for 2 min twice.
10. Incubate the samples with amplification solution containing polymerase at 37 °C for 100 min.
11. Wash the samples with 1× wash buffer B for 10 min twice.
12. Wash the samples with 0.01× wash buffer B for 1 min twice.
13. Put the samples in the dark.
14. Incubate the samples with Duolink II mounting medium with DAPI.
15. The red spot and blue nuclear, which indicate RARα-RARα heterodimer and cell nuclei, respectively, in cells are examined and photographed by fluorescence microscopy.

3.6 siRNA or miR Transfection

3.6.1 siRNA Transfection

1. 70–80% confluence ECs are cultured in 8 mL medium without antibiotics in 10 cm dish.
2. Dilute siRARα, siRXRα, siGATA6, or siCL in 2 mL Opti-MEM I Reduced Serum Medium, and mix well.
3. Add 25 μL Lipofectamine RNAiMAX into Opti-MEM I Reduced Serum Medium containing siRNA.
4. Mix well and incubate in the room temperature for 20 min.
5. Add the siRNA-Lipofectamine RNAiMAX mixture into the dish containing ECs to make the final concentration of siRNA at 5–40 nM.
6. Rocking the dish back and forth.

7. Incubate the ECs at 37 °C for 24–48 h in a CO_2 incubator.
8. After transfection, ECs are subjected to static or shear condition.

3.6.2 MiR transfection

For miR transfection, 70–80% confluence ECs are transfected with PreR-10a, AMR-10a, or CL-miR at 1–25 nM for 24–48 h using similar procedure.

3.7 Treatment of ECs with RARα- and RXRα-Specific Agonists in Response to OS or PS

1. 80–90% confluence ECs are cultured on glass slide with growth medium.
2. ECs are pre-treated with vehicle control DMSO, RARα-specific agonists (AM580, 10 μM), or RXRα-specific agonists (CD3254, 10 μM), or their combination for 1 h before the exposure to OS, PS, or static condition.
3. ECs are then subjected to static or flow conditions whose mediums contain DMSO, RARα-specific agonist (AM580, 10 μM), or RXRα-specific agonist (CD3254, 10 μM), or their combination, respectively.
4. The experimental ECs are collected for further examination.

3.8 ChIP Assay

ChIP assay is performed according to protocols of the EZ ChIP TM assay kit. In brief,

Prepare before the experiments: SDS lysis buffer is put in the room temperature; prepare cold PBS containing proteinase inhibitor and SDS lysis buffer containing proteinase inhibitor.

1. ECs subjected to static or shear condition are washed with PBS.
2. ECs on glass slides are fixed with growth medium containing 1% formaldehyde in the room temperature for 10 min.
3. ECs are incubated with 2 mL 10× glycine in the room temperature for 5 min to quench un-reacted formaldehyde.
4. ECs are washed with cold PBS containing proteinase inhibitor twice.
5. Add 200 μL cold PBS containing proteinase inhibitor to collect ECs.
6. Spin at 700 × *g* at 4 °C for 5 min.
7. Remove the supernatant and add 130 μL SDS lysis buffer containing proteinase inhibitor to resuspend the pellet to get cell lysates of ECs.
8. Put on ice for 5 min.
9. The cell lysates of ECs are subjected to sonication with sonicator.
10. Centrifuge at 14,000 × *g* for 10 min and collect the supernatant to the new tube.

11. Add 900 μL dilution buffer containing proteinase inhibitor.
12. Pre-clean: incubate the samples with 60 μL protein G agarose rotationally at 4 °C for 1 h.
13. Centrifuge at 5000 × *g* for 1 min.
14. Collect the supernatant containing chromatin DNA to new tube. In addition, 10 μL (1%) of the chromatin DNA used for immunoprecipitation is collected as input.
15. Incubate chromatin complex with 3 μL anti-RARα antibody rotationally at 4 °C for overnight.
16. Incubate chromatin complex with 60 μL protein G agarose rotationally at 4 °C for 2 h.
17. Centrifuge at 5000 × *g* for 1 min, remove the supernatant, and get the protein G agarose.
18. Wash the protein G agarose-antibody/chromatin complex with 1 mL following wash buffer, including low salt wash buffer (one wash), high salt wash buffer (one wash), LiCl wash buffer (one wash), and TE buffer (two washes). Each wash for 5 min and then centrifuges at 5000 × *g* for 1 min to get the agarose.
19. Take 1 M $NaHCO_3$ in the room temperature and vortex to prevent clumps; make the water bath for 65 °C; prepare elution buffer for each tube.
20. For input tube, add 190 μL elution buffer and set in the room temperature.
21. Incubate protein G agarose-antibody/chromatin complex with 100 μL elution buffer rotationally in the room temperature for 15 min, and then centrifuge at 5000 × *g* for 1 min to get the supernatant to the new tube.
22. Repeat **step 21** to get supernatant to the same tube.
23. To all tubes (IPs or Input), add 8 μL 5 M NaCl to the samples and incubate at 65 °C for 5 h or overnight.
24. To all tubes, add 1 μL RNase A to the samples and incubate at 37 °C for 30 min.
25. Add 4 μL 0.5 M EDTA, 8 μL 1 M Tris–HCl, and 1 μL proteinase K to the samples and incubate at 45 °C for 1–2 h.
26. Take the spin filter/collection tube.
27. Add 1 mL binding reagent A to all samples (IPs and Inputs), and mix well.
28. Get 600 μL sample/binding reagent A mixture from **step 27** to the spin filter/collection tube, centrifuge at 12,000 × *g* for 30 s, and discard the solution in the collection tube.

29. Get the other 600 μL sample/binding reagent A mixture from **step 27** to the spin filter/collection tube and repeat **step 28**.
30. Add 500 μL wash reagent B to the spin filter/collection tube, centrifuge at 12,000 × *g* for 30 s, and discard the solution in the collection tube.
31. Centrifuge at 12,000 × *g* for 30 s again and discard the liquid.
32. Put the spin filter to the new collection tube.
33. Add 50 μL elution buffer C to the spin filter/collection tube.
34. Centrifuge at 12,000 × *g* for 30 s, remove spin filter, and get the collection tube containing eluate to −20 °C.
35. The immunoprecipitated DNA fragments are used as templates for PCR with the following PCR primers: human DR5 type 10 RARE, 5′-ACAGACAAGGTGGACTG-ATGCAG-3′ and 5′-CATTCAGGTCGGCTGCAGAGCCC-3′ [15] (the PCR procedure is as Subheading 3.11, **steps 10–13**).
36. 1% of the chromatin DNA used for immunoprecipitation is also subjected to PCR analysis and defined as input (the PCR procedure is as Subheading 3.11, **steps 10–13**).

3.9 RNA Isolation

1. Rinse EC with cold PBS and add 1 mL Trizol reagents per 10 cm dish to lysis cell and transfer the cell lysates to new tube.
2. Incubate at 15–30 °C for 5 min.
3. Add 200 μL chloroform to tube and inverse the tube 10–15 times.
4. Incubate at 15–30 °C for 3 min.
5. Centrifuge at 12,000 × *g* at 4 °C for 15 min.
6. The mixture contains lower red phenol-chloroform phase, an interphase, and a colorless upper aqueous phase.
7. Take the upper aqueous phase carefully into the new tube (*see* **Note 8**).
8. Add 500 μL isopropanol to the tube.
9. Inverse the tube 10–15 times.
10. Incubate samples at 15–30 °C for 10 min, and centrifuge at 12,000 × *g* for 15 min at 4 °C to get the pellet.
11. Wash the pellet with 1 mL 75% ethanol.
12. Votex and then centrifuge at 7500 × *g* for 5 min at 4 °C.
13. Collect the pellet and discard the liquid.
14. Air-dry RNA pellet for 5 min.
15. Dissolve the RNA pellet with 12–15 μL RNase-free water and store at −80 °C (*see* **Note 9**).

3.10 RT-miR Quantitative Real-Time PCR

Quantitative stem-loop PCR method [19, 20] is used to detect miR expression. In brief,

1. The same amounts of total RNA are extracted from ECs subjected to static or shear condition.
2. Mix the 1 μg RNA with 1 μL 10 mM dNTP Mix, 1 μL 2 μM miR-specific RT primer (miR-10a RT-stem loop) or oligo dT, and add RNase-free water to 12 μL.
3. Incubate RNA mixture at 65 °C for 5 min, and then puts on the ice for 1 min.
4. Make the buffer mixture containing 4 μL 5× FS buffer, 2 μL 0.1 M DTT, and 1 μL RNase out. Mix the RNA mixture and buffer mixture, and spin down.
5. Incubate at 42 °C for 2 min.
6. Add 1 μL Superscript II reverse transcriptase into RNA/buffer mixture (*see* **Note 10**).
7. Incubate at 42 °C for 50 min.
8. Incubate at 70 °C for 15 min.
9. Put the tubes on ice and keep at 4 °C.
10. Add 30 μL RNase-free water into the tube (final volume 50 μL).
11. The level of miR-10a is detected by quantitative real-time PCR with iQ SYBR Green master mix.
12. Mix 2.5 μL of cDNA, 0.5 μL 10 μM miR-10a forward primer, 0.5 μL 10 μM universal primer, 2.5 μL SYBR Green PCR master mix, and RNase-free water to total 10 μL.
13. Amplification curves are performed on initial denaturing and enzyme activation step at 95 °C for 10 min, followed by denaturing, annealing, and extension at 45 cycles of 95 °C for 5 s, 60 °C for 10 s, and 70 °C for 1 s.
14. MiR-10a expression is normalized to the expression of the snU6.

3.11 RT-PCR

1. Incubate 4 μg RNA at 65 °C for 5 min, put on ice, and then add RNase-free water to 10 μL.
2. Mix the RNA solution with 1 μL oligo dT and 1 μL 10 mM dNTP Mix.
3. Incubate RNA mixture at 65 °C for 5 min, and then put on ice for 1 min.
4. Make the buffer mixture containing 4 μL 5× FS buffer, 2 μL 0.1 M DTT, and 1 μL RNase-free water. Mix the RNA mixture and buffer mixture and spin down.
5. Incubate at 42 °C for 2 min.

6. Add 1 μL Superscript II reverse transcriptase into RNA/buffer mixture (*see* **Note 10**).
7. Incubate at 42 °C for 50 min.
8. Incubate at 70 °C for 15 min.
9. Add 130 μL RNase-free water into the tube.
10. 3 μL (3 μg) cDNA is amplified through PCR with the use of Taq DNA polymerase.
11. Mix the cDNA with buffer mixture containing 3 μL 10× Standard Taq Reaction Buffer, 2.4 μL 2.5 mM dNTP Mix, 1 μL 10 μM forward primer, 1 μL 10 μM reverse primer, 0.5 μL Taq DNA polymerase, and 19.1 μL RNase-free water.
12. The PCR reactions are carried out in PCR machine.

    PCR reaction condition is as follows: heat denaturation at 94 °C for 5 min; PCR cycle with 35 cycles: heat denaturation at 94 °C for 30 s, primer annealing at 60 °C for 45 s, and primer extension at 72 °C for 45 s; primer extension at 72 °C for 5 min.
13. The amplified cDNAs are analyzed by 1% agarose gel electrophoresis and ethidium bromide staining.

3.12 Administration of ApoE$^{-/-}$ Mice with RARα/RXRα Agonists and AMR-10a [18]

The design of animal experiment is shown in Fig. 4. The detailed procedures are as follows:

1. 8–12-week-old ApoE$^{-/-}$ mice are prepared.
2. ApoE$^{-/-}$ mice are divided to three groups, i.e., DMSO+CL-miR group, RARα−/RXRα-specific agonists+CL-miR group, and RARα−/RXRα-specific agonists+AMR-10a group and treated with WD for 12 weeks.

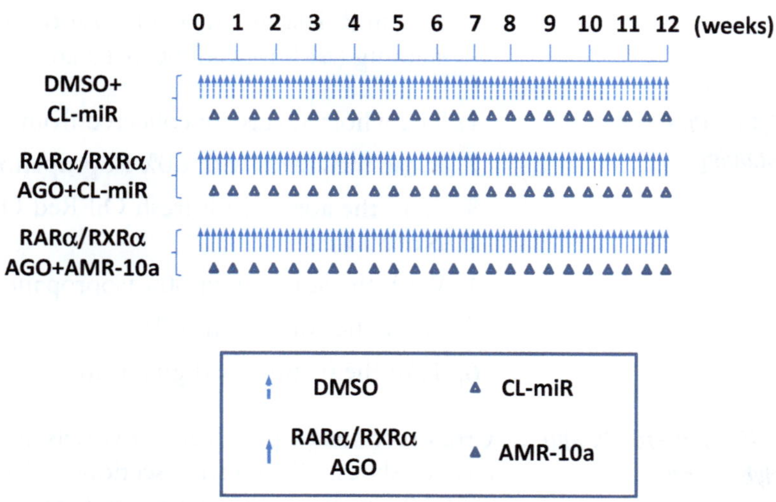

Fig. 4 Schematic diagram of the design of ApoE$^{-/-}$ mice experiment. *CL-miR* control miR, *AGO* agonist, *AMR-10a* antagomiR-10a

3. During the period of WD treatment, ApoE$^{-/-}$ mice receive intraperitoneal injections with the combination of RARα- and RXRα-specific agonists (1 mg/kg body weight each) for 6 days per week (except Sunday) and tail-vein injections of miR-Invivofectamine mixture of CL-miR (RARα/RXR-α-specific agonists+CL-miR group) or AMR-10a (RARα/RXRα-specific agonists+AMR-10a group) (12.5 μg/mouse body) twice per week.
4. Control mice (DMSO+CL-miR group) receive intraperitoneal injections of olive oil plus 5% DMSO for 6 days per week (except Sunday) and tail-vein injections of miR-Invivofectamine mixture of CL-miR twice per week.
5. The experiments proceed for 12 weeks, and the mice are sacrificed for further experiment.

3.13 ApoE$^{-/-}$ Mice Tissue Preparation

1. ApoE$^{-/-}$ mice are euthanized with CO_2 and transcardially perfused with 100 mL of lukewarm saline, followed by 50 mL of 10% neutral-buffered zinc-formalin.
2. After perfusion, the aortic arch from ApoE$^{-/-}$ mice are collected and fixed in the fixative solution overnight.
3. Whole aortas are collected from experimental mice and fixed in the fixative solution overnight. Oil Red O staining is used to examine atherosclerotic lesions (*see* Subheading 3.14).
4. Post-fixed aortas are rinsed with PBS, immersed in 70% ethanol, dehydrated in increasing concentrations of ethanol, cleared in xylol, and embedded in paraffin.
5. The microtome is used to make paraffin serial cross-sections (4 μm thick).
6. The cross-sections of aortic arch are examined by H&E staining, miR-10a in situ hybridization, or immunohistochemical staining (*see* Subheadings 3.15 and 3.16).

3.14 Oil Red O Staining

1. The whole aortas are collected from ApoE$^{-/-}$ mice.
2. Rinse the aortas with 60% isopropanol for 3 min.
3. Stain the aortas with fresh Oil Red O solution for 15 min (*see* **Note 11**).
4. Wash the aortas with 60% isopropanol for 5 min three times.
5. Rinse the aortas with PBS.
6. Take the picture by digital camera.

3.15 miR-10a In Situ Hybridization

Cross-sections (4 μm thick) of vessels are placed on poly-L-lysine-coated slides. Two serial sections of vessel are performed by miR-10a in situ hybridization and immunohistochemical staining with the EC marker (i.e., vWF), respectively (*see* Subheading 3.16).

1. The cross-sections are incubated in the oven at 70 °C for 1 h.
2. The cross-sections are incubated in xylene for 10 min three times, and each uses fresh xylene.
3. The cross-sections are incubated in 100% ethanol for 10 min twice, and each uses fresh ethanol.
4. The cross-sections are sequentially incubated in 95%, 75%, or 50% ethanol. Each for 10 min.
5. Wash the cross-sections with PBS for 5 min twice.
6. Boil the cross-sections with 0.01 M sodium citrate (PH 6.0) at 100 °C for 15 min.
7. Cool the cross-sections in the room temperature for 30 min.
8. Wash the cross-sections with sterile ddH_2O under shaking for 5 min twice.
9. Wash the cross-sections with PBS under shaking for 5 min twice.
10. Incubate the cross-sections with 1 mL 0.5% Triton X-100 in PBS containing 25 μL RNase out.
11. Incubate the cross-sections with the pre-hybridization buffer at hybridization temperature for 20 min (*see* **Note 12**).
12. Pre-warm hybridization buffer in hybridization temperature.
13. Incubate the cross-sections in the hybridization buffer containing 5'DIG-labeled locked-nucleic acid probes (probe: buffer = 1:40) at hybridization temperature for 1.5 h (*see* **Note 12**).
14. Pre-warm Wash buffer I, II, III, at the wash temperature (*see* **Note 13**).
15. Wash the cross-sections with Wash buffer I at wash temperature for 5 min four times, and each uses the fresh Wash buffer I.
16. Wash the cross-sections with Wash buffer II at wash temperature for 5 min.
17. Wash the cross-sections with Wash buffer III at wash temperature for 5 min.
18. Wash the cross-sections with PBS at room temperature for 5 min.
19. Incubate the cross-sections with fresh 3% H_2O_2 in PBS at room temperature for 20 min (*see* **Note 14**).
20. Wash the cross-sections with TN buffer for 5 min.
21. Incubate the cross-sections with TNB blocking buffer at room temperature for 30 min.
22. Incubate the cross-sections with the HRP-conjugated anti-DIG antibody in the TNB blocking buffer (1:50) at 4 °C overnight.

23. Wash the cross-sections with TNT buffer for 5 min three times.
24. The following steps (**steps 25–31**) are examined in the dark condition.
25. Incubate the cross-sections with TSA solution at room temperature for 15 min.
26. Wash the cross-sections with TNT buffer for 5 min in the room temperature three times.
27. Wash the cross-sections with PBS for 5 min twice.
28. Stain the nuclear with DAPI (1:1000 diluted in PBS) for 10 min.
29. Wash the cross-sections with PBS three times, each for 5 min.
30. The slides are mounted by mounting medium.
31. The images are photographed with an epifluorescence microscope.

3.16 Immunohistochemical Staining

Cross-sections (4 μm thick) of aortic arch from the experimental mice are placed on poly-L-lysine-coated slides. Two serial sections of vessel are performed for staining pro-inflammatory molecules, i.e., GATA6 and VCAM-1, and the EC marker, i.e., vWF, respectively. For detecting the infiltration of inflammatory cells, the cross-sections of vessels are immunostained for CD45, which is a common biomarker for leukocytes and CD45-positive cells consisting of neutrophils, macrophages, and monocytes.

1. The cross-sections placed on poly-L-lysine-coated slides are incubated in the oven at 70 °C for 1 h.
2. Put the cross-sections in the container containing xylene for 10 min three times, each uses the fresh xylene.
3. Incubate the cross-sections with 100% ethanol for 10 min twice, each uses fresh ethanol.
4. Incubate the cross-sections with 95%, 75%, or 50% ethanol in the sequential manner. Each for 10 min.
5. Wash the cross-sections with PBS for 5 min twice.
6. Boil the cross-sections in 0.01 M sodium citrate (PH 6.0) containing Tween 20 at 100 °C for 15 min.
7. Cool the cross-sections at room temperature for 30 min.
8. Wash the cross-sections with PBS three times, each for 5 min.
9. Block the background of the cross-section with blocking solution for 1 h.
10. Wash the cross-sections with PBS three times, each for 5 min.
11. Incubate the cross-sections with primary antibody diluted in the working solution (1:100) at 4 °C overnight (One cross-section is incubated with an antibody against pro-inflammatory

molecule, i.e., GATA6 or VCAM-1, and the other cross-section is incubated with an antibody against EC marker, i.e., vWF). On the other hand, the cross-section is incubated with an antibody against CD45 for detecting the infiltration of inflammatory cells.

12. Wash the cross-sections with PBS three times, each for 5 min.
13. The following steps (**steps 14–19**) are examined in the dark condition.
14. Incubate the cross-sections with the secondary antibody (i.e., rhodamine- or FITC- conjugated secondary antibody) diluted in the working solution (1:1000) at room temperature for 1 h.
15. Wash the cross-sections with PBS three times, each for 5 min.
16. Stain the cell nuclei with DAPI diluted in PBS (1:1000) for 10 min.
17. Wash the cross-sections with PBS three times, each for 5 min.
18. The slides are mounted by mounting medium.
19. The images are taken by an epifluorescence microscope.

4 Notes

1. Ligase is added to ligation solution immediately before the reaction of samples.
2. Polymerase is added to amplification solution immediately before the reaction of samples.
3. In physiological condition, the mean flow rate of OS developed in branches of arterial tree is closed to zero. The fluid shear stress with 0.5 dynes/cm^2 is generated to provide the basal nutrient and oxygen delivery.
4. The parameters of PS and OS are based on the mean flow rate developed on the physiological condition in vivo [1].
5. The volume of lysis buffer is based on the amount of collected cells.
6. 70–80% cellular confluence is optional for the in situ PLA experiment.
7. For detecting protein-protein interaction, two primary antibodies used in the PLA must be against different and non-competing epitopes on two different target proteins, raised in different species, i.e., mouse and rabbit, and examined under the same condition.
8. Take the upper aqueous phase carefully without getting any other phase of the mixture.

9. Do not use DEPC-treated water to dissolve the RNA pellet because DEPC-treated water disrupts the RT reaction.
10. Superscript II reverse transcriptase is added to solution immediately before the reaction of samples.
11. Each time uses fresh Oil Red O solution for staining.
12. Hybridization temperature is 22–25 °C lower than Tm of probes.
13. Wash temperature is 5 °C higher than hybridization temperature.
14. This step needs to use fresh H_2O_2 to quench endogenous peroxidase activity.

References

1. Chiu JJ, Chien S (2011) Effects of disturbed flow on vascular endothelium: pathophysiological basis and clinical perspectives. Physiol Rev 91:327–387
2. Chien S (2008) Effects of disturbed flow on endothelial cells. Ann Biomed Eng 36:554–562
3. Kumar S, Kim CW, Simmons RD et al (2014) Role of flow-sensitive microRNAs in endothelial dysfunction and atherosclerosis: mechanosensitive athero-miRs. Arterioscler Thromb Vasc Biol 34:2206–2216
4. Hahn C, Schwartz MA (2009) Mechanotransduction in vascular physiology and atherogenesis. Nat Rev Mol Cell Biol 10:53–62
5. Nagy L, Szanto A, Szatmari I et al (2012) Nuclear hormone receptors enable macrophages and dendritic cells to sense their lipid environment and shape their immune response. Physiol Rev 92:739–789
6. Rhee EJ, Nallamshetty S, Plutzky J (2012) Retinoid metabolism and its effects on the vasculature. Biochim Biophys Acta 1821:230–240
7. Mendelsohn C, Lohnes D, Decimo D et al (1994) Function of the retinoic acid receptors (RARs) during development (II). Multiple abnormalities at various stages of organogenesis in RAR double mutants. Development 120:2749–2771
8. Kastner P, Grondona JM, Mark M et al (1994) Genetic analysis of RXR alpha developmental function: convergence of RXR and RAR signaling pathways in heart and eye morphogenesis. Cell 78:987–1003
9. Lee DY, Lee CI, Lin TE et al (2012) Role of histone deacetylases in transcription factor regulation and cell cycle modulation in endothelial cells in response to disturbed flow. Proc Natl Acad Sci U S A 109:1967–1972
10. Perissi V, Aggarwal A, Glass CK et al (2004) A corepressor/coactivator exchange complex required for transcriptional activation by nuclear receptors and other regulated transcription factors. Cell 116:511–526
11. Kagechika H, Kawachi E, Hashimoto Y et al (1988) Retinobenzoic acids. 1. Structure-activity relationships of aromatic amides with retinoidal activity. J Med Chem 31:2182–2192
12. le Maire A, Teyssier C, Erb C et al (2010) A unique secondary-structure switch controls constitutive gene repression by retinoic acid receptor. Nat Struct Mol Biol 17:801–807
13. Minucci S, Leid M, Toyama R et al (1997) Retinoid X receptor (RXR) within the RXR-retinoic acid receptor heterodimer binds its ligand and enhances retinoid-dependent gene expression. Mol Cell Biol 17:644–655
14. Condorelli G, Latronico MV, Cavarretta E (2014) MicroRNAs in cardiovascular diseases: current knowledge and the road ahead. J Am Coll Cardiol 63:2177–2187
15. Weiss FU, Marques IJ, Woltering JM et al (2009) Retinoic acid receptor antagonists inhibit miR-10a expression and block metastatic behavior of pancreatic cancer. Gastroenterology 137:2136–2145
16. Fang Y, Shi C, Manduchi E et al (2010) MicroRNA-10a regulation of proinflammatory phenotype in athero-susceptible endothelium *in vivo* and *in vitro*. Proc Natl Acad Sci U S A 107:13450–13455
17. Lee DY, Lin TE, Lee CI et al (2017) MicroRNA-10a is crucial for endothelial response to different flow patterns via interaction of retinoic acid receptors and histone

deacetylases. Proc Natl Acad Sci U S A 114:2072–2077

18. Lee DY, Yang TL, Huang YH et al (2018) Induction of microRNA-10a using retinoic acid receptor-α and retinoid x receptor-α agonists inhibits atherosclerotic lesion formation. Atherosclerosis 271:36–44

19. Wu RM, Wood M, Thrush A et al (2007) Real time quantfication of plant miR using universal probelibrary technology. Biochemica 2:12–15

20. Leucht C, Bally-Cuif L (2007) The universal probelibrary- a versatile tool for quantitative expression analysis in the zebrafish. Biochemica 2:16–18

Chapter 12

Evaluating the Role of RARβ Signaling on Cellular Metabolism in Melanoma Using the Seahorse XF Analyzer

Christina Dahl, Per Guldberg, and Cecilie Abildgaard

Abstract

Dysregulation of retinoic acid signaling is implicated in several human cancer types, including melanoma where the gene encoding retinoic acid receptor beta (RARβ) is frequently silenced by promoter hypermethylation. In this chapter, we describe some of the experimental procedures that we have used to characterize the role of RARβ signaling on the regulation of cellular metabolism in melanoma. Central to these studies is the use of the Seahorse XF Analyzer, which allows real-time assessment of the oxygen consumption rate (OCR) and extracellular acidification rate (ECAR) in cultured cells as readouts for oxidative phosphorylation and glycolysis, respectively. The levels of RARβ signaling can be modulated using RARβ agonists (e.g., all-*trans* retinoic acid) and antagonists (e.g., LE135). The bioenergetic profiles of melanoma cells in response to RARβ modulators and other metabolic modifiers can be the basis for defining new therapeutic strategies.

Key words Retinoic acid receptor β, Seahorse XF Analyzer, Cancer metabolism, Melanoma, Mitochondrial function, Glycolytic activity, Warburg effect, Oxygen consumption rate, Extracellular acidification rate, Metabolic profile

1 Introduction

Reprogramming of the metabolic network in cancer cells was discovered already in the 1920s when Otto Warburg demonstrated an increased consumption of glucose in tumors, shifting the cells from mitochondrial respiration toward aerobic glycolysis (the Warburg effect) [1, 2]. Metabolic reprogramming is now considered a hallmark of cancer independent of the tissue of origin [3] and has lately received increased attention, not least due to the prospect of exploiting the metabolic vulnerabilities of cancer cells for therapeutic interventions [4, 5].

A major substrate for energy production in both normal and cancer cells is glucose, which is catabolized to pyruvate through glycolysis in the cytoplasm, with a net production of two molecules of ATP per glucose molecule. In the presence of oxygen, pyruvate

Swapan K. Ray (ed.), *Retinoid and Rexinoid Signaling: Methods and Protocols*, Methods in Molecular Biology, vol. 2019, https://doi.org/10.1007/978-1-4939-9585-1_12,

enters the mitochondria and is converted into acetyl-CoA, which is then further oxidized in the tricarboxylic acid (TCA) cycle and by oxidative phosphorylation (OXPHOS), increasing the yield of energy to more than 30 ATP molecules per glucose molecule.

Assessment of energy metabolism at the cellular level requires simultaneous quantification of glycolysis and OXPHOS. A widely used platform designed for this purpose is the Seahorse XF Analyzer, which builds on the correlation between changes in pH and oxygen tension and the activity of glycolysis (lactate production) and mitochondrial respiration, respectively. This system uses an array of sensors to measure metabolic fluxes from cells in microplates. The sensor cartridge is designed to create a temporary micro chamber with limited diffusion, allowing measurements of small changes in oxygen tension and pH within minutes. Assessing the oxygen consumption rate (OCR) before and after serial injections of inhibitors of different components of the mitochondrial electron transport chain (ETC) allows characterization of key features of mitochondrial function, including basal respiration, ATP production-coupled respiration, maximal capacity, spare capacity, and non-mitochondrial respiration (*see* Fig. 1). Information on glycolytic activity can be obtained by measuring the extracellular acidification rate (ECAR) during sequential injection of metabolic modulators. Non-glycolytic acidification can be measured in medium without glucose, or after the injection of the glycolysis inhibitor 2-deoxy-D-glucose (2-DG). Basal glycolysis can be assessed after the injection of glucose, while the maximal glycolytic

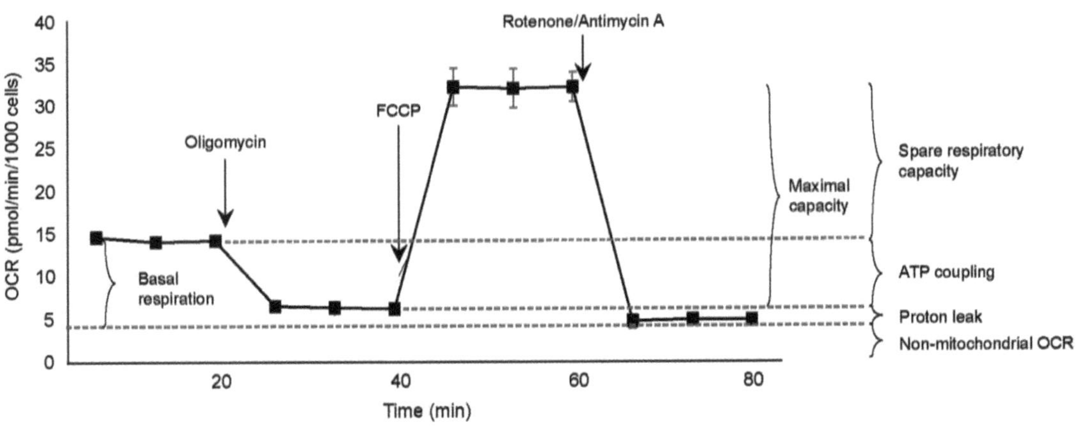

Fig. 1 Mitochondrial profile. Evaluation of OCR by serial injections of specific inhibitors targeting different components of the ETC in the mitochondria. Initially, baseline cellular OCR is measured. Addition of oligomycin inhibits the ATP synthase, and OCR decreases proportionally to the mitochondrial respiration associated with cellular ATP production. FCCP disrupts the mitochondrial membrane and allows ETC to function at its maximal rate. Finally, the mix of rotenone and antimycin A effectively shuts down the ETC function, enabling the non-mitochondrial respiration to be determined. In addition, it is possible to calculate proton leak (defined as the difference between ATP production and non-mitochondrial respiration) and spare respiratory capacity (defined as the difference between maximal respiration and basal respiration)

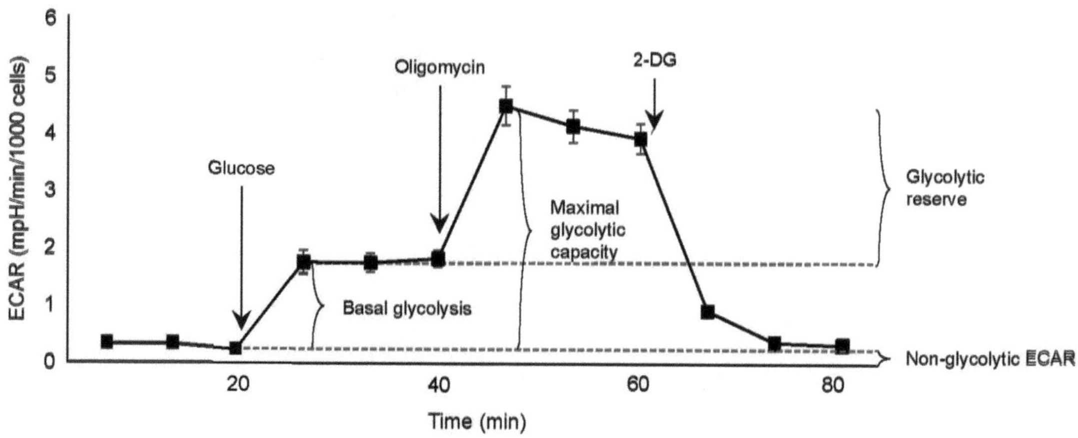

Fig. 2 Glycolytic profile. ECAR is an indicator of glycolysis. At the outset, glucose starvation provides information on the intracellular non-glycolytic acidification. Addition of saturating amounts of glucose stimulates glycolysis and allows measurement of basal glycolysis. Injection of oligomycin blocks mitochondrial ATP production and drives the cell to its maximal glycolytic capacity. The glycolytic reserve indicates the difference between maximal glycolytic capacity and the basal rate. Non-glycolytic acidification is the extracellular acidification derived from sources different from glycolysis and can be measured after inhibition of glycolysis with 2-DG

capacity and glycolytic reserve can be estimated after the injection of oligomycin, an inhibitor of the mitochondrial ATP synthase (*see* Fig. 2). Subsequent integration of these data reveals the interplay between OXPHOS and glycolysis. Plotting basal OCR against basal ECAR provides a metabolic profile and indicates how cells respond to metabolic modifiers (*see* Fig. 3).

Development of melanoma, the most lethal form of skin cancer, is associated with metabolic reprogramming induced by genetic and epigenetic alterations targeting regulators of cellular metabolism [6, 7]. A prominent example is DNA methylation-dependent silencing of the gene encoding retinoic acid receptor beta (RARβ) in a large proportion of melanomas [8]. Recent work has shown that abrogation of the RARβ-signaling pathway shifts the metabolism of melanoma cells toward glycolysis and therefore may act as a component of the Warburg effect in this cancer [9, 10]. Importantly, targeting RARβ signaling in RARβ-expressing melanoma cells led to energetic stress and increased sensitivity to other metabolic modifiers, including the pyruvate dehydrogenase kinase inhibitor dichloroacetate (DCA), offering new opportunities for therapeutic intervention [10].

In this chapter, we describe the protocol for how metabolic profiling can be used to identify RARβ signaling as a regulator of cellular metabolism and potential therapeutic target in melanoma. For a more general description of the assay, please refer to Agilent's website. We have used a model system of RARβ-expressing melanoma cell lines, and we have modulated the levels of RARβ

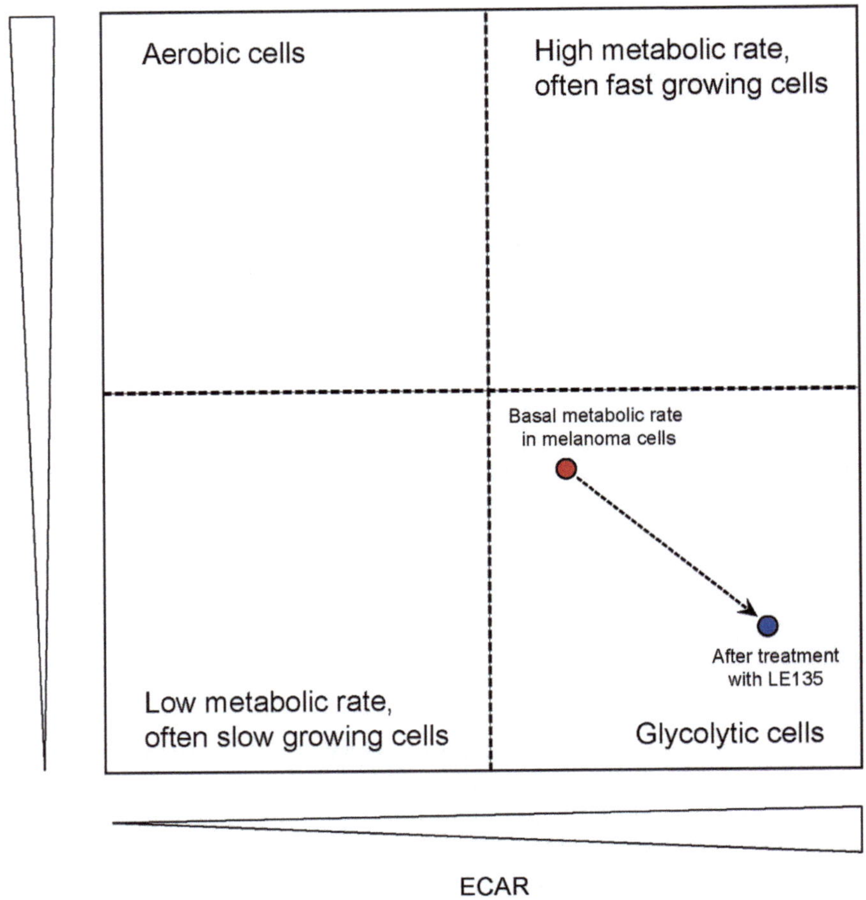

Fig. 3 Metabolic profile. Four different phenotypes can be defined by plotting basal OCR against ECAR. The switch in metabolic phenotype of melanoma cells after treatment with the RARβ antagonist LE135 is exemplified by the red (before treatment) and blue (after treatment) dots

signaling in these cells using the RARβ agonist all-*trans* retinoic acid (ATRA) and the RARβ antagonist LE135 [11]. The assay can be performed within 2 days.

2 Materials

2.1 Culture of Melanoma cells

1. Melanoma cell lines.
2. Tissue culture treated flasks.
3. RPMI-1640 supplemented with 10% (v/v) fetal bovine serum (FBS), pH regulated to 7.4.
4. Phosphate Buffered Saline (PBS).
5. Dissociation reagent like TrypLE Express (1×), no phenol red.
6. 37 °C incubator with humidified atmosphere of 5% CO_2.

2.2 Metabolic Profiling of Melanoma Cells

1. Seahorse XFe96 Analyzer with controller (Agilent).
2. Non-CO_2 incubator, 37 °C.
3. Calibrated pH meter.
4. Seahorse XFe96 sensor cartridges.
5. Seahorse XF calibrant for sensor cartridge.
6. Seahorse XFe96 cell culture microplates.
7. Melanoma cell lines.
8. RPMI-1640 supplemented with 10% (v/v) FBS, pH regulated to 7.4.
9. TrypLE Express (1×), no phenol red (or similar).
10. LE135 dissolved in DMSO.
11. ATRA dissolved in DMSO.
12. Oligomycin.

2.2.1 Specific for Testing Mitochondrial Function

1. Carbonyl cyanide-4 (trifluoromethoxy)phenylhydrazine (FCCP) dissolved in medium (100 μM).
2. Rotenone/antimycin A dissolved in medium (50 μM).

2.2.2 Specific for Testing Glycolytic Activity

1. L-Glutamine (100 mM),
2. Glucose (100 mM).
3. 2-DG dissolved in medium (1 M).
4. Seahorse XF Base Medium.

3 Methods

3.1 Culturing Melanoma Cells

We used three melanoma cell lines from the European Searchable Tumour line Database (ESTDAB); ESTDAB-007 (FM-3), ESTDAB-117 (Mel-NT3-00), and ESTDAB-196 (Ma-Mel-51), all of which express RARβ [8, 10]. All cell lines were tested negative for mycoplasma contamination and were cultured without the addition of antibiotics.

1. Remove and discard the cell culture media from the adherent melanoma cells, when they are in log-phase growth.
2. Gently wash the cells with PBS to remove any traces of serum and salts.
3. Remove and discard the PBS.
4. Add pre-warmed dissociation reagent (like TrypLE). Make sure that the cell layer is completely covered.
5. Incubate at room temperature and visually inspect the cells under microscope for detachment.

6. Tilt the vessel and add the equivalent of 2 volumes of pre-warmed RPMI-1640 supplemented with 10% (v/v) FBS.
7. Transfer the cells to a 15 mL conical tube and centrifuge at 200 × *g* for 7 min. Centrifuge speed and time vary depending on the cell line.
8. Resuspend the cell pellet in pre-warmed RPMI-1640 supplemented with 10% (v/v) FBS.
9. Dilute the cell suspension, transfer into new cell culture vessels and incubate the cells at 37 °C with humidified atmosphere of 5% CO_2.

3.2 Metabolic Profiling of Melanoma Cells

Day 1

3.2.1 Cell Seeding and Cartridge Hydration

1. Harvest the melanoma cells as described in Subheading 3.1, and plate them in the Seahorse cell culture microplate in 80 μL of RPMI-1640 medium supplemented with 10% (v/v) FBS per well (*see* **Note 1**). The number of cells plated should be optimized, to ensure 100% confluency at the day of the experiment (*see* **Note 2**). The number of replicates should be at least six because some variation can be expected.
2. Incubate the microplate at 37 °C with humidified atmosphere of 5% CO_2 for 4 h.
3. Add 100 μL of additional media containing drug(s) of interest or vehicle (*see* **Note 3**).
4. Incubate the microplate at 37 °C with humidified atmosphere of 5% CO_2 overnight.
5. Hydrate the sensor cartridge by adding 200 μL of pre-warmed Seahorse XF calibrant to each well. Incubate overnight in a non-CO_2 incubator at 37 °C.
6. Design an assay template using the Wave software.
7. Turn on the Seahorse XFe96 Analyzer to warm up to 37 °C and let it stabilize overnight (or at least 5 h).

Day 2

3.2.2 Cartridge Loading

On the day of assay, the cartridge loading protocol will differ for mitochondrial function and glycolytic activity, as outlined below:

For Assessment of Mitochondrial Function

1. Heat the RPMI-1640 medium supplemented with 10% (v/v) FBS to 37 °C (*see* **Note 4**).
2. Adjust pH to 7.4 ± 0.05 using 0.1 N NaOH, sterile filter (0.2 μm), and keep at 37 °C until use.

3. Equilibrate oligomycin, FCCP, and rotenone/antimycin A to room temperature for approximately 15 min.
4. Dilute each solution in assay medium.
   - Oligomycin, 20 μM.
   - FCCP, 10 μM.
   - Rotenone/antimycin A, 10 μM.
5. Load the sensor cartridge with oligomycin, FCCP, and rotanone/antimycin A into the appropriate injection ports (*see* **Note 5**).
   - Port A: 20 μL of oligomycin (final concentration 2 μM). The concentration of oligomycin should be optimized for each cell line (*see* **Note 6**).
   - Port B: 22 μL of FCCP (final concentration 1 μM). The concentration of FCCP should be optimized for each cell line (*see* **Note 6**).
   - Port C: 25 μL of rotenone/antimycin A (final concentration 1 μM).
6. Proceed to Subheading "For Assessment of Glycolytic Activity" if the analysis should also include assessment of glycolytic activity; otherwise proceed to Subheading 3.2.3.

For Assessment of Glycolytic Activity

1. Supplement the Seahorse XF Base Medium with 2 mM L-glutamine (the medium composition can be varied depending on cell type (*see* **Note 7**)), and heat the assay medium to 37 °C.
2. Adjust pH to 7.4 ± 0.05 using 0.1 N NaOH, sterile filter, and keep at 37 °C until use.
3. Equilibrate glucose, oligomycin, and 2-DG to room temperature for approximately 15 min.
4. Dilute each solution in assay medium.
   - Glucose, 100 mM.
   - Oligomycin, 20 μM.
   - 2-DG, 1 M.
5. Load the sensor cartridge with the different compounds into the appropriate injection ports (*see* **Note 8**).
   - Port A: 20 μL of glucose (final concentration 10 mM).
   - Port B: 22 μL of oligomycin (final concentration 2 μM).
   - Port C: 25 μL of 2-DG (final concentration 0.1 M).
6. Proceed to Subheading 3.2.3.

3.2.3 Pre-incubation and Plate Reading

1. Examine the cells seeded in the Seahorse XF cell culture microplate under microscope to confirm seeding uniformity and confluence.
2. Change the medium in the microplate with 180 μL of pre-warmed RPMI-1640 medium supplemented with 10% (v/v) FBS (mitochondrial function) or Seahorse XF Base Medium supplemented with 2 mM L-glutamine (glycolytic activity).
3. Incubate the cells at 37 °C in a non-CO_2 incubator for 45–60 min (*see* **Note 9**).
4. To run the assay, place the calibration plate (without lid) with the loaded sensor cartridge on the instrument tray and start the calibration (takes approximately 20 min).
5. After calibration, replace the calibration plate with the cell culture microplate and run the assay.
6. Optional: Normalize to the number of cells or protein concentration using standard methods (*see* **Note 10**).
7. Export XF data to Excel and calculate variables, or analyze using the Seahorse XF Mito Stress Test Report Generator and the Seahorse XF Glycolysis Test Report Generator for mitochondrial function and glycolysis, respectively.

4 Notes

1. The Seahorse XF cell culture microplates may give rise to edge effects resulting in incorrect measurements, particularly around the perimeter. To reduce well-to-well variability, it may be necessary to avoid using the outer wells of the plate and instead fill them with medium or PBS.
2. It is important that the layer of cells is as even as possible. Gaps in the cell layer can result in a large variation in the measurements of OCR and ECAR.
3. Addition of agonists or antagonists of RARβ can be added to the Seahorse XF cell culture microplates in the desired concentration either the day before the experiment or as the initial injection during the Seahorse experiment. We have used the specific RARβ antagonist LE135 at a final concentration of 5 μM. However, optimization is usually required for each cell line to achieve the concentration that gives the greatest effect on metabolism without affecting cell growth.
4. For measurements of OCR, the assay medium should contain carbon sources for optimal function of the mitochondrial respiration, including glucose, pyruvate, and L-glutamine. The medium used for normally culturing the cells can be used.

5. It is possible to modulate the cellular metabolism by adding up to four different reagents through the four injection ports (A, B, C, and D). To include an additional injection prior to oligomycin, use port A for the desired compound, port B for oligomycin, port C for FCCP, and port D for rotenone/antimycin A and adjust the volumes correspondingly.
6. In order to characterize cellular metabolic function properly, it is necessary to determine the optimal concentration of oligomycin required for maximal inhibition of the ATP synthase for each cell line. Likewise, optimization of the FCCP concentration required for maximal oxygen consumption should be determined empirically. In our laboratory, we have found that 2 μM oligomycin and 1 μM FCCP are optimal for melanoma cells.
7. For measurements of ECAR, the medium composition can be varied depending on the cell lines investigated, as long as the medium does not contain pH buffers or serum.
8. It is possible to modulate the cellular metabolism by addition of up to four different reagents through the four injection ports (A, B, C, and D). To include an additional injection prior to glucose use port A for the desired compound, port B for glucose, port C for oligomycin, and port D for 2-DG and adjust the added volumes corresponding.
9. This incubation step is required to stabilize cellular metabolism after changing the media and permits outgassing from the plate, which is a prerequisite for accurate ECAR measurements.
10. Treatment with LE135 at a concentration of up to 5 μM does not affect cell growth or viability during the incubation period. If other treatments are applied, it may be necessary to normalize the results to the number of living cells or protein levels.

References

1. Warburg O, Wind F, Negelein E (1927) The metabolism of tumors in the body. J Gen Physiol 8(6):519–530
2. Vander Heiden MG, Cantley LC, Thompson CB (2009) Understanding the Warburg effect: the metabolic requirements of cell proliferation. Science 324(5930):1029–1033
3. Hanahan D, Weinberg RA (2011) Hallmarks of cancer: the next generation. Cell 144 (5):646–674
4. Vander Heiden MG (2011) Targeting cancer metabolism: a therapeutic window opens. Nat Rev Drug Discov 10(9):671–684
5. Amoedo ND, Obre E, Rossignol R (2017) Drug discovery strategies in the field of tumor energy metabolism: Limitations by metabolic flexibility and metabolic resistance to chemotherapy. Biochim Biophys Acta 1858 (8):674–685
6. Ratnikov BI, Scott DA, Osterman AL, Smith JW, Ronai ZA (2017) Metabolic rewiring in melanoma. Oncogene 36(2):147–157
7. Abildgaard C, Guldberg P (2015) Molecular drivers of cellular metabolic reprogramming in melanoma. Trends Mol Med 21(3):164–171
8. Dahl C, Christensen C, Jönsson G, Lorentzen A, Skjødt ML, Borg A, Pawelec G, Guldberg P (2013) Mutual exclusivity analysis of genetic and epigenetic drivers in melanoma identifies a link between p14ARF and RARbeta signaling. Mol Cancer Res 11(10):1166–1178

9. Hoon DS, Spugnardi M, Kuo C, Huang SK, Morton DL, Taback B (2004) Profiling epigenetic inactivation of tumor suppressor genes in tumors and plasma from cutaneous melanoma patients. Oncogene 23(22):4014–4022
10. Abildgaard C, Dahl C, Abdul-Al A, Christensen A, Guldberg P (2017) Inhibition of retinoic acid receptor ß signaling confers glycolytic dependence and sensitization to dichloroacetate in melanoma cells. Oncotarget 8(48):84210–84223
11. Li Y, Hashimoto Y, Agadir A, Kagechika H, Zhang X (1999) Identification of a novel class of retinoic acid receptor beta-selective retinoid antagonists and their inhibitory effects on AP-1 activity and retinoic acid-induced apoptosis in human breast cancer cells. J Biol Chem 274 (22):15360–15366

Chapter 13

Highly Sensitive Quantitative Determination of Retinoic Acid Levels, Retinoic Acid Synthesis, and Catabolism in Embryonic Tissue Using a Reporter Cell-Based Method

Leo M. Y. Lee, Selina T. K. Tam, Peter J. McCaffery, and Alisa S. W. Shum

Abstract

The effect of all-*trans* retinoic acid (RA) on embryogenesis is tissue specific and highly concentration dependent. Using a liquid chromatography/mass spectrometry-based method to quantify trace amounts of RA in embryonic tissue requires expensive specialist facilities. Here, we describe the use of a RA response element (RARE)-lacZ reporter cell-based method, which is simple and cost effective, to measure RA levels in small pieces of tissue from the embryo. We further apply this method to quantitatively assay activities of RA-synthesizing and RA-catabolizing enzymes, the key regulators of RA bioavailability in tissues and developing organs of the embryo.

Key words Retinoic acid, Retinoic acid reporter cell line, Retinoic acid synthesis, Retinoic acid metabolism, Embryo, RALDH, CYP26, Cell culture

1 Introduction

All-*trans* retinoic acid (RA) differs greatly in concentration between varied tissues of the embryo [1] and its effect is highly concentration dependent. Thus, to study the role of RA during normal and abnormal embryogenesis, it is important to examine tissue-specific RA levels. One of the most commonly used techniques to identify and quantitate RA is high-performance liquid chromatography (HPLC) coupled with spectrophotometric detection. Although this technique can be employed to assay RA levels in embryonic tissue, it requires pooling of tissues from large numbers of embryos [1]. Thus, it may not be suitable for measuring RA in localized regions of the embryo, especially if the tissue is limited in availability. Tandem mass spectrometry (MS) coupled to HPLC has greater sensitivity for RA and less tissue is needed for the analysis [2]. However, it requires expensive specialist equipment and reagents. Various RA-sensitive reporter cell lines have been developed to detect

Swapan K. Ray (ed.), *Retinoid and Rexinoid Signaling: Methods and Protocols*, Methods in Molecular Biology, vol. 2019, https://doi.org/10.1007/978-1-4939-9585-1_13,

RA in embryonic tissue [1, 3–5]. Here we describe the use of a F9 RARE-lacZ cell line generated by Wagner and colleagues [3, 6] for quantitative measurement of RA levels in very small pieces of embryonic tissue [7]. This cell line was derived from the stable transfection of mouse F9 teratocarcinoma cells with the *E. coli lacZ* gene under the control of a retinoic acid response element (RARE) from the promoter of the human RA receptor *β* (*RARβ*) gene. In the presence of RA, the RARE is activated and drives the expression of *lacZ*, which encodes the enzyme β-galactosidase. After processing for β-galactosidase-mediated conversion of X-gal to a blue reaction product, the staining intensity in these RA-sensitive reporter cells can be measured by spectrophotometry. This cell line is most sensitive to RA (all-*trans*) with a detection threshold of 10^{-12} M, while its sensitivity to 13-*cis* and 9-*cis* isomers being at least 100-fold lower [8].

An alternative approach to study the bioavailability of RA in tissue is via the analysis of the activity of the enzymes that mediate RA biosynthesis and catabolism. RA is synthesized via irreversible oxidation of all-*trans* retinaldehyde (RAL) by retinaldehyde dehydrogenases (RALDHs). On the other hand, RA is metabolically inactivated by cytochrome P450 family 26 (CYP26) enzymes. The RA-synthesizing and RA-catabolizing enzymes express in complementary non-overlapping domains to precisely regulate RA levels in specific tissues [9]. Here, we describe a protocol, modified from McCaffery and Dräger [10], and Yamamoto et al. [11], for quantitative assay of RA synthesis and catabolism in small pieces of embryonic tissue [12]. This protocol quantifies the activity of RA-synthesizing enzymes in tissue homogenates by measuring the conversion of exogenous RAL to RA, or alternatively quantifies the activity of RA-catabolizing enzymes by measuring the inactivation of exogenous RA. The amount of RA being synthesized, or alternatively remaining in the reaction mixture after incubation with the homogenate, is determined using the RA reporter cells. This RA reporter cell-based method is simple, easy to set up, and cost effective. Yet, it allows quantification of RA using a standard curve constructed from serially diluted RA standard solutions, with a linear response in the range of 10^{-10} to 10^{-7} M RA (Fig. 1).

2 Materials

Prepare solutions using ultrapure water, and store solutions and medium at 4 °C (unless indicated otherwise).

2.1 Cell Culture

1. Gelatin stock solution (10×): 2% gelatin in water. Sterilize by filtering through a 0.22 μm pore size membrane filter. Dilute to 0.2% gelatin (1×) with autoclaved water.

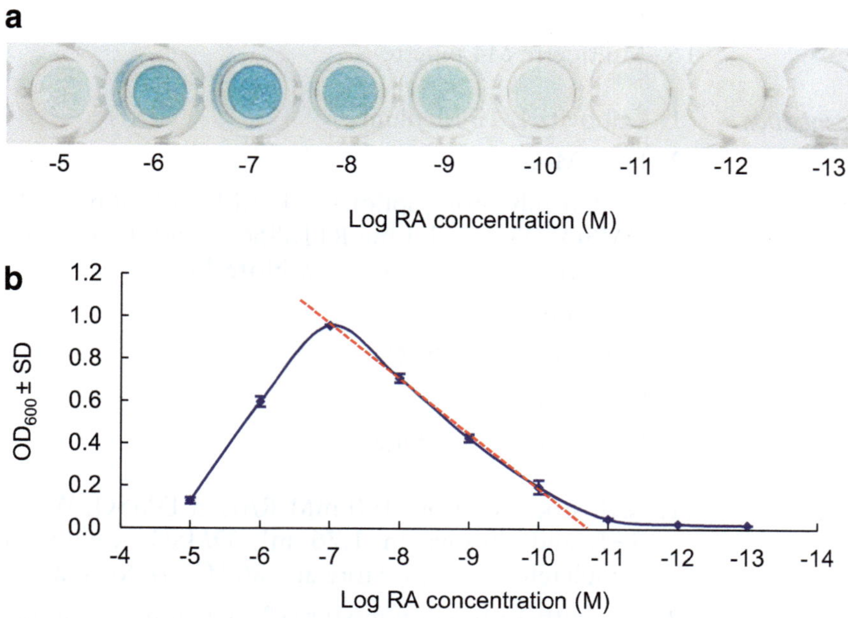

Fig. 1 Staining intensity of RA reporter cells in response to varying concentrations of RA. (**a**) Image of the X-gal stained RA reporter cells on the 96-well plate that had been incubated with serially diluted RA standard solutions for 24 h. (**b**) A standard curve showing the staining intensity at optical density 600 nm plotted against log RA concentrations from 10^{-5} to 10^{-13} M. There is linear response between 10^{-7} and 10^{-10} M RA

2. Hank's balanced salt solution (HBSS).
3. F9 RARE-lacZ cells (RA reporter cells) stored in liquid nitrogen.
4. Dulbecco's Modified Eagle's Medium plus (DMEM+): DMEM containing 2 mg/mL G418, 100 U/mL Penicillin-Streptomycin. Sterilize by filtering through a 0.22 μm pore size membrane filter.
5. Culture medium: DMEM+ containing 10% fetal bovine serum (FBS).
6. Trypsin-EDTA (0.25%).
7. Freezing medium: Culture medium containing 5% dimethyl sulfoxide (DMSO). Prepare freshly before use.
8. Cell culture flask (25 cm^2).
9. Centrifuge tube (15 mL).
10. Culture plate (96-well).
11. Water bath.
12. Centrifuge.
13. CO_2 incubator.
14. Hemocytometer.

15. Inverted microscope.
16. Multipette M4 pipette.

2.2 RA Concentration Assay

1. Leibovitz L-15 Medium.
2. DMEM+.
3. CYP26 inhibitor (optional): 1 mM R115866 (10,000×) in DMSO. Dissolve 1.1 mg R115866 in DMSO to a final volume of 3 mL. Store at −20 °C (*see* **Note 1**).
4. Bradford reagent.
5. FBS stored at −20 °C.
6. Sonicator.
7. Refrigerated centrifuge.

2.3 RA Synthesis Assay

1. RAL stock solution: 100 mM RAL in DMSO. Weigh 50 mg RAL and dissolve in 1.76 mL DMSO. Gently mix until completely dissolved. Store at −80 °C (*see* **Note 2**).
2. RA synthesis reaction mixture (2×): Culture medium containing 1.2 mg/mL DL-Dithiothreitol (DTT), 1.6 mg/mL β-Nicotinamide adenine dinucleotide hydrate (NAD) (*see* **Note 3**), 100 nM RAL (*see* **Note 4**), 0.2 μM R115866 and 2× protease inhibitor cocktail (*see* **Note 5**). Prepare freshly before use.
3. RALDH inhibitor (optional): Citral. Store at room temperature.

2.4 RA Catabolism Assay

1. RA stock solution: 100 mM RA in DMSO. Weigh 50 mg RA and dissolve in 1.66 mL DMSO. Gently mix until completely dissolved. Store at −80 °C (*see* **Note 2**).
2. RA catabolism reaction mixture (2×): Culture medium containing 0.6 mg/mL DTT, 3.2 mg/mL β-Nicotinamide adenine dinucleotide 2′-phosphate reduced tetrasodium salt hydrate (NADPH) (*see* **Note 3**), 100 nM RA (*see* **Note 4**) and 2× protease inhibitor cocktail (*see* **Note 5**). Prepare freshly before use.
3. CYP26 inhibitor (optional): 1 mM R115866 (10,000×) in DMSO. Store at −20 °C.

2.5 X-Gal Staining

1. PBS.
2. Fixative: 1% glutaraldehyde. Mix 0.24 mL of 25% glutaraldehyde with 5.76 mL phosphate buffered saline (PBS). Prepare freshly before use.
3. 5-Bromo-4-chloro-3-indolyl-β-D-galactopyranoside (X-gal): 40 mg/mL X-gal in *N*,*N*-dimethylformamide. Dissolve 0.5 g X-gal in *N*,*N*-dimethylformamide to a final volume of 12.5 mL. Store at −20 °C protected from light.

4. Magnesium chloride: 1 M $MgCl_2$. Dissolve 5.08 g $MgCl_2 \cdot 6H_2O$ in water to a final volume of 25 mL. Store at room temperature.
5. Potassium ferrocyanide (100×): 0.33 M $K_4[Fe(CN)_6]$. Dissolve 3.48 g $K_4[Fe(CN)_6] \cdot 3H_2O$ in water to a final volume of 25 mL. Store protected from light.
6. Potassium ferricyanide (100×): 0.33 M $K_3[Fe(CN)_6]$. Dissolve 2.72 g $K_3[Fe(CN)_6]$ in water to a final volume of 25 mL. Store protected from light.
7. Staining solution: 2 mg/mL X-gal, 4 mM $MgCl_2$, 3.3 mM $K_4Fe(CN)_6$ and 3.3 mM $K_3Fe(CN)_6$ in PBS (*see* **Note 6**). Prepare freshly before use.
8. Microplate spectrophotometer.

3 Methods

Conduct in vitro culture experiments under sterile conditions in a class II biological safety cabinet. Perform all procedures at room temperature unless indicated otherwise.

3.1 Cell Line Maintenance

1. Coat a 25 cm^2 cell culture flask with 3 mL of 0.2% gelatin solution for 30 minutes (min). Remove excess gelatin solution and rinse the coated flask twice with 3 mL of HBSS (*see* **Note 7**).
2. Place the cryogenic vial containing the RA reporter cells in a 37 °C water bath until thawed. Transfer the cell suspension into a 15 mL centrifuge tube and dilute tenfold with HBSS pre-warmed at 37 °C.
3. Spin at 400 × *g* for 3 min. Discard the supernatant and resuspend the cell pellet in 5 mL of pre-warmed culture medium.
4. Add the cell suspension to the gelatin-coated flask and culture at 37 °C in a 5% CO_2 incubator. Change the medium the next day. Subculture/plating when cells reach 80–90% confluency (around 1×10^7 cells per flask) (*see* **Note 8**).

3.2 Subculture and Freezing

1. Remove the culture medium and wash the cells with 3 mL of pre-warmed HBSS. Remove HBSS and incubate the cells with 2 mL of pre-warmed trypsin-EDTA at 37 °C in a 5% CO_2 incubator for 1 min to detach the cells (*see* **Note 9**).
2. Inactivate trypsin-EDTA by adding 4 mL of culture medium. Transfer the cell suspension into a 15 mL centrifuge tube and spin at 400 × *g* for 3 min.
3. To subculture, discard the supernatant and resuspend the cell pellet in 2 mL of pre-warmed culture medium. Mix 400 μL of

cell suspension (around 2×10^6 cells) with 5 mL of pre-warmed culture medium and then transfer to a gelatin-coated 25 cm^2 cell culture flask.

4. To freeze the cells, discard the supernatant and resuspend the cell pellet in 3 mL of freezing medium. Divide the cell suspension into three aliquots (around 3×10^6 cells per mL) and store in a cryogenic vial in liquid nitrogen.

3.3 Plating of Cells

1. Coat a 96-well plate with 50 μL per well of 0.2% gelatin for 30 min. Remove excess gelatin solution and wash with HBSS (*see* **Note 7**).
2. Detach cells by trypsinization and estimate the number of cells in the suspension using a hemocytometer. Adjust the density to 5×10^5 cells per mL of suspension.
3. Apply 100 μL per well of cell suspension to the gelatin-coated 96-well plate using a Multipette M4 pipette (*see* **Note 10**). Incubate at 37 °C in a 5% CO_2 incubator for 24 hours (h) or until the cells reach 80–90% confluency.

3.4 Preparation of Sample for Measurement of RA Content

To prevent photoisomerization and photodegradation of RA, perform all procedures under a dim yellow light and sterile conditions. Apply also to Subheadings 3.5 and 3.6.

1. Dissect the target tissue from the embryo in Leibovitz L-15 Medium and pool the tissue from an appropriate number of embryos. Store at −80 °C.
2. Homogenize the sample by sonication on ice or by freeze-thaw (*see* **Note 11**) in DMEM+ (*see* **Note 12**) containing 0.1 μM R115866 (CYP26 inhibitor) [13] to inhibit RA degradation by CYP26 enzymes that may be present in the tissue sample (optional) (*see* **Note 13**). Take a few microliters of the homogenate for measurement of protein concentration using Bradford assay.
3. Spin down large debris at 3000 × *g* for 3 min at 4 °C (optional) (*see* **Note 14**).
4. Add FBS to the remaining homogenate to a final concentration of 10%.
5. The sample is ready for loading onto the RA reporter cells to determine the RA content.

3.5 Preparation of Sample for Measurement of RA Synthesis

1. Homogenize the tissue sample in DMEM+ by sonication on ice or by freeze-thaw (*see* **Note 15**). Keep the homogenate on ice and take a few microliters for measurement of protein concentration by Bradford assay. Then, equalize the protein concentration of all sample homogenates by adjusting with culture medium.

2. Add 50 μL of adjusted sample homogenate containing a fixed amount of protein to 50 μL of RA synthesis reaction mixture in a microfuge tube (*see* **Note 16**).
3. Incubate the mixture at 37 °C in a 5% CO_2 incubator for 2–4 h to allow in vitro conversion of RAL to RA by the RA-synthesizing enzymes present in the sample (*see* **Note 17**).
4. To confirm that the conversion of RAL in the reaction mixture to RA is mediated via RALDH enzymes in the sample, a negative control can be included by addition of 0.01% citral (RALDH inhibitor) [14] (*see* **Note 18**).
5. At the end of the culture period, mix thoroughly and then dilute the mixture 30-fold with culture medium (*see* **Note 19**) before loading onto the RA reporter cells to determine the amount of RA synthesized in the mixture.

3.6 Preparation of Sample for Measurement of RA Catabolism

1. Homogenize the tissue sample in DMEM+ by sonication on ice or by freeze-thaw (*see* **Note 15**). Keep the homogenate on ice and take a few microliters for measurement of protein concentration by Bradford assay. Then, equalize the protein concentration of all sample homogenates by adjusting with culture medium.
2. Add 50 μL of adjusted sample homogenate containing a fixed amount of protein to 50 μL of RA catabolism reaction mixture in a microfuge tube (*see* **Note 16**).
3. Incubate the mixture at 37 °C in a 5% CO_2 incubator for 2–4 h, during which time RA in the reaction mixture will be degraded by RA-catabolizing enzymes present in the sample (*see* **Notes 17** and **20**).
4. To confirm that degradation of RA in the reaction mixture is mediated via CYP26 enzymes in the sample, a negative control can be included by addition of 0.1 μM R115866 (CYP26 inhibitor).
5. At the end of the culture period, mix thoroughly and then dilute the mixture 30-fold with culture medium (*see* **Note 19**) before loading onto the RA reporter cells to determine the amount of RA remaining in the mixture.

3.7 RA Standard Solutions

1. Dilute the 100 mM RA stock solution stepwise with DMSO and culture medium following Table 1 to obtain RA standard solutions in tenfold dilution from 10^{-6} to 10^{-11} M (*see* **Note 4**).

3.8 Loading Samples onto RA Reporter Cells

1. Add 90 μL per well of the RA standard solutions in duplicate, and 90 μL per well of the sample (*see* Subheadings 3.4–3.6) in single measurement or in replicates, to the 96-well plate seeded with RA reporter cells (*see* **Notes 10** and **21**).

Table 1
Stepwise dilution from 100 mM to 1 μM RA or RAL stock solution to avoid precipitation

Volume of RA or RAL	Volume of DMSO (μL)	Volume of culture medium (μL)	Final RA or RAL concentration
10 μL of 100 mM RA or RAL	90		10 mM (100% DMSO)
10 μL of 10 mM RA or RAL	90		1 mM (100% DMSO)
50 μL of 1 mM RA or RAL		50	0.5 mM[a] (50% DMSO)
20 μL of 0.5 mM RA or RAL		980	10 μM[a] (1% DMSO)
100 μL of 10 μM RA or RAL		900	1 μM[b] (0.1% DMSO)

[a]The RA or RAL solution at 1 mM in DMSO is first diluted 2-fold and then 50-fold in culture medium to prevent precipitation

[b]RA standard solutions between 10^{-7} and 10^{-11} M can then be made from tenfold serial dilution of 1 μM RA solution with culture medium and are at a retinoid concentration sufficiently low where precipitation is not a problem

2. Cover the plate loosely with aluminum foil and incubate at 37 °C in a 5% CO_2 incubator for 24 h (*see* **Note 22**).
3. Measure β-galactosidase activity by X-gal staining assay.

3.9 X-Gal Staining

1. Wash the cells in the 96-well plate with PBS once and then fix with 50 μL per well of 1% glutaraldehyde for 10 min (*see* **Notes 23** and **24**).
2. Discard the fixative and rinse cells three times with PBS. Apply 100 μL per well of staining solution (*see* **Notes 23** and **24**).
3. Seal the plate with Parafilm to prevent evaporation of the staining solution and wrap with aluminum foil to protect from light. Incubate at 37 °C for 24 h to allow color development (Fig. 1a).
4. Measure the optical density at a wavelength of 600 nm (OD_{600}) using a microplate spectrophotometer. Determine the amount of RA in the sample using the standard RA response curve constructed from the RA standard solutions (Fig. 1b).

4 Notes

1. To avoid repeated freeze-thaw, store Rl15866 stock solution (10,000×) in aliquots. Each aliquot will be used only once. First dilute the stock solution 100-fold with culture medium, and then add to the reaction mixture at 1:100.

2. To prevent photoisomerization and photodegradation of RA and RAL, perform all procedures under a dim yellow light. To prevent oxidation, air bubbles should be avoided during preparation of the stock solution, which should be stored in aliquots in microfuge tubes and the tubes gassed with nitrogen or other inert gas. Seal the tube with Parafilm and wrap with aluminum foil. Store at −80 °C and use within 2 months. Each aliquot will be used only once.
3. DTT together with NAD in the RA synthesis reaction mixture or NADPH in the RA catabolism reaction mixture are required for optimal activity of RALDH or CYP26 enzymes, respectively. Their activities will be greatly reduced if omitted [12]. Prepare a 10x solution in culture medium on the day of use and then add to the reaction mixture at 1:10.
4. RA or RAL stock solutions at 100 mM should be dissolved stepwise with DMSO and culture medium (*see* Table 1) to prevent precipitation. The reaction mixture contains 50 nM RA or RAL in 0.005% DMSO, which will further be diluted 30-fold with culture medium before loading onto the RA reporter cells. We have tested to confirm that DMSO at a concentration of 0.1% or lower will not affect cell viability and LacZ expression.
5. We use commercially available Protease Inhibitor Cocktail (*Sigma*), which contains AEBSF at 104 mM, Aprotinin at 80 μM, Bestatin at 4 mM, E-64 at 1.4 mM, Leupeptin at 2 mM, and Pepstatin A at 1.5 mM in DMSO.
6. X-gal is light sensitive. The staining solution should be prepared under dim light.
7. The gelatin-coated culture flask/plate should not be allowed to dry, otherwise the gelatin will lose its adherent qualities. Thus, when the excess gelatin solution is removed, the flask/plate should be washed and cell suspension added immediately after washing.
8. The RA reporter cells will reach 80–90% confluency after 2–3 days of culture.
9. Avoid over incubation with trypsin.
10. Mix the cell suspension thoroughly and ensure even plating of cells. We do not use the outer wells of the plate for seeding cells to avoid edge effect. We add PBS to these unused outer wells. We only seed cells in the 60 inner wells. Samples and RA standard solutions can be added to the plate following Fig. 2. According to our experience, the number of cells at 90% confluency from a 25 cm^2 flask is enough to seed three 96-well plates. Each plate can measure 16 samples in triplicate, or up to 48 samples in a single assay.

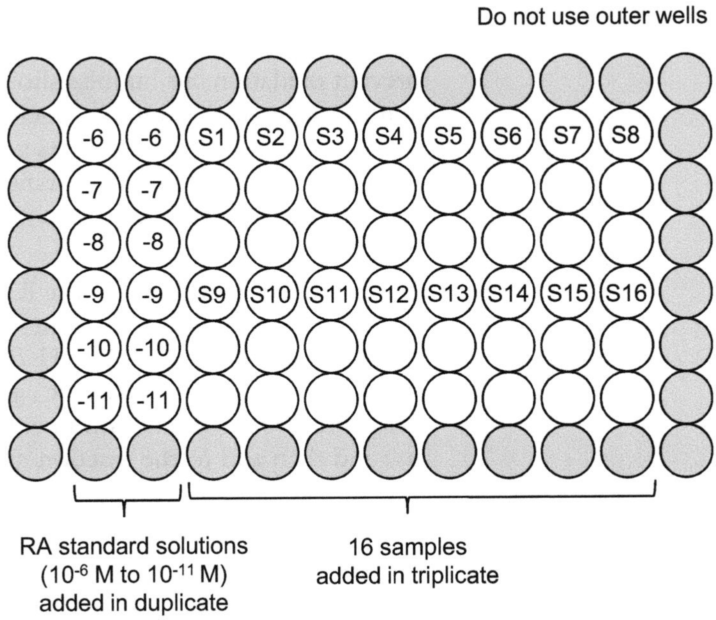

Fig. 2 An example of how RA standard solutions and samples are loaded onto the RA reporter cells seeded in the inner 60 wells of a 96-well plate. The outer wells are filled with PBS

11. The sonication amplitude or the number of freeze (in liquid nitrogen for 1 min)-thaw (in 37 °C water bath for 1 min) cycles will depend on the size of the tissue. As a reference, for a sample with 2 metanephroi of mouse embryos at gestational day (GD) 12, we conduct 5 freeze-thaw cycles.
12. Adjust the volume of medium according to the size of the tissue, RA levels in the tissue and the number of replicates in measurement required. For the metanephroi of mouse embryos at GD11-12, we pool 2 metanephroi into one sample and add 100 μL of medium for a single measurement of RA content in the sample [7].
13. Add the CYP26 inhibitor to tissue that expresses high levels of CYP26, such as the tailbud of the embryo [12].
14. If the homogenate appears cloudy, large debris should be spun down to prevent them from sticking to the RA reporter cells and killing the cells.
15. Adjust the volume of medium according to the size and protein content of the tissue, and the expression level (activity) of the RA-synthesizing/catabolizing enzymes in the tissue. For example, to determine RA-synthesizing activity in the metanephros of the mouse embryo at GD12, we pool 5 metanephroi into one sample. To determine RA-catabolizing activity in the

tailbud of the mouse embryo at GD9, we pool 4 tailbuds into one sample [12]. The sample is homogenized in 50 μL of medium.

16. The amount of protein used for the reaction will depend on the expression levels of RA-synthesizing/catabolizing enzymes in the tissue being studied. For example, *Raldh2* is expressed at high levels in the metanephroi of GD11-12 mouse embryos [7], and *Cyp26a1* is expressed at high levels in the tailbud of GD9 mouse embryo [12]. To determine the RA-synthesizing activity in the metanephros or RA-catabolizing activity in the tailbud, 10 μg protein is required.

17. The incubation time will depend on the expression level (activity) of the RA-synthesizing/catabolizing enzymes in the tissue being studied. Fix the length of time for all samples. It is suggested that a preliminary study be conducted on varying the length of incubation time.

18. Dilute citral (*Sigma*) 100-fold with culture medium, and then add to the reaction mixture at 1:100.

19. A 30-fold dilution is necessary to minimize the toxic effect of DTT and NAD/NADPH. Less than 30-fold is not recommended.

20. We have previously shown that there is minimal natural (non-enzymatic) degradation of RA during the incubation period at 37 °C [12].

21. Pre-load 100 μL of samples and RA standard solutions into individual wells of an empty 96-well plate following Fig. 2. Turn the 96-well plate seeded with RA reporter cells upside down and shake downward to drain off the medium. Immediately, use a multichannel pipette to transfer 90 μL of samples and RA standard solutions from the pre-loaded plate to this 96-well plate to avoid drying out of the cells. Add 90 μL of PBS to the outer wells.

22. We have tested various incubation times and found that 24 h is the optimal length of time, which gives the maximal staining intensity (β-galactosidase activity). Further increase in incubation time will result in reduction of staining intensity.

23. To prevent the cells from dislodging, do not add solution directly onto the cells. Apply the solution gently down the wall of the well.

24. Turn the plate upside down and shake downward to drain off the PBS before adding the fixative or staining solution. The plate should be placed upside down on tissue paper to soak up residual PBS before applying the staining solution. This will prevent dilution of the staining solution, which will affect the staining intensity.

Acknowledgments

This work was supported by the Hong Kong Research Grants Council General Research Fund Ref No. 14113314 and 14174017 to A.S.W.S. and P.J.M. We thank Dr. Michael Wagner (SUNY Downstate Medical Center) for the gift of the F9 RARE-lacZ cell line, and *Johnson & Johnson* for the gift of R115866.

References

1. Maden M, Sonneveld E, van der Saag PT, Gale E (1998) The distribution of endogenous retinoic acid in the chick embryo: implications for developmental mechanisms. Development 125:4133–4144
2. Kane MA, Chen N, Sparks S, Napoli JL (2005) Quantification of endogenous retinoic acid in limited biological samples by LC/MS/MS. Biochem J 388:363–369
3. Wagner M, Han B, Jessell TM (1992) Regional differences in retinoid release from embryonic neural tissue detected by an in vitro reporter assay. Development 116:55–66
4. Colbert MC, Linney E, LaMantia AS (1993) Local sources of retinoic acid coincide with retinoid-mediated transgene activity during embryonic development. Proc Natl Acad Sci U S A 90:6572–6576
5. Chen Y, Huang L, Solursh M (1994) A concentration gradient of retinoids in the early *Xenopus laevis* embryo. Dev Biol 161:70–76
6. Wagner M (1998) Detection and measurement of retinoic acid production by isolated tissues using retinoic acid-sensitive reporter cell lines. Methods Mol Biol 89:41–53
7. Lee LM, Leung CY, Tang WW, Choi HL, Leung YC, McCaffery PJ, Wang CC, Woolf AS, Shum AS (2012) A paradoxical teratogenic mechanism for retinoic acid. Proc Natl Acad Sci U S A 109:13668–13673
8. McCaffery P, Dräger UC (1994) Hot spots of retinoic acid synthesis in the developing spinal cord. Proc Natl Acad Sci U S A 91:7194–7197
9. Romand R, Niederreither K, Abu-Abed S, Petkovich M, Fraulob V, Hashino E, Dollé P (2004) Complementary expression patterns of retinoid acid-synthesizing and -metabolizing enzymes in pre-natal mouse inner ear structures. Gene Expr Patterns 4:123–133
10. McCaffery P, Dräger UC (1997) A sensitive bioassay for enzymes that synthesize retinoic acid. Brain Res Brain Res Protoc 1:232–326
11. Yamamoto M, Dräger UC, McCaffery P (1998) A novel assay for retinoic acid catabolic enzymes shows high expression in the developing hindbrain. Brain Res Dev Brain Res 107:103–111
12. Lee LM, Leung MB, Kwok RC, Leung YC, Wang CC, McCaffery PJ, Copp AJ, Shum AS (2017) Perturbation of retinoid homeostasis increases malformation risk in embryos exposed to pregestational diabetes. Diabetes 66:1041–1051
13. Stoppie P, Borgers M, Borghgraef P, Dillen L, Goossens J, Sanz G, Szel H, Van Hove C, Van Nyen G, Nobels G, Vanden Bossche H, Venet M, Willemsens G, Van Wauwe J (2000) R115866 inhibits all-trans-retinoic acid metabolism and exerts retinoidal effects in rodents. J Pharmacol Exp Ther 293:304–312
14. Kikonyogo A, Abriola DP, Dryjanski M, Pietruszko R (1999) Mechanism of inhibition of aldehyde dehydrogenase by citral, a retinoid antagonist. Eur J Biochem 262:704–712

Chapter 14

A Behavioral Assay to Study Effects of Retinoid Pharmacology on Nervous System Development in a Marine Annelid

M. Handberg-Thorsager, V. Ulman, P. Tomançak, D. Arendt, and M. Schubert

Abstract

Autonomous animal locomotion, such as swimming, is modulated by neuronal networks acting on cilia or muscles. Understanding how these networks are formed and coordinated is a complex scientific problem, which requires various technical approaches. Among others, behavioral studies of developing animals treated with exogenous substances have proven to be a successful approach for studying the functions of neuronal networks. One such substance crucial for the proper development of the nervous system is the vitamin A-derived morphogen retinoic acid (RA). In the larva of the marine annelid *Platynereis dumerilii*, for example, RA is involved in the specification and differentiation of individual neurons and responsible for orchestrating the swimming behavior of the developing larva. Here, we report a workflow to analyze the effects of RA on the locomotion of the *P. dumerilii* larva. We provide a protocol for both the treatment with RA and the recording of larval swimming behavior. Additionally, we present a pipeline for the analysis of the obtained data in terms of swimming speed and movement trajectory. This chapter thus summarizes the methodology for analyzing the effects of a specific drug treatment on larval swimming behavior. We expect this approach to be readily adaptable to a wide variety of pharmacological compounds and aquatic species.

Key words 13-*cis* and all-*trans* retinoic acid, Behavioral analysis, Live imaging, Marine invertebrate larvae, Movement trajectory, Pharmacological treatments, *Platynereis dumerilii*, Swimming speed

1 Introduction

Autonomous animal locomotion by swimming is the outcome of a series of interactions between neurons and their projections to and coordination of effector cells, such as ciliary and muscle cells. In the course of development, marine organisms often alternate between ciliary- and muscle-based swimming movements. For example, in the young larva of the annelid *Platynereis dumerilii* swimming

---

Electronic supplementary material: The online version of this chapter (https://doi.org/10.1007/978-1-4939-9585-1_14) contains supplementary material, which is available to authorized users.

Swapan K. Ray (ed.), *Retinoid and Rexinoid Signaling: Methods and Protocols*, Methods in Molecular Biology, vol. 2019, https://doi.org/10.1007/978-1-4939-9585-1_14, © Springer Science+Business Media, LLC, part of Springer Nature 2019

depends solely on the activity of equatorial bands of ciliated cells, which are directly connected to a simple nervous system [1]. Active ciliary beating keeps the larva steady in the water column, whereas ciliary arrest leads to the sinking of the larva. Later during development, in advanced larval stages with differentiated muscle cells, movement is accomplished by an alternation of ciliary beating and muscle activity [2, 3].

Studies into the molecular mechanisms that coordinate the activity of the nervous system, the ciliary band, and the muscles during larval movement require the development of behavioral assays applicable to a wide array of candidate molecules. Previous studies have shown that larvae of *P. dumerilii* are suitable models for studying the effects of signaling molecules during locomotion. A detailed characterization of the developing neuro-ciliary system, for instance, has revealed that the hormone melatonin is crucial for the nocturnal sinking of *P. dumerilii* larvae by increasing the frequency of ciliary arrest [4]. Additionally, it has already been established that neuropeptides control the vertical up and down movements of young larvae through the neuro-ciliary system [5, 6]. Regarding larval movements mediated by muscles, we have previously shown that the signaling molecule retinoic acid (RA), a derivate of vitamin A and crucially required for orchestrating animal development [7–9], is involved in specification and differentiation of the nervous system in older *P. dumerilii* larvae, where locomotion relies both on the neuro-ciliary and neuro-muscular systems [10]. An increase of RA signaling activity by exogenous treatments with RA thus resulted in morphological malformations of the larval nervous system, which were accompanied by a decrease in larval swimming speed and changes in larval swimming patterns [10]. The treatment-induced alterations of nervous system topology and connectivity were thus reflected by changes in the swimming behavior of *P. dumerilii* larvae.

Here, making use of this interconnection between nervous system development and swimming behavior, we present a detailed protocol for analyzing the effects of pharmacological compounds on the swimming speed and movement trajectory of aquatic larvae. Based on the example of RA treatments of developing *P. dumerilii* larvae, we explain how to apply the drug and how to perform live imaging of larval swimming. We further describe a data analysis pipeline for determining the swimming speed and movement trajectory of individual larvae from the movies. To optimize the utility of this protocol, the methodology is based exclusively on commercially available material and open-source software.

2 Materials

RA isomers, such as all-*trans*, 13-*cis*, and 9-*cis* RA (respectively referred to here as ATRA, 13cRA, and 9cRA) [11], are light-sensitive compounds and, when working with these drugs, care should be taken to avoid exposure to light. Reconstitution and aliquoting of stock solutions and dilution to work concentrations are best performed in a dark room under red light. In addition, RA solutions need to be stored in amber, light-tight Eppendorf tubes at −20 °C or −80 °C to optimize drug preservation. Note that the use of freshly prepared and rapidly frozen RA dilutions is highly recommended, as drug activity tends to decrease by repeated freeze-thaw cycles (*see* **Note 1**).

RA isomer choice as well as final treatment concentrations vary between organisms [10, 12, 13] and thus need to be assessed individually for each species (*see* **Note 2**). Treatments of *P. dumerilii* embryos and larvae with ATRA and 13cRA, for example, were based on the observation that ATRA and 13cRA, but not 9cRA, are present at measurable concentrations in tissues of both developing and adult annelids [10].

As RA is involved in various biological processes in humans, special caution needs to be taken when handling the drug. Wear gloves and work under a fume hood. Residues should be discarded following validated safety protocols.

2.1 Drug Preparation

1. Dissolve ATRA (Sigma-Aldrich) and 13cRA (Sigma-Aldrich) in dimethyl sulfoxide (DMSO) or ethanol at a final concentration of 10 mM (stock solution) by vortexing (*see* **Note 3**). Aliquot the stock solution into amber Eppendorf tubes (Sigma-Aldrich) and store at −20 °C or −80 °C until use. Thaw a new aliquot for each experiment and discard the leftover.

2.2 Incubation Solution

1. Filter, through a 0.22 μm mesh (Corning), the solution to be used for incubation of embryos and larvae during the drug treatment. For marine organisms, either natural seawater (NSW) or artificial seawater (ASW) is used. The water temperature needs to match the culturing conditions (*see* **Note 4**).

3 Methods

RA treatment experiments are carried out in 6 mL NSW in 6-well plates in three biological replicates. For longer drug treatments, keep the number of embryos to 100–200 per well. High embryo density negatively affects well-being and development of embryos and larvae. Perform treatment experiments at the culturing temperature of embryos and larvae (*see* **Note 4**).

3.1 Drug Treatment

1. Prepare a 6-well plate with 2.5 mL, per well, of filtered (0.22 μm) NSW. Add embryos in 0.5 mL NSW to the wells (*see* **Note 5**) to reach a 3 mL volume per well and keep the plates in the culturing incubator. An additional 3 mL will be added to each well to establish the different treatment conditions (*see* Subheading 3.1, **step 2**). Use two of the wells for ATRA treatment, two for 13cRA treatment, one for a control group treated with DMSO (or ethanol if used for dissolving the retinoids) to check for potential effects of DMSO (or ethanol) on animal development, and one for wild-type embryos to confirm normal development of the embryo batch used for the experiment.
2. Prepare ATRA and 13cRA at the final working concentrations of 0.5 μM and 1.0 μM right before use (*see* **Note 2**). Add 3 mL of filtered (0.22 μm) NSW to 15 mL Falcon tubes, preparing one 15 mL Falcon tube per treatment. Subsequently, add 0.3 μL ATRA stock solution (at 10 mM) or 0.3 μL 13cRA stock solution (at 10 mM) to prepare the 0.5 μM final working concentrations, 0.6 μL ATRA stock solution (at 10 mM) or 0.6 μL 13cRA stock solution (at 10 mM) to prepare the 1.0 μM final working concentrations or 0.6 μL DMSO (or ethanol) to prepare the final working concentration for the treatment control. Vortex the Falcon tubes immediately to dissolve the drug and transfer, with a Pasteur pipette, the solution in each Falcon tube to the corresponding well of the 6-well plate. This step thus yields a final treatment volume of 6 mL per well. Gently mix embryos and drug solution with the Pasteur pipette. Make sure to use a clean Pasteur pipette for each experimental condition (i.e., for each Falcon tube).
3. Cover the 6-well plate with aluminum foil to avoid light exposure and incubate at the appropriate culturing temperature. Our RA drug treatments were performed for 24 hours on developing *P. dumerilii* at 48 hours-post-fertilization (hpf).

3.2 Live Recording of Larval Swimming

Use a stereomicroscope with a fast-speed camera (25 frames per second), e.g., a Nikon Digital Sight DS-Fi1 or a DMK 42BUC03 (The Imaging Source). This setup allows the recording of the swimming behavior of the treated larvae (with a size of about 200 μm) both in a horizontal plane (covering a swimming area of 11.34 cm^2) and, albeit to a lesser extent, in a vertical plane (covering 0.53 cm of the water column). For more detailed measurements of vertical swimming behavior in the water column, home-built vertical columns need to be used, as described, for example, by Conzelmann et al. [5].

1. Remove the aluminum foil and record the swimming behavior of the larvae in the 6-well plate. Record one well at a time and

change between plates of the biological replicates for each recording to ensure the same recording conditions (i.e., incubation in the dark before recording). Record for 1 min at maximum speed. Cover the plate with aluminum foil after recording.

2. Save the files in a file format that can be opened by the program Fiji [14], such as TIFF or JPEG. Make sure to save the metadata to extract the exact time frames, especially if these are not constant. This is important for subsequent speed calculations (*see* Subheading 3.3, **steps 16–18**).

3.3 Analysis of Larval Swimming Behavior

Analysis of the swimming behavior can be divided into two parts: measuring the swimming speed and assessing the swimming trajectory. In both cases, tracking of embryos and larvae over time is needed. Different open-source tracking programs are available under Fiji [14], including TrackMate [15] and MTrackJ [16]. As TrackMate [15] offers automated segmentation and tracking of individual specimens, it was used here (*see* **Note 6**).

1. Download and install the required software. Fiji [14] can be downloaded from https://fiji.sc and TrackMate [15] from https://imagej.net/TrackMate. Download also the additional "Track analysis" plugins under "5.1 Downloadable jars" at http://imagej.net/TrackMate#Downloadable_jars to be able to extract the "TOTAL_DISTANCE_TRAVELED" parameter (*see* Subheading 3.3, **step 18**). Place the track analysis jar file in the Fiji.app/jars folder and restart Fiji.
2. Open the previously recorded movie in Fiji [14] and select an area that completely excludes the margins of the 6-well plate (*see* **Note 7**). Save the resulting file as TIFF.
3. Launch TrackMate [15]. All the operations are performed from the TrackMate window, which appears next to the movie (*see* Fig. 1). Start by adjusting the image and movie properties, if needed ("Calibration Settings"), and select the region and time points of interest ("Crop Settings"). When saving the TrackMate file (bottom "Save"), an XML file will be created containing all the processed data.
4. Navigate to the next window to select a detector for segmentation of the specimens (i.e., embryos or larvae). Use the LoG detector for a diameter of the specimen between 5 and 20 pixels. Indicate the estimated specimen size in pixels in the next window ("Estimated blob diameter"). To provide an example, a typical annelid larva in the RA treatment experiment was close to 10 pixels in size. "Preview" to inspect the segmentation result. Selected spots are highlighted with a purple circle (*see* Fig. 2a). Application of a threshold value (here 1.0) and "Use median filter" will remove many of the falsely detected spots

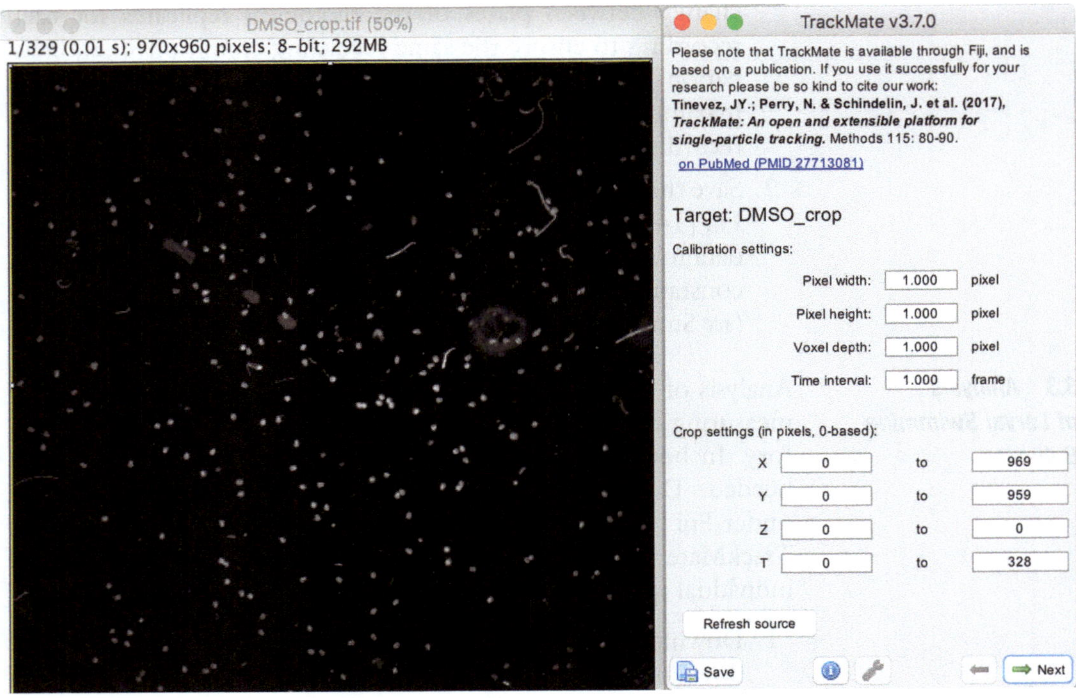

Fig. 1 Launching TrackMate. Defining image properties in the initial TrackMate window

(*see* Fig. 2b). In the following windows, application of different filtering criteria will further lower the number of segmented spots.

5. The first filtering option allows the definition of a quality value, above or below which the spots will be included in the tracking analysis. This value, calculated by the segmentation software, reflects the probability of a spot to represent a real biological specimen. Of note, segmented spots removed in this step will be removed from the metadata, which will speed up the processing time of the calculations. In the current analysis, assessing the effects of RA treatments on the swimming behavior of *P. dumerilii* larvae, we used a quality value above 4.4.
6. Select a displayer for the visualization of specimen spots and tracks on the live imaging recordings. For our 2D recordings, we chose the option "HyperStack Displayer." Press "Next."
7. In the next panel, various filtering options are available and can be applied to the data by clicking on the "+" button. Select a filtering feature in the scroll down menu and determine the filtering criteria by manually adjusting the histogram or by pressing "Auto." Included spots are highlighted in the image. For our movies, "Contrast" (>0.5) or "Signal/Noise" (>1.0) were used to ensure unique segmentation of the larvae. This step is critical to lower the number of selected spots before

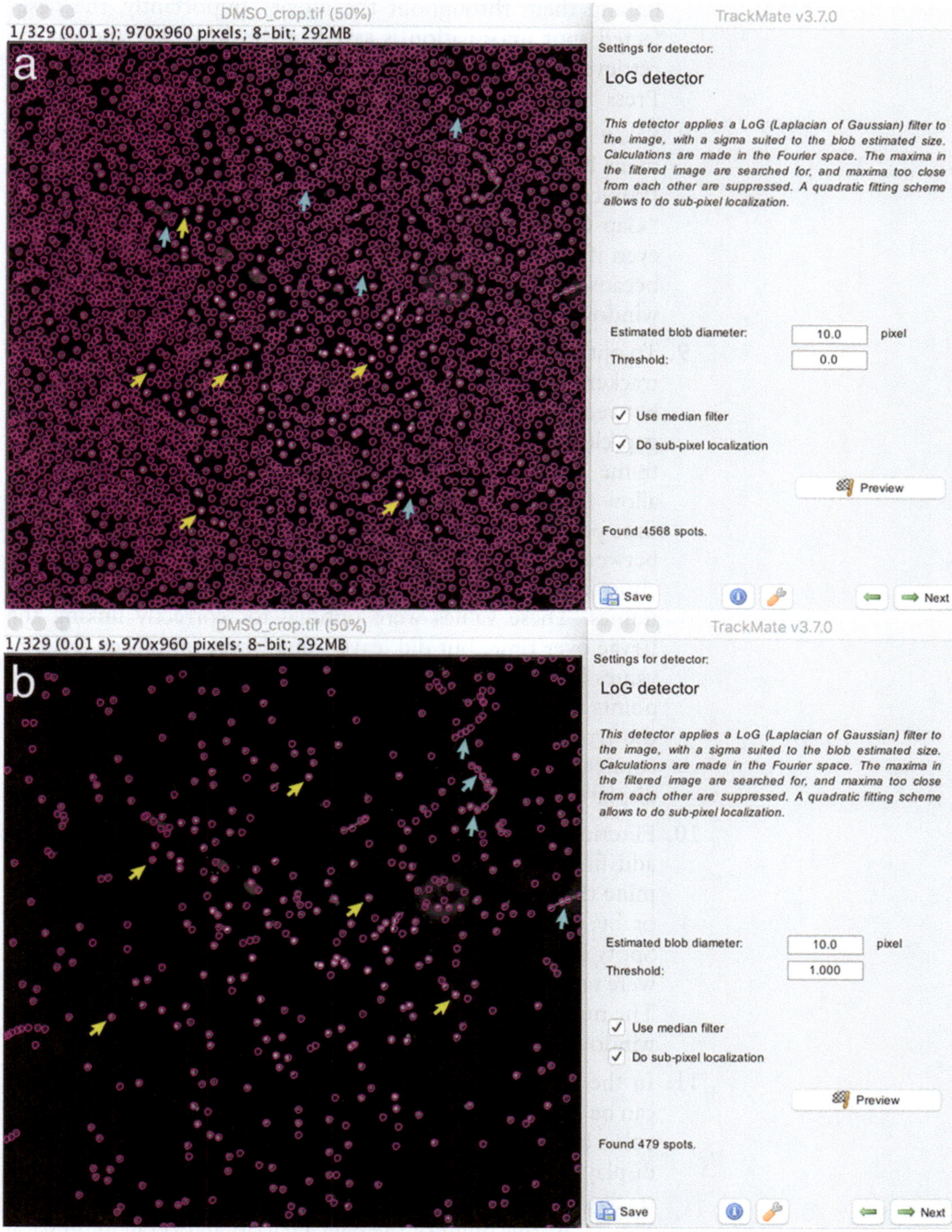

Fig. 2 Detection of embryos and larvae by TrackMate. (**a**) Define the diameter of the specimen and inspect by pressing "Preview." Here, many of the segmented spots are larvae (yellow arrows), but TrackMate also detected many other objects (blue arrows). (**b**) Introducing a threshold can eliminate falsely labeled spots (compare yellow and blue arrows between (**a**) and (**b**))

linking them throughout the movie. Importantly, the deselected spot information is saved in the XML file and can be retrieved again by navigating back in the TrackMate window. Press "Next."

8. Select a tracking algorithm to connect the segmented specimens between time points. We chose the "Linear Assignment Problem" (or "LAP") tracker, which includes the feature "Gap-Closing Events" to connect the track of a specimen, even if it disappears in some of the time frames (for example, because the specimen is out of focus or outside the imaging window).
9. To optimize the tracking, apply the following parameters to the tracker: (1) the maximal allowed linking distance of a specimen between two time frames, (2) the maximal distance for gap-closing of a specimen, and (3) the allowed number of frame gaps in the gap-closing feature. The latter two values allow for the connection of a specimen between two separate time points, even if the specimen is undetectable in the period between these two time points. For our movies, we respectively chose 30 pixels, 25 pixels, and a maximum of 2 gap-closing frames. These values were optimal for correctly linking the larvae over time, but did leave some larvae untracked. Higher values increasing the allowed specimen distance between time points increased the number of tracked larvae, but also caused the erroneous linking of larvae (i.e., different specimens were connected over time). The number of calculated tracks will be displayed when pressing "Next."
10. Filtering of the tracks is also necessary. Use the "+" button to add filtering parameters from the scroll down menu. Determine the filtering criteria by manually adjusting the histogram or by pressing "Auto." For our purpose, both "Number of Spots in a Track" (>35) and "Track Displacement" (>100) were used for incrementing the ratio of real tracks (*see* **Note 8**). The number of remaining tracks is displayed in the TrackMate window as the filtering options are applied to the dataset.
11. In the next window, the visualization of the spots and tracks can be adjusted. We selected "Uniform color" for the spots and color by "Track start" for the tracks. Adjust also the "Track display mode" to show all tracks or only a selection of the tracks, according to the options in the scroll down menu (*see* Fig. 3). At this point of the data analysis, an overall impression of swimming paths and behaviors under different experimental conditions can be displayed (*see* Fig. 4).
12. In the same window, the tracks can be visualized, inspected, and curated with "TrackScheme." For manual curation, in "Display tracks," set "Show local tracks, backwards" and

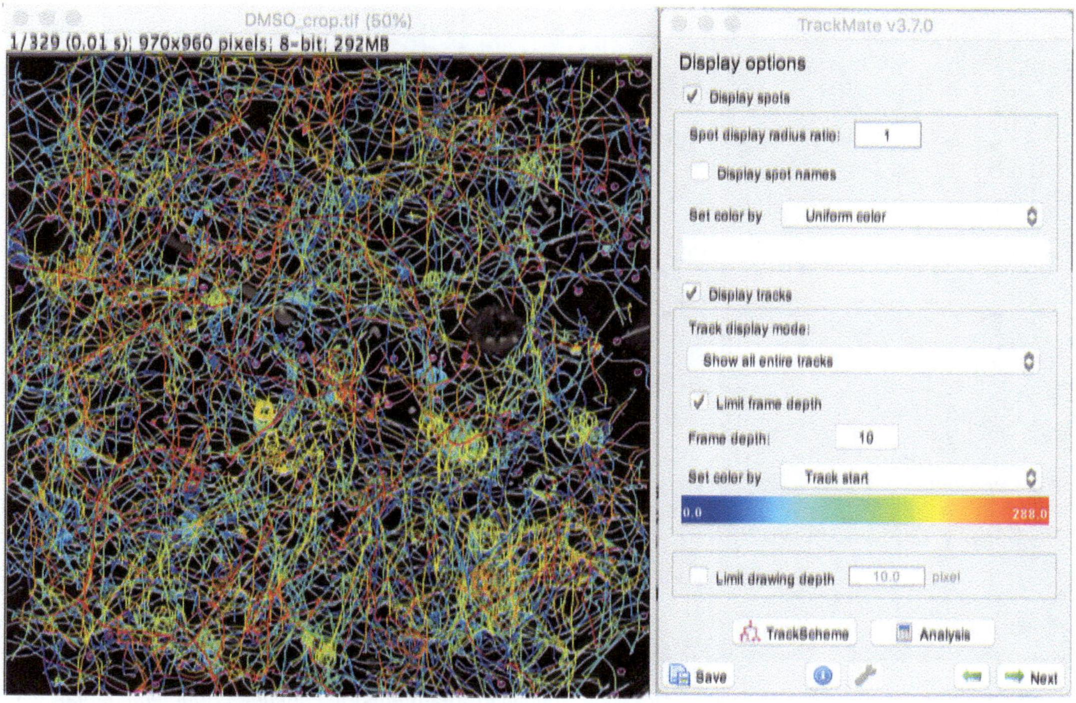

Fig. 3 TrackMate display options of segmented larvae (i.e., spots) and tracks. Scroll down menus offer various options for differential visualization. Here, spots were labeled with "Uniform colors" (pink) and tracks by "Track start" (indicated by a progressive color change from the beginning to the end of the movie)

"Limit frame depth" to 20, and then open "TrackScheme." When selecting a spot in "TrackScheme," this spot will be highlighted in green in the movie (*see* Fig. 5). A dragon tail displays the track of the specimen in the last 20 frames of the movie (*see* **Note 9**). Use the arrows to move through the movie over time in "TrackScheme" to inspect the path of the specimen. Perform manual curation in "TrackScheme" or the displayer window, for example, to move a spot along the *x*- and/or *y*-axis (change the position of the spot with the mouse), to delete a spot (with the "D" key), to add a spot (with the "A" key), or to link different tracks (with the "L" key). In our current experiment, we manually curated a minimum of 30 larval tracks per experiment. Save the resulting XML file.

13. Non-corrected tracks can be deleted by selecting them in "TrackScheme" (click the right mouse button and delete with "Remove spots and links"). Save the project as XML file under a different name, since the deleted information will be irretrievably lost.
14. Visualize the curated swimming tracks by displaying "Show all entire tracks" in the "Display options" window of TrackMate.

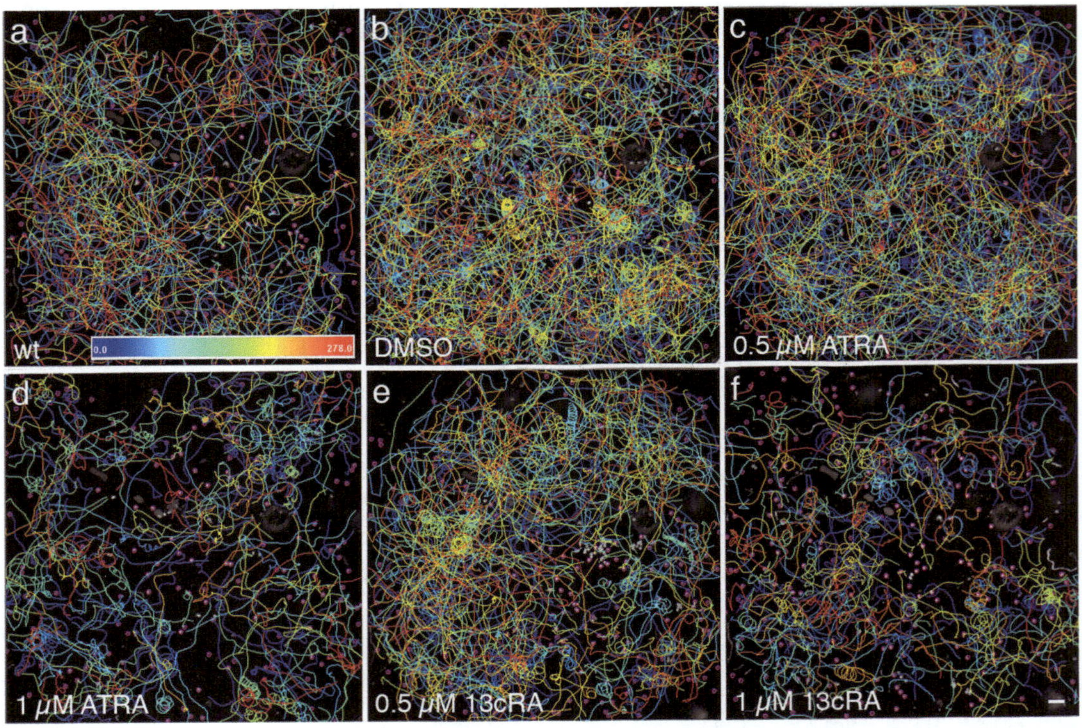

Fig. 4 Tracking results, obtained with TrackMate, following treatments of *Platynereis dumerilii* larvae with retinoic acid (RA). (**a**) Wild-type (wt) larvae. (**b**) Control larvae treated with dimethyl sulfoxide (DMSO). (**c**) Larvae treated with 0.5 μM all-*trans* RA (ATRA). (**d**) Larvae treated with 1 μM all-*trans* RA (ATRA). (**e**) Larvae treated with 0.5 μM 13-*cis* RA (13cRA). (**f**) Larvae treated with 1 μM 13-*cis* RA (13cRA). The color of the track indicates the starting time point in the movie, following the color code indicated in (**a**). Scale bar: 500 μm

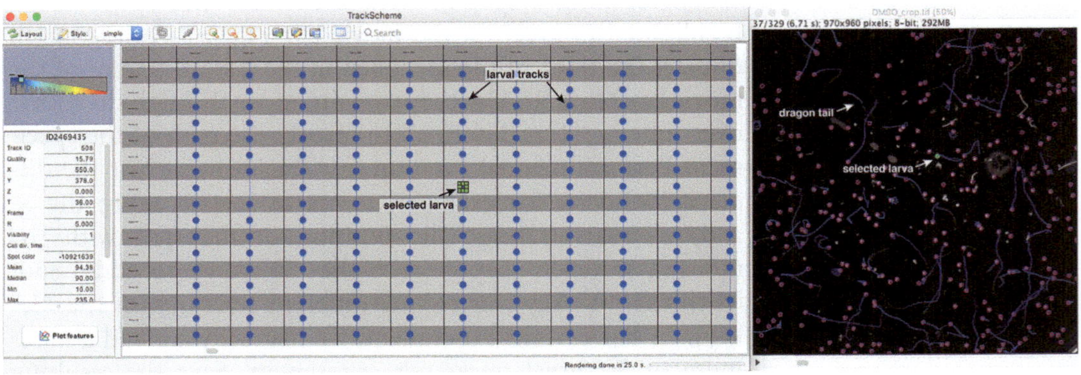

Fig. 5 Manual curation of TrackMate larval tracks. To inspect a larval track, select a spot in "TrackScheme" and follow its track over time in the display window. The selected larva is highlighted in green in both "TrackScheme" and display window. Here, the larval track is visualized with a dragon tail consisting of the last 20 time frames. Larval tracks can easily be split or linked in "TrackScheme." For splitting, select the linker between two spots, click the right mouse button, and select "Remove spots and links." For connecting two tracks, select the end spot of one track and the first spot of the subsequent track and press the "Toogle Linking" button in the "TrackScheme" menu (alternatively, press the "L" key). For deleting multiple spots in a larval track, highlight the spots in "TrackScheme," click the right mouse button, and select "Remove spots and links." Spots can be added by pressing the "A" key in the display window. Press "Layout" in the "TrackScheme" menu to update the "TrackScheme" window

Create a picture (or take a screenshot) of the completed swimming tracks to compare the swimming patterns between experimental conditions (*see* Fig. 6a–f).

15. Visualize the swimming behavior directly by overlaying the specimen tracks on the recorded movie. To do this, set "Show local tracks, backwards" in the "Display options" window of TrackMate. Navigate forward two windows and execute the "Capture overlay" action (*see* Movie 1) (*see* **Note 10**).
16. For calculation of the swimming speed, navigate two windows back to "Display options." Select "Analysis" to extract all the spot and tracking data. Three files will open: "Track statistics," "Spots in tracks statistics," and "Links in tracks statistics." Save these three files in CSV format for further analysis.
17. The larval swimming speed (i.e., the "TRACK_MEAN_SPEED" values) can be extracted directly from the "Track statistics" file, if the time intervals between time frames in the recordings are identical (*see* **Note 11**). However, if the time intervals vary between time frames, as in our case, the swimming speed has to be calculated manually.
18. For manual calculation of the swimming speed, extract the time at each time frame of the movie. This information is available in the metadata of the recordings. Annotate also the time frames covering each of the manually curated specimen tracks ("FRAME" values from the file "Spots in tracks statistics"). Determine the time points of the first and the last time frame of the specimen track to obtain the total duration of the track. Thereafter, extract the corresponding length of the specimen track ("TOTAL_DISTANCE_TRAVELED" value from the file "Tracks statistics"). This value is in pixels and can be converted into a metric value by multiplying the pixel value with the image pixel size (11.33 μm, when using the Nikon Digital Sight DS-Fi1 camera, a 0.6× projection lens, and a 0.5× objective lens) (*see* **Note 12**). Calculate the swimming speed as total distance traveled divided by total time. In our case, this resulted in values measured in μm/s.
19. Display the swimming speeds in a box plot for comparison between the different experimental conditions (*see* Fig. 6g). Free online tools, such as BoxPlotR (http://shiny.chemgrid.org/boxplotr/), can be used for generating the box plots.
20. Perform statistical analyses, like a Student's *t*-test, to assess the statistical significance of the effects on larval swimming behavior observed after different drug treatments (*see* Fig. 6g). Free statistical analyses can be carried out, for example, by making use of the QuickCalcs program suite provided online by GraphPad Software (https://www.graphpad.com/quickcalcs/).

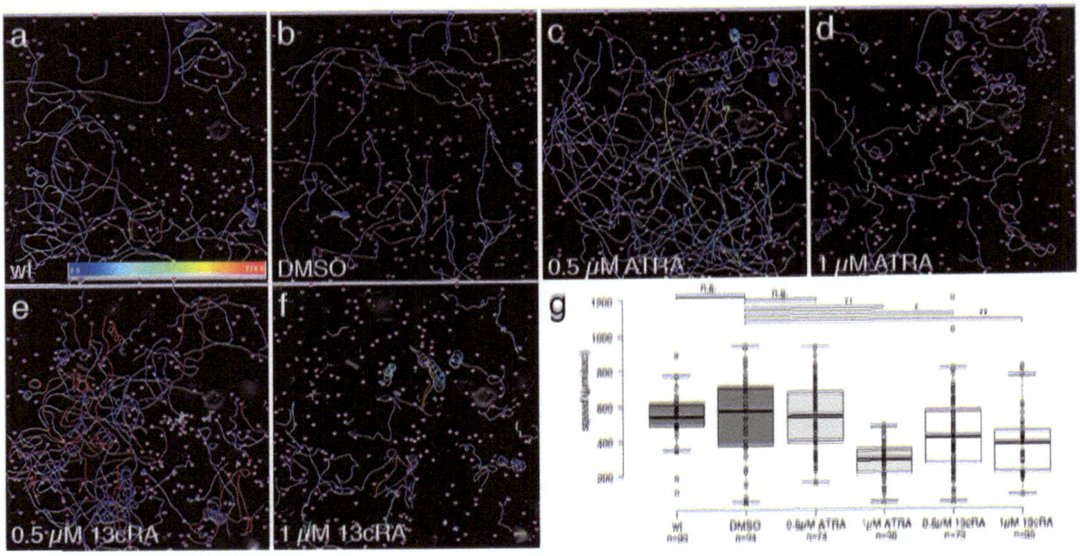

Fig. 6 Analysis of curated larval tracking results following retinoic acid (RA) treatments of *Platynereis dumerilii* larvae. (**a–f**) Curated larval tracks using TrackMate. (**a**) Wild-type (wt) larvae. (**b**) Control larvae treated with dimethyl sulfoxide (DMSO). (**c**) Larvae treated with 0.5 μM all-*trans* RA (ATRA). (**d**) Larvae treated with 1 μM all-*trans* RA (ATRA). (**e**) Larvae treated with 0.5 μM 13-*cis* RA (13cRA). (**f**) Larvae treated with 1 μM 13-*cis* RA (13cRA). The color of the track indicates the starting time point in the movie, following the color code indicated in (**a**). Scale bar: 500 μm. (**g**) Exogenous RA treatments slow down *P. dumerilii* larval swimming. Box plots showing the speed of *P. dumerilii* larval swimming following different RA treatments. Data distribution (circles), median values (bold line), and Tukey whiskers are shown. The number of curated larval tracks is indicated (n). An unpaired Student's *t*-test on the mean value was used for statistical analysis (n.s., non-significant; $*$, $P < 0.05$; $**$, $P < 0.01$). $P = 0.9330$ for DMSO versus wild-type (wt); $P = 0.8810$ for DMSO versus 0.5 μM ATRA; $P = 0.0001$ for DMSO versus 1 μM ATRA; $P = 0.0416$ for DMSO versus 0.5 μM 13cRA; $P = 0.0038$ for DMSO versus 1 μM 13cRA

4 Notes

1. To ensure reproducibility of treatment-based experiments, strictly respect handling and care guidelines of the compounds to be used and note that the stability of the drug may be altered once dissolved. Generally avoid repeated freeze-thaw cycles once the compound is in solution. If possible, aliquot stock solutions into small volumes for single use.
2. It is advisable to assess the optimal treatment concentrations with small trial experiments before initiating the actual live recording experiment. As a rule of thumb, RA treatments in vertebrates are carried out at concentrations between 1 nM and 0.1 μM, whereas higher concentrations, between 0.1 μM and 5 μM, are generally used in invertebrates, due to lower ligand binding affinity of most invertebrate RA receptors [10, 12, 13].

3. ATRA can be ordered in aliquots of 50 mg (Sigma-Aldrich). To obtain a 10 mM ATRA stock solution from 50 mg ATRA, the powder has to be dissolved in 16.66 mL DMSO (or ethanol). 13cRA can be purchased as aliquots of 100 mg (Sigma-Aldrich), and a 10 mM 13cRA stock solution is prepared by dissolving the powder in 33.32 mL DMSO (or ethanol).
4. Temperature influences development. *P. dumerilii* embryos, larvae, and adults are kept at 18 °C in the laboratory, but can grow at temperatures between 14 °C and 30 °C [2]. Development generally accelerates at higher temperatures and decelerates at lower temperatures [2]. Therefore, in order to guarantee comparability between experiments, it is crucially important to keep all cultures at the same, constant, temperature.
5. Embryos can easily be transferred to individual wells of a 6-well plate, if they are transferred before they start to swim. If the embryos are already swimming, they can be concentrated into a smaller volume of NSW by passing the seawater with the embryos through a 50 μm mesh (Fisher Scientific). Since *P. dumerilii* embryos measure between 160 and 200 μm, they will be retained by the mesh and can be transferred into a small volume of NSW. Keep the mesh in seawater at all times to avoid desiccation of the embryos. To ensure the appropriate dilution of the drug, 0.5 mL of concentrated embryos should be pipetted into a well containing 2.5 mL of NSW. The final treatment volume of 6 mL will then be reached once the 3 mL drug solution is added (*see* Subheading 3.1, **step 2**).
6. A detailed guide for the use of TrackMate is available at https://imagej.net/Getting_started_with_TrackMate.
7. Any particle, dirt, or shiny part of the 6-well plate can mistakenly be interpreted as a specimen and thus be segmented during the detection process. It is therefore highly recommended to define the movie area in a way that it only contains information related to the specimens to be tracked. Since filtered NSW is used as incubation medium (*see* Subheading 2.2), the amount of particles and dirt in the culture should be minimal.
8. Note that "Track Displacement" is a measurement of the displacement of a tracked specimen between the first and the last time point and does thus not indicate the length of the track itself. This parameter is useful to spot and remove segmentation and tracking of immobile embryos and larvae.
9. At this step, the consequences of the defined track filtering options will become evident, since some spots will have a dragon tail (i.e., these spots were included and thus tracked),

while other spots do not have a dragon tail (i.e., these spots were discarded during the track filtering process).

10. For comparisons of the movies under different experimental conditions, use the "Combine..." option in Fiji [14] to organize the movies next to each other. The number of frames should be the same between movies to use this option. Since the movies are generally recorded for 1 min at maximum speed (and thus not at a predefined speed), the number of frames can differ between experimental conditions. If this is the case, individual movies need to be duplicated to obtain the desired frame number by using the "Duplicate..." option in Fiji [14]. Use "Label..." to add the experimental conditions and the time stamps to each movie. This is possible if a fixed time interval is used. In our case, however, the time intervals vary between time points. We hence wrote a script, which uses the time frame information from the supplementary metadata of each movie to label the individual time stamps of each movie in Fiji [14]. The experimental condition can be added as well. The script is available from the "TomancakLab" update site in Fiji [14]—click "Manage update sites" in the Fiji Updater, then click "Close" and "Apply changes." Alternatively, it can be downloaded from https://github.com/xulman/TomancakLab/tree/master/7_putLabels. The web page indicates how to present the time frame information in the supplementary metadata and how to use the script.
11. Make sure that proper time intervals are used. This information can be obtained from the metadata of the movies and should be applied to the image properties and/or the initial TrackMate window.
12. The pixel size of a given image can be determined by dividing the camera pixel size by the magnification of the objective used. In our setup, we used a Nikon Digital Sight DS-Fi1 camera with a camera pixel size of 3.4 μm, a 0.6× projection lens, a 0.5× objective lens, and an optical zoom of 1. The image pixel size was thus $3.4\ \mu m/(0.6 \times 0.5 \times 1) = 11.33\ \mu m/\text{pixel}$.

Acknowledgments

The authors are indebted to Miquel Vila Farré and Jochen Rink for help with microscopy recordings. Mette Handberg-Thorsager is supported by the Deutsche Forschungsgemeinschaft (DFG, grant number TO563/7-1), Vladimir Ulman by the German Federal Ministry of Research and Education (BMBF) under the code 031L0102 (de.NBI), and Detlev Arendt by the European Molecular Biology Laboratory and the European Research Council (BrainEvoDevo No. 294810).

References

1. Jékely G, Colombelli J, Hausen H et al (2008) Mechanism of phototaxis in marine zooplankton. Nature 456:395–399
2. Fischer AH, Henrich T, Arendt D (2010) The normal development of *Platynereis dumerilii* (Nereididae, Annelida). Front Zool 7:31
3. Randel N, Asadulina A, Bezares-Calderón LA (2014) Neuronal connectome of a sensory-motor circuit for visual navigation. elife 3:e02730
4. Tosches MA, Bucher D, Vopalensky P et al (2014) Melatonin signaling controls circadian swimming behavior in marine zooplankton. Cell 159:46–57
5. Conzelmann M, Offenburger S, Asadulina A et al (2011) Neuropeptides regulate swimming depth of *Platynereis* larvae. Proc Natl Acad Sci U S A 108:E1174–E1183
6. Conzelmann M, Williams EA, Tunaru S et al (2013) Conserved MIP receptor-ligand pair regulates *Platynereis* larval settlement. Proc Natl Acad Sci U S A 110:8224–8229
7. Niederreither K, Dollé P (2008) Retinoic acid in development: towards an integrated view. Nat Rev Genet 9:541–553
8. Cunningham TJ, Duester G (2015) Mechanisms of retinoic acid signalling and its roles in organ and limb development. Nat Rev Mol Cell Biol 16:110–123
9. Zieger E, Schubert M (2017) New insights into the roles of retinoic acid signaling in nervous system development and the establishment of neurotransmitter systems. Int Rev Cell Mol Biol 330:1–84
10. Handberg-Thorsager M, Gutierrez-Mazariegos J, Arold ST et al (2018) The ancestral retinoic acid receptor was a low-affinity sensor triggering neuronal differentiation. Sci Adv 4:eaao1261
11. Kane MA, Chen N, Sparks S et al (2005) Quantification of endogenous retinoic acid in limited biological samples by LC/MS/MS. Biochem J 388:363–369
12. Escriva H, Bertrand S, Germain P et al (2006) Neofunctionalization in vertebrates: the example of retinoic acid receptors. PLoS Genet 2:e102
13. Gutierrez-Mazariegos J, Nadendla EK, Studer RA et al (2016) Evolutionary diversification of retinoic acid receptor ligand-binding pocket structure by molecular tinkering. R Soc Open Sci 3:150484
14. Schindelin J, Arganda-Carreras I, Frise E et al (2012) Fiji: an open-source platform for biological-image analysis. Nat Methods 9:676–682
15. Tinevez JY, Perry N, Schindelin J et al (2017) TrackMate: an open and extensible platform for single-particle tracking. Methods 115:80–90
16. Meijering E, Dzyubachyk O, Smal I (2012) Methods for cell and particle tracking. Methods Enzymol 504:183–200

Chapter 15

Retinoic Acid as a Modulator of Proximal-Distal Patterning and Branching Morphogenesis of the Avian Lung

Rute S. Moura

Abstract

Retinoic acid modulates numerous cellular events, namely, proliferation, differentiation, apoptosis, and patterning, hence influencing both embryo development and adult homeostasis. In vitro explant culture is a valuable technique for studying the impact of growth factors and signaling molecules, such as retinoic acid, in organ development since tissue architecture is maintained. This technique allows controlled supplementation of culture medium and straightforward analysis of its effect on morphogenesis. This chapter describes the detailed protocol for culturing embryonic chick lung explants and testing the impact of retinoic acid in branching and patterning, based on morphometric and molecular analysis.

Key words Chicken embryo, Lung branching, Proximal-distal patterning, Explant culture, In situ hybridization, Morphometric analysis

1 Introduction

Retinoic acid (RA), the bioactive form of vitamin A, is key to vertebrate embryo development and its levels need to be tightly controlled, otherwise congenital malformations will appear. Through the regulation of gene expression at the transcriptional level, RA mediates cellular processes such as proliferation, differentiation, and patterning of several organs [1]. Molecular targets of RA signaling range from transcription factors (such as *sox*) to homeodomain genes (*hox*), among many others [1]. During embryonic development, the lung is one of the organs that rely on RA signaling. In fact, impairment of this pathway leads to critical developmental defects that compromise normal lung function [2].

Model organisms are crucial to understanding the molecular and genetic circuits that control animal physiology, reproduction, development, and have served to explore cell and tissue interactions in both normal and disease context. The avian lung has recently emerged as an alternative to study branching morphogenesis. Despite the structural differences between mammalian and avian

Swapan K. Ray (ed.), *Retinoid and Rexinoid Signaling: Methods and Protocols*, Methods in Molecular Biology, vol. 2019, https://doi.org/10.1007/978-1-4939-9585-1_15, © Springer Science+Business Media, LLC, part of Springer Nature 2019

adult lung, the embryonic lung, at early developmental stages, presents molecular and morphological similarities between the two species [3–5]. Moreover, the chick model (*Gallus gallus*) presents some advantages when compared to mammalian models, namely: availability, accessibility for surgical manipulations, low cost with no need of special housing facilities, and external development (outside the mother) avoiding progenitor death to obtain the embryos.

This chapter outlines the basic procedure for performing in vitro (ex vivo) embryonic chick lung culture, in an air-liquid interface, for up to 2 days. This method is adapted to the avian species, nonetheless, it can be used for mammalian models with the corresponding amendments [6]. In this system, lung structure is preserved allowing to study the complex interactions between epithelial and mesenchymal compartments that ultimately control growth, differentiation, and patterning through specific signaling pathways [3–5]. With this technique, it is possible to manipulate a chemically defined culture medium by adding different compounds (e.g., growth factors, inhibitors, drugs, recombinant proteins, antibodies) and then examine their impact on lung morphogenesis, by morphometric analysis. Additionally, explants can be analyzed at the molecular level, to identify molecular targets and unveil potential crosstalks by, for instance, whole mount in situ hybridization. This technique reveals mRNA spatial distribution and expression levels, uncovering abnormal or ectopic expression sites/levels, thus exposing molecular interactions that may clarify the outcome in lung organogenesis.

The combination of the two techniques has proved very useful to assess the impact of RA in lung branching and on the expression of *sox2* and *sox9*, that are crucial for establishing the proximal-distal lung fate [7]. Nonetheless, this is a valuable approach to study tissue organogenesis and its underlying molecular mechanisms.

2 Materials

Prepare all solutions using ultrapure water (prepared by purifying deionized water, to attain a sensitivity of 18.2 MΩ-cm at 25 °C) and molecular biology/cell culture reagents.

Always wear disposable gloves when manipulating reagents, solutions, glass material, and RNA.

Carry out all procedures at room temperature, and on the bench, unless otherwise indicated.

Prepare and, then, store all reagents/solutions at 4 °C unless otherwise stated.

Glass material should be baked at 250 °C, overnight, to eliminate RNases. Wrap the top of beakers, graduated cylinders, graduated bottles, etc. with aluminum foil to prevent contamination after baking. Plastic caps must be wrapped with aluminum foil and autoclaved (121 °C, 20 min).

Magnetic stir bars should be washed with ethanol absolute and, afterward, extensively rinsed with ultrapure water before use.

Microsurgical dissection instruments should be sterilized by dry heating, at 130 °C under normal air pressure, for 8 h.

Always use filter tips and tubes that have been tested and certified RNase-free.

Handle all the biological waste safely in accordance with waste disposal regulations.

Discard the culture medium in a container with a diluted solution of domestic bleach and leave it inactivating for a couple of hours. Subsequently, pour it down in the sink with an excess of water. Solid waste such as culture plates, falcon tubes, and tips should be autoclaved.

2.1 Tissue Collection

1. Fertilized eggs.
2. Forced air humidified egg incubator set to 38 °C.
3. Glass bowl.
4. Plastic disposable petri dish (60 mm).
5. Dissection tools including Fine scissors-sharp (straight), Moria perforated spoon, and Dumont Fine forceps.
6. Flat bottom cell culture plate with lid, 24-well plate.
7. Stereomicroscope with an image acquisition system.
8. Sterile transfer pipettes.
9. PBS 10× (2 L): add 160 g NaCl, 4 g KCl, 4 g KH_2PO_4, 23 g $Na_2HPO_4{\cdot}2H_2O$ to an RNase-free graduated bottle with approximately 1.8 L of ultrapure water. Mix with a magnetic stir bar. Adjust the pH to 7.2–7.4 with NaOH (*see* **Note 1**). Make up the final volume in a 2 L RNase-free graduated cylinder. Autoclave in the RNase-free graduated bottle (121 °C, 20 min) and store at room temperature.
10. PBS 1× (1 L): dilute 100 mL of PBS 10× to 900 mL of ultrapure water to a 1 L RNase-free graduated cylinder. Store in an RNase-free graduated bottle at room temperature.

2.2 Lung Culture Medium

Medium preparation should be performed in the laminar flow hood.

1. Heat-inactivated fetal calf serum: inactivate the serum at 56 °C, for 30 min with shaking. Aliquot and store at −20 °C.
2. L-Ascorbic acid: prepare a 50 mg/mL solution with filtered (0.22 μm) ultrapure water in the laminar flow hood. Aliquot and store at 4 °C protected from light.

3. Lung medium preparation: medium 199 supplemented with 10% (V/V) chick serum, 5% (V/V) heat-inactivated fetal calf serum, 1% (V/V) L-glutamine (*see* **Note 2**), 1% (V/V) penicillin/streptomycin (5000 IU/mL each), and 0.25 mg/mL of ascorbic acid (*see* **Note 3**).

2.3 Retinoic Acid Solution

Retinoic acid should be manipulated in the laminar flow hood.

1. Stock solution (yellow): dissolve Retinoic Acid (R2625; Sigma, USA) in DMSO at its maximum solubility (40 mg/mL = 133.1 mM). Aliquot and store at −20 °C, protected from light.
2. Working solution: prepare 10 mM solution by diluting from the stock solution in DMSO. Prepare 1 mM solution by diluting from the 10 mM solution with DMSO. Aliquot and store at −20 °C, protected from light.

2.4 Lung Explant Culture

1. Medium 199.
2. Lung culture medium.
3. DMSO.
4. RA working solution (1 and 10 mM).
5. Nuclepore Track-Etched Polycarbonate Membrane Filter (8 μm pore size, 13 mm, Whatman).
6. Fine forceps.
7. Flat bottom cell culture plate with lid, 24-well plate.
8. Stereomicroscope with an image acquisition system.
9. Plastic disposable petri dish (60 mm).
10. Laminar flow hood.
11. Bunsen's burner.
12. CO_2 incubator.
13. 30G needles.
14. PBS 1×.
15. EGTA 0.5 M, pH 8 (100 mL): add 19.02 g to a graduated bottle with approximately 80 mL of ultrapure water. Make up the final volume in a 100 mL RNase-free graduated cylinder. Store at room temperature.
16. PBS 1× +EGTA (250 mL): 1 mL EGTA 0.5 M (pH 8) in PBS 1×.
17. In situ hybridization fixation solution (50 mL): 5 mL of formaldehyde 37% in PBS 1× +EGTA, pH 7.5 (adjusted with NaOH 1 M). Prepare and use on the same day.

2.5 Probe Synthesis

1. Refrigerated benchtop centrifuge.
2. 500 ng of purified DNA fragment (amplified by conventional PCR).
3. DTT 0.1 M.
4. Digoxigenin RNA labeling mix (Roche Applied Sciences, Germany).
5. RNA polymerase and transcription buffer.
6. RNase inhibitor (40 U/μL).
7. Thermoblock.
8. Dnase-RNase free (1 U/μL).
9. Agarose.
10. TAE 50× (1 L): add 242 g of Tris base and 18.61 g EDTA to 800 mL of distilled water. Stir until it dissolves. Add 57 mL of glacial acetic acid in the chemical hood. Adjust the volume to 1 L, autoclave (121 °C, 20 min) and store at room temperature.
11. TAE 1× (2 L): dilute 40 mL of TAE 50× to 2 L of distilled water in a volumetric flask. Store at room temperature.
12. LiCl 4 M (10 mL): add 1.69 g LiCl to 10 mL of ultrapure water. Store at room temperature.
13. Tris 1 M, pH 8 (100 mL): add 12.11 g Tris base to a graduated bottle with approximately 60 mL of ultrapure water. Adjust the pH with HCl. Make up the final volume in an RNase-free graduated cylinder. Store at room temperature.
14. EDTA 0.5 M, pH 8 (500 mL): add 95.05 g EDTA to a graduated bottle with approximately 400 mL of ultrapure water. Adjust the pH with NaOH pellets and then with NaOH solution (*see* **Note 1**). Make up the final volume in a 500 mL RNase-free graduated cylinder. Store at room temperature (*see* **Note 4**).
15. TE (100 mL): 1 ml Tris 1 M pH 8, 0.2 mL EDTA 0.5 M pH 8. Make up the final volume with ultrapure water.
16. Ethanol absolute: keep an aliquot at −20 °C.
17. Ethanol 70% (50 mL): add 35 mL of ethanol absolute to 15 mL of ultrapure water. Store at −20 °C.
18. EDTA 10 mM (10 mL): dilute 200 μL of EDTA 0.5 M, pH 8 to 10 mL of ultrapure. Store at room temperature.

2.6 In Situ Hybridization

1. Stereomicroscope with an image acquisition system.
2. 70 °C incubator.
3. 37 °C incubator.
4. Bunsen's burner.

5. Vacuum filtration system.
6. Parafilm M®.
7. Horizontal rocking platform.
8. Aluminum foil.
9. PBS 1×.
10. Methanol.
11. Proteinase K, stock solution: 20 mg/mL in ultrapure water. Aliquot and store at −20 °C.
12. t-RNA solution: 20 mg/mL solution of t-RNA from baker's yeast (Roche Applied Sciences) with ultrapure water. Aliquot and store at −20 °C.
13. Heparin solution: dissolve heparin sodium salt from porcine intestinal mucosa (180 U/mg) with ultrapure water to obtain a final concentration of 50 mg/mL. Aliquot and store at −20 °C.
14. Blocking reagent (Roche Applied Sciences).
15. Goat serum.
16. Anti-Digoxigenin-AP, Fab fragments (Roche Applied Sciences).
17. NBT (Roche Applied Sciences).
18. BCIP (Roche Applied Sciences).
19. Azide.
20. PBT (1 L): to 1000 mL of PBS 1× add with 1 mL of Tween 20.
21. PBT/Methanol (50/50) (500 mL): 250 mL of PBT plus 250 mL methanol.
22. PBT/Methanol (25/75) (500 mL): 125 mL of PBT plus 375 mL methanol.
23. PBT/Methanol (75/25) (500 mL): 375 mL of PBT plus 125 mL methanol.
24. Proteinase K, working solution (50 mL): 25 μL Proteinase K, stock solution, in PBT. Prepare and use on the same day (*see* **Note 5**).
25. Post-fixation solution (50 mL): 5 mL of formaldehyde 37%, 200 μL of glutaraldehyde 25% in PBT. Prepare and use on the same day.
26. Hybridization mixture (500 mL): 250 mL formamide, 32.5 mL 20× SSC, 5 mL EDTA 0.5 M pH 8, 1.25 mL t-RNA solution, 1 mL Tween 20, 1 mL Heparin solution, 2.5 g CHAPS. Store at −20 °C.

27. PBT/hybridization mixture (250 mL): 125 mL of PBT plus 125 mL hybridization mixture.
28. Probe solution: 10 μL of antisense RNA probe per mL of hybridization mixture.
29. MAB 5× (2 L): add 116 g $C_4H_4O_4$, 87 g NaCl, and 70 g NaOH to a graduated bottle with approximately 1.6 L of ultrapure water. Mix with a magnetic stir bar. Adjust the pH with concentrated NaOH solution (*see* **Note 1**). Make up the final volume in a 2 L RNase-free graduated cylinder. Autoclave in the RNase-free graduated bottle (121 °C, 20 min) and store at 4 °C.
30. MAB 1× (1 L): dilute 200 mL of PBS 10× to 800 mL of ultrapure water to a 1 L RNase-free graduated cylinder. Filter through a vacuum filtration system, near a Bunsen's burner, to avoid contamination. Store at 4 °C.
31. MABT (1 L): add 10 mL of Tween 20 to 1 L of filtered MAB 1×, near a Bunsen's burner.
32. MABT/hybridization solution mixture (100 mL): 50 mL of MABT plus 50 mL hybridization mixture. Prepare and use on the same day.
33. Blocking solution 10% (100 mL): add 10 g of blocking reagent to 100 mL of MAB 1×. Autoclave (121 °C, 20 min), aliquot, and store at −20 °C.
34. MABT/blocking solution (50 mL): 10 mL blocking solution plus 40 mL of MABT. Prepare and use on the same day.
35. MABT/blocking solution/goat serum (50 mL): 10 mL blocking solution, 10 mL goat serum plus 30 mL of MABT. Prepare and use on the same day.
36. Antibody solution (1 mL): add 1 μL of Anti-Digoxigenin antibody to 2 mL of MABT/blocking solution/goat serum.
37. NaCl 5 M (500 mL): add 146.3 g NaCl to a graduated bottle with approximately 450 mL of ultrapure water. Make up the final volume in a 500 mL RNase-free graduated cylinder. Store at room temperature.
38. Tris–HCl 2 M, pH 9.5 (500 mL): add 121.1 g Tris-base to a graduated bottle with approximately 400 mL of ultrapure water. Adjust the pH with HCl. Make up the final volume in a 500 mL RNase-free graduated cylinder. Store at room temperature.
39. $MgCl_2$ 2 M (250 mL): add 101.65 g $MgCl_2 \cdot 6H_2O$ to a graduated bottle with approximately 200 mL of ultrapure water. Make up the final volume in a 250 mL RNase-free graduated cylinder. Store at room temperature.

40. NTMT (50 mL): 1 mL NaCl 5 M, 2.5 mL Tris–HCl 2 M pH 9.5, 1.25 mL $MgCl_2$ 2M, 500 μL Tween 20. Make up the final volume with ultrapure water. Prepare and use on the same day.
41. Developing solution (20 mL): 67.5 μL NBT, 70 μL BCIP in NTMT. Protect from light.
42. Azide 10×: add 5 g of sodium azide to 50 mL of ultrapure water. Store at room temperature.
43. PBT/azide (50 mL): 500 μL of azide 10×, 50 mL PBT.

3 Methods

3.1 Ethical Issues

Experiments carried out in the chicken embryo (*Gallus gallus*) were performed in accordance with the Directive 2010/63/EU of the European Parliament and of the Council of 22 September 2010 on the protection of animals used for scientific purposes. The Portuguese Directive 113/2013 of 7 August 2013 does not contain any kind of restriction to the use of non-mammalian embryos.

3.2 Tissue Collection

Do the tissue collection at room temperature, on the bench, but as fast as possible. Fertilized chicken eggs may be obtained from local commercial sources.

1. Incubate fertilized chicken eggs in a 38 °C incubator, with 49% humidity, for 4.5–5.5 days.
2. After the incubation period, carefully open a window in the egg shell (with Fine scissors) to access the embryo.
3. Gently peel off the egg shell membrane with Fine forceps.
4. Remove the embryo from the egg with the help of the Moria perforated spoon and cut the vitelline blood vessels with Fine scissors.
5. Briefly wash the embryo with PBS 1× in a glass bowl, to remove the egg yolk and the allantoic fluid.
6. Transfer the embryo, with the Moria perforated spoon, to a Petri dish with PBS 1× and place it under the stereomicroscope.
7. Remove the amniotic sac with Fine forceps, to release the whole embryo.
8. Stage the embryo according to Hamburger and Hamilton developmental table [8] (*see* **Note 6**).
9. Section the embryo along the spinal axis to separate head, spine, and limbs from the ventral tissue that comprises the heart, lung, and gastrointestinal system (esophagus, stomach, liver, etc.).

10. Gradually remove the surrounding tissue, with Fine forceps, until you pinpoint the lung.
11. Finally, separate the trachea from the esophagus with extreme care, to not damage the lung (*see* **Note 7**).
12. Stage the embryonic lung according to the number of secondary branches that emerge from the main bronchus: b1, b2, or b3 if they exhibit 1, 2, or 3 secondary buds/bronchus.
13. Transfer the lung, with a sterile transfer pipette, to a 24-well plate filled with PBS 1×.
14. Repeat this procedure until you obtain 12 b2-stage lungs.

3.3 Lung Explant Culture

1. In the laminar flow hood, prepare a 24-well plate by filling the wells with 400 μL of medium 199.
2. Pick one polycarbonate membrane with a Fine forceps and place it, bright side up, floating on top of the medium.
3. Let it stand at room temperature for, at least, 1 h.
4. After this period, replace the medium 199 by 200 μL of lung culture medium, in the laminar flow hood.
5. Carefully transfer dissected lungs on top of the previously soaked membrane, one per membrane, using a micropipette with a cut tip, next to a Bunsen's burner to maintain sterile conditions (*see* **Note 8**).
6. Incubate at 37 °C in a 5% CO_2 incubator for 30 min.
7. Photograph the lungs, by using a camera coupled to the stereomicroscope, and confirm lung stage (*see* **Note 9**).
8. Randomly assign lungs to the three experimental groups.
9. Substitute the medium with fresh lung culture medium supplemented with 1 mM or 10 mM retinoic acid to achieve 1 μM or 10 μM final concentration in the well, respectively; or with DMSO (vehicle control) to achieve a final concentration of 0.1% (*see* **Note 10**).
10. Maintain the floating culture at 37 °C in a 5% CO_2 incubator, for 24 h.
11. Photograph the lungs, and replace the medium with fresh lung culture medium as described in 9.
12. Maintain the culture at 37 °C in a 5% CO_2 incubator for additional 24 h.
13. Photograph the lungs. An example of chick lung explants cultured for 48 h is presented in Fig. 1.
14. Put the membrane, with the explant, on the top of a 60 mm diameter petri dish and pinch the secondary buds with a 30G needle (*see* **Note 11**).

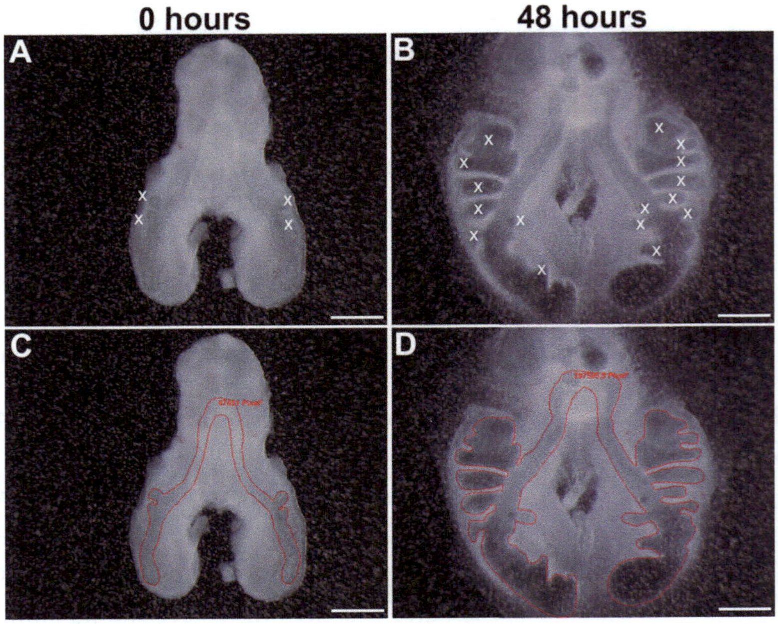

Fig. 1 In vitro chick lung explants and morphometric analysis. Representative example of a DMSO-treated b2 lung at 0 h (**a**, **c**) and 48 h (**b**, **d**) of culture. (**a**, **b**) Branching analysis: *x* identifies the secondary buds. (**c**, **d**) Morphometric analysis: the red line shapes the limits of the epithelium and reveals the epithelial area. The D48/D0 ratio is used as a measure of lung growth, provided that the images have the same dimension. Scale bar: 500 μm

15. Wash the explants three times with PBS 1× to remove the excess of medium and to release the tissue from the membrane.
16. Replace the PBS 1× with in situ hybridization fixation solution and keep at 4 °C, overnight.
17. Dehydrate the explants the following day (Subheading 3.5, **steps 1–4**).

3.4 Probe Synthesis by In Vitro Transcription

This protocol has been adapted from the datasheets of the different reagents and has been proved to be efficient when working with embryonic tissues. Under standard conditions, approximately 10 μg of full-length digoxigenin-labeled RNA is transcribed from 1 μg of template. Mix the reagents in the following order, on ice, and calculate the volume of ultrapure water needed to attain a final volume of 35 μL.

1. In an RNase-free microtube add the volume corresponding to 500 ng of amplified DNA fragment from *sox2* [9] or *sox9* [10] plasmid.
2. Add ultrapure water.
3. Add 4 μL of DTT, 2 μL of Digoxigenin RNA labeling mix.

4. Add the RNA polymerase transcription buffer to achieve a final concentration of 1× (*see* **Note 12**).
5. Add 2 μL of T7 RNA polymerase (for both probes), 2 μL of RNase inhibitor.
6. Incubate for 3 h, at 37 °C in a thermoblock.
7. Prepare a 0.8% agarose gel with TAE 1×.
8. Run 1 μL of probe as a checkpoint of the reaction.
9. Add 2 μL of Dnase RNase-free, 2 μL of RNase inhibitor and incubate for another 30 min, 37 °C.
10. Precipitate the probe with 20 μL of LiCl 4 M, 200 μL of TE, 600 μL of ice-cold ethanol absolute.
11. Leave 45 min at −80 °C or overnight at −20 °C.
12. Spin at 4 °C, 16200 × *g*, for 30 min.
13. Remove the supernatant and wash the pellet with ice–cold 70% ethanol. Be sure to detach the pellet from the bottom of the tube.
14. Spin at 4 °C, 16200 × *g*, for 15 min.
15. Remove the supernatant thoroughly, without disturbing the pellet.
16. Leave it to dry, on ice, for 15 min.
17. Resuspend the pellet in 60 μL of EDTA 10 mM and keep it at −20 °C until needed.

3.5 Proximal-Distal Patterning Assessment, by In Situ Hybridization

This protocol is based on previously described method [11], and it has been adapted to chicken lung tissue. All the steps should be carried out with gentle shaking, in a horizontal rocking platform, at room temperature and for 15 min, unless otherwise specified.

1. Rinse the explants in PBT.
2. Wash with PBT/methanol (50/50).
3. Wash with methanol, twice.
4. Replace with fresh methanol, wrap the plate with Parafilm M® and store it at −20 °C, minimum overnight.
5. Rehydrate the explants through a methanol series: PBT/methanol (25/75), PBT/methanol (50/50), PBT/methanol (75/25).
6. Rinse with PBT, twice.
7. Incubate the lungs, for 2 min with proteinase K working solution, without shaking (*see* **Note 13**).
8. Immediately rinse with PBT, three times.
9. Incubate the tissues with post-fixation solution, for 20 min.
10. Rinse with PBT, twice.

11. Wash with PBT/hybridization mixture.
12. Rinse with hybridization mixture, twice.
13. Replace with fresh hybridization mixture and incubate at 70 °C for, at least, 1 h (no shaking) (*see* **Note 14**).
14. Replace with heated probe solution and incubate at 70 °C, overnight (no shaking) (*see* **Note 15**).
15. Rinse with heated hybridization mixture, twice (no shaking and no incubation time).
16. Replace with heated hybridization mixture and incubate at 70 °C for 30 min (no shaking).
17. Rinse with heated MABT/hybridization mixture, at 70 °C, for 15 min (no shaking).
18. Wash twice with MABT, no shaking, and with no incubation.
19. Wash with MABT, 15 min.
20. Incubate the tissues with MABT/blocking solution, for at least 2 h.
21. Replace the solution with MABT/blocking solution/goat serum, for at least 2 h.
22. Incubate overnight, at room temperature and with gentle shaking, with antibody solution.
23. Rinse, briefly, three times with MABT.
24. Incubate with MABT for at least 2 h.
25. Repeat **step 24** as many times possible during the day.
26. Finally, let the tissues incubating with MABT, overnight.
27. Wash twice with NTMT.
28. Incubate the tissues with developing solution and protected from light (cover the plate with aluminum foil), for 15 min, at room temperature.
29. Check by the stereomicroscope the aspect of the tissues.
30. Continue the incubation at 37 °C, but every 30 min check by the stereomicroscope until the tissue presents a dark blue precipitate that identifies the expression site (*see* **Note 16**).
31. Wash the tissues three times with PBT, and replace by PBT/azide (*see* **Note 17**).
32. Photograph the explants with stereomicroscope with an image acquisition system. An example of chick lung explants followed by *sox2* and *sox9* in situ hybridization is displayed in Fig. 2.

3.6 Morphometric Analysis

1. Branching analysis is evaluated by counting the total number of secondary buds at 0 h and after 48 h of culture. Results are expressed as D48/D0 ratio and reflect lung growth (Fig. 1a, b).

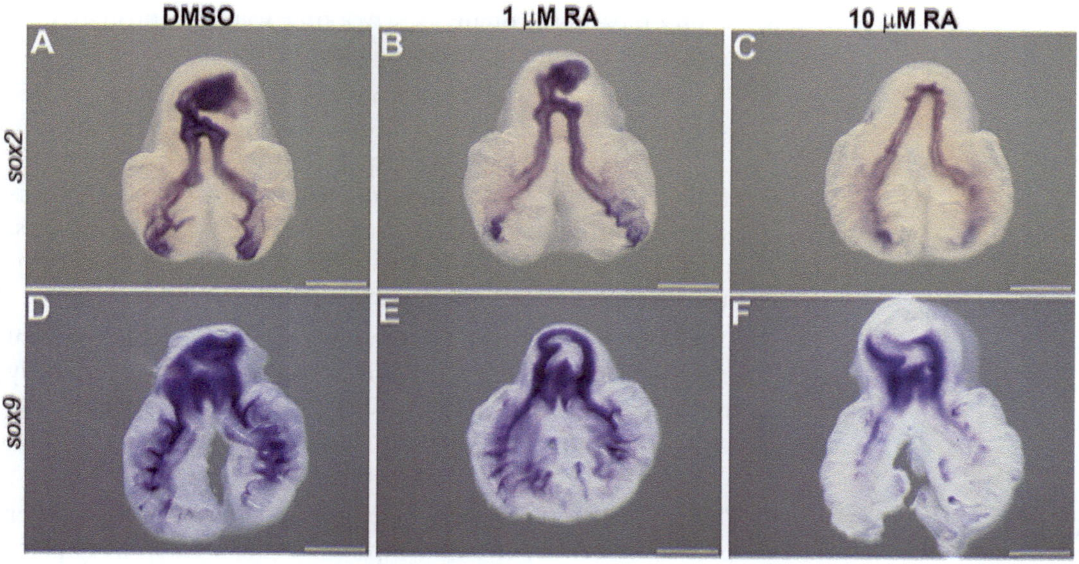

Fig. 2 In vitro chick lung explants, DMSO and RA-treated, followed by *sox2* and *sox9* expression analysis. Representative examples of stage b2 lung explant culture at 48 h of culture, treated with DMSO (**a**, **d**), 1 μM RA (**b**, **e**) or 10 μM RA (**c**, **f**), and probed with *sox2* (**a–c**) and *sox9* (**d–f**). *sox2* expression remains unaffected; *sox9* expression decreases and becomes more proximalized pointing to a role for RA in controlling lung patterning. Scale bar: 500 μm. Adapted from [5]

2. Morphometric analysis is assessed by outlining the internal perimeter of the lung (epithelium) and the outer perimeter of the lung (mesenchyme) at 0 h and 48 h, using a standard imaging analysis software. Results are expressed as D48/D0 ratio and reveal development of both compartments, independently (Fig. 1c, d).

4 Notes

1. To adjust the pH please make sure to use NaOH solution prepared with ultrapure water. Additionally, wash the pH meter several times with ultrapure water before usage (to try to minimize RNA contamination).
2. When handling glutamine solution confirm that there is no solid in suspension before supplementing lung culture medium.
3. Lung medium should be prepared on a weekly basis. Moreover, if the ascorbic acid solution presents a yellowish aspect, discard it and prepare a fresh solution from solid.
4. EDTA will not dissolve until the pH of the solution is approximately 8.0, with the addition of NaOH.

5. When preparing proteinase K working solution, proteinase K stock solution should be crystal clear before use.
6. The Hamburger and Hamilton developmental table is frequently used for chicken embryo staging. After incubating 4.5–5.5 days, embryos are expected to be in the range of HH24–27. The identification of stages relies on external features as, in this case, the toe plate, the elbow and knee joints, and toes. This staging will be useful to determine if development occurred within the expected range.
7. The esophagus and the trachea have the same developmental origin and, in these embryonic stages, they are completely attached along their proximal-distal axis. For this reason, the separation of these two structures, without damaging the lung, is the most challenging part of the dissection.
8. This procedure is not performed in the laminar flow chamber since it is necessary to have a top view of the membrane to place the lungs carefully in the central part of the floating membrane and avoid sliding off the membrane.
9. Photograph the lungs as quickly as possible with the lid open (to avoid vapor condensation) to obtain clear images.
10. DMSO final concentration in the well, in control and treated-explants, should be kept to the minimum possible (0.1%). To avoid pipetting errors of measuring very small volumes (0.2 μL/200 μL lung culture medium/well) it is advisable to prepare a larger volume that can be used for several wells (for example, 1500 μL lung culture medium/1.5 μL DMSO). Taking this into consideration, to maintain 0.1% DMSO, and obtain the desired RA concentration in the well, RA stock solution needs to be 1000× more concentrated (1 mM for 1 μM; 10 mM for 10 μM).
11. This step is critical to avoid colorimetric substrate accumulation during in situ hybridization procedure.
12. Depending on the RNA polymerase brand, transcription buffer may be 5× or 10× concentrated. Accordingly, pipette the volume to achieve a final concentration of 1×.
13. Proteinase K is an endopeptidase that cleaves peptide bonds. This treatment aims to increase the permeability of the tissue and thus facilitate the access of the probe to its target. Treatment with proteinase K is extremely time sensitive and incubation time should be precisely controlled to avoid damaging the structural integrity tissue completely.
14. At this point, probe solution should also be placed at 70 °C. This step elicits the elimination of RNA secondary structures that will impede probe binding to the tissue mRNA. The presence of formamide in the hybridization mixture will also

help to eliminate secondary structures (in both probes and tissues). Additionally, hybridization mixture and MABT/hybridization mixture should be placed at 70 °C, since they are needed for the first steps of the following day.

15. Probes can be retrieved, kept at −20 °C, and reused for, at least, one more time; some probes may even be reused more than twice but they should be tested. If the probe solution is found in a solid state (frozen) at −20 °C, discard it.
16. This step must be carefully monitored under the dissection microscope. The time needed to stop the reaction will depend on the amount of mRNA present in the tissue. The reaction must be stopped, at the same time, for all lung explants conditions incubated with the same probe. Stopping reaction at the same time allows a direct comparison between the experimental groups, and allows to draw a conclusion about treated vs. control explants.
17. If probes take longer to develop, it is possible to continue the process the following day. If so, leave the tissues in NTMT solution, overnight, at 4 °C. Resume the process, the following morning, in **step 28** with fresh NTMT and developing solution.

Acknowledgments

This work has been funded by FEDER funds, through the Competitiveness Factors Operational Programme (COMPETE), and by National funds, through the Foundation for Science and Technology (FCT), under the scope of the Project UID/Multi/50026/2019; and by the Project NORTE-01-0145-FEDER-000013, supported by the Northern Portugal Regional Operational Programme (NORTE 2020), under the Portugal 2020 Partnership Agreement, through the European Regional Development Fund (FEDER).

References

1. Cunningham TJ, Duester G (2015) Mechanisms of retinoic acid signalling and its roles in organ and limb development. Nat Rev Mol Cell Biol 16(2):110–123. https://doi.org/10.1038/nrm3932
2. Marquez HA, Cardoso WV (2016) Vitamin A-retinoid signaling in pulmonary development and disease. Mol Cell Pediatr 3(1):28. https://doi.org/10.1186/s40348-016-0054-6
3. Moura RS, Coutinho-Borges JP, Pacheco AP, daMota PO, Correia-Pinto J (2011) FGF signaling pathway in the developing chick lung: expression and inhibition studies. PLoS One 6(3):e17660. https://doi.org/10.1371/journal.pone.0017660
4. Moura RS, Carvalho-Correia E, daMota P, Correia-Pinto J (2014) Canonical Wnt signaling activity in early stages of chick lung development. PLoS One 9(12):e112388. https://doi.org/10.1371/journal.pone.0112388
5. Fernandes-Silva H, Vaz-Cunha P, Barbosa VB, Silva-Gonçalves C, Correia-Pinto J, Moura RS (2017) Retinoic acid regulates avian lung branching through a molecular network. Cell

Mol Life Sci 74(24):4599–4619. https://doi.org/10.1007/s00018-017-2600-3

6. Yeganeh B, Bilodeau C, Post M (2018) Explant culture for studying lung development. Methods Mol Biol 1752:81–90. https://doi.org/10.1007/978-1-4939-7714-7_8
7. Zhu Y, Li Y, Jun Wei JW, Liu X (2012) The role of sox genes in lung morphogenesis and cancer. Int J Mol Sci 13(12):15767–15783. https://doi.org/10.3390/ijms131215767
8. Hamburger V, Hamilton HL (1951) A series of normal stages in the development of the chick embryo. J Morphol 88(1):49–92
9. Dady A, Blavet C, Duband JL (2012) Timing and kinetics of E-N-cadherin switch during neurulation in the avian embryo. Dev Dyn 241(8):1333–1349. https://doi.org/10.1002/dvdy.23813
10. Domowicz MS, Henry JG, Wadlington N, Navarro A, Kraig RP, Schwartz NB (2011) Astrocyte precursor response to embryonic brain injury. Brain Res 1389:35–49. https://doi.org/10.1016/j.brainres.2011.03.006
11. Henrique D, Adam J, Myat A, Chitnis A, Lewis J, Ish-Horowicz D (1995) Expression of a delta homologue in prospective neurons in the chick. Nature 375:787–790. https://doi.org/10.1038/375787a0

Chapter 16

Translation of Effects of Retinoids and Rexinoids: Extraction and Quality Assessment of RNA from Formalin-Fixed Tissues

Iván P. Uray and Loretta László

Abstract

Retinoids and rexinoids directly and selectively activate their nuclear receptors, resulting in changes in the transcript levels of their target genes. Consequently, quantitating mRNA levels transcribed from cognate target genes is the most accurate measure of retinoid action. These changes can serve as relevant endpoints in biomarker trials, as well as in vivo preclinical studies. In gene expression analyses of archival material such as formalin-fixed paraffin-embedded (FFPE) tissues, assessing the quality of the extracted RNA is essential for the validation of the studies. With next generation sequencing (NGS) becoming the method of choice for gene expression profiling, RNA quality has become a critical aspect of study feasibility. In this chapter, we describe a method to extract RNA and to assess the intactness of RNA samples extracted from paraffin-embedded tissues.

Key words FFPE, RNA degradation, Quantitative PCR, Next generation sequencing, Staggered primers

1 Introduction

Long-term treatment of experimental animals and humans with retinoids results in modulation of retinoid-responsive genes in target tissues (*see* **Note 1**) [1, 2]. Specimen from tissues reflecting these changes can often be accessed from archival materials in a formalin-fixed condition [3]. RNA in formalin-fixed tissue is degraded to varying degrees, depending on the length of storage [4], the length of time between removal of active blood supply and fixation, and additional factors that are difficult to assess [5]. The accuracy and reliability, as well as the feasibility of downstream applications, depend on the investigator's assessment of what portion of the cellular transcripts still remain intact.

The analysis of gene expression from partially degraded total RNA is feasible, but the level of degradation may predetermine the chance for success. As an alternative to an RNA integrity number

Swapan K. Ray (ed.), *Retinoid and Rexinoid Signaling: Methods and Protocols*, Methods in Molecular Biology, vol. 2019, https://doi.org/10.1007/978-1-4939-9585-1_16,

(RIN) algorithm [6], which relies on electrophoretic detection, we present a simple and customizable method to assess the level of intactness of RNA samples from formalin-fixed tissues.

2 Materials

- Prepare all solutions in sterile, nuclease-free, or Diethylpyrocarbonate (DEPC)-treated water (*see* **Note 2**).
- Microtome used in histology for sectioning paraffin-embedded tissue blocks (*see* **Note 3**).
- Solvent and clearing agent for the removal of paraffin: Xylene substitute solution, commercially available as Hemo-De or Histo-Clear (*see* **Note 4**).
- Tissue lysis solution: 10% SDS solution prepared in 10 mM phosphate buffer containing 100 mM NaCl and 10 mg/mL proteinase K.
- Proteinase K: to make a stock solution from powder dissolve in 10 mM sterile Tris–HCl at pH 8 and 50 vol% glycerol, 3 mM $CaCl_2$. Store at −20 °C. Working concentration may range from 0.1 to 1 mg/mL. Optimize activation of the enzyme for more effective RNA (or DNA) release. Working solution should contain at least 0.5% SDS, 100 mM NaCl, and 1 mM EDTA. Enzyme activation may be optimized by adding calcium chloride or calcium acetate to 1–5 mM final concentration.
- Trichloroacetic acid (TCA): prepared in 10 mM phosphate buffer (pH 7.0) and used at 10% w/v concentration.
- Guanidinium thiocyanate-phenol-chloroform (a.k.a. TRIzol or Tri-reagent) extraction reagent.
- *Alternatively, silica column-based purification with a vacuum or centrifuge setup.*
- Primer and probe sets for real-time detection of quantitative PCR reactions. Reagents and enzymes for reverse transcription and PCR.

3 Methods

3.1 RNA Isolation Protocol

The isolation of RNA from paraffin-embedded tissues depends on four major chemical alterations to the specimen: the removal of hydrophobic substances, the release of RNA from the highly cross-linked protein matrix, the precipitation and removal of all proteinaceous materials, and the breakdown of genomic DNA. With these changes accomplished, RNA can be separated from the remaining solution with relative purity. Deparaffinization and protein digestion is done manually, while the separation of the nucleic acid components may be done either by solubility-based applications, or may be assisted by resin-based RNA extraction columns included in various kits. Both options will be described here.

The tissue structure, density, and the amount of connective elements in tissue will have a great impact on the yield of RNA resulting from the process. Therefore, depending on the size and density of the tissue samples, use at least 3 microtome-cut sections not thicker than 10 μm.

1. Using a microtome cut 5 μm sections from every tissue block included in the study.
2. Place individual tissue sections in separate 1.5 mL tubes and add 1 mL of xylene substitute solution to each sample. *If you use xylene, perform this step under a fume hood!* Shake the tube vigorously several times and let sit for 10 min at room temperature (*see* **Note 5**). Spin down microfuge tubes at 10,000 rpm (10,680 × *g*) for 1 min and discard the supernatant. Repeat this step once (*see* **Note 6**).
3. Wash deparaffinized tissue sections with 1 mL of 95% ethanol. Vortex. Spin down at 10,000 rpm (10,680 × *g*) for 1 min and discard the supernatant. Repeat this step once.
4. Add 0.5 mL of the tissue lysis buffer containing 10% SDS and 0.5 mg/mL proteinase K. Vortex. If a shaker heating block is available incubate for 30 min at +85 °C shaking at 600 rpm. When using a stationary heating block vortex samples every 10 min.
5. Add fresh proteinase K solution to amount to 1 mg/mL final concentration. Incubate overnight (a minimum of 3 h for loose tissues) at 55 °C. If a shaker heating block is available, perform digestion at a constant 600 rpm.
6. Add trichloroacetic acid (TCA) to a final concentration of 10% w/v (**Note 7**).
7. Centrifuge at 12,000 × *g* for 5 min, transfer the supernatant to a new tube, and discard the pellet. At this point you may choose to purify RNA using resin-based spin columns—if so continue on **step 11**.
8. Extract the RNA by adding 0.5 mL of a guanidinium thiocyanate-phenol-chloroform-containing reagent (i.e., TRIzol) to each tube (*see* **Note 8**). Invert tube several times gently, let stand for 10–15 min at room temperature. The mixture will separate into an upper and a lower phase, separated by an interphase. Transfer the colorless upper, aqueous phase to a fresh microcentrifuge tube.
9. Add 0.5 mL of 2-propanol per mL of TRIzol used. Let stand for 5–15 min at RT. Spin down at 12,000 × *g* for 10 min at 4 °C in a refrigerated centrifuge (*see* **Note 9**).

10. Remove the supernatant and wash the RNA pellet by adding 1 mL of cold 75% ethanol (*see* **Note 10**).
11. Vortex the sample and centrifuge at 7500 × *g* for 5 min at 4 °C.
12. Remove the ethanol and air-dry the pellet for 5–10 min (*see* **Note 11**). Resolubilize the RNA by vortexing or suspending with a pipette, in a small volume of RNAse-free water or formamide (*see* **Note 12**).

    The subsequent steps include a simple spin column-based RNA isolation protocol that works with kits from various manufacturers, to replace **steps 7–11** of the base protocol.
13. Prepare RNA for binding: add RNA binding buffer provided in the kit of your choice and absolute ethanol equal to the combined volume of buffer and sample.
14. RNA binding to the column (filtration tube): pipet the RNA lysate into the reservoir of the column and centrifuge for 30 s at 6000 × *g*. In order to dry the filter resin centrifuge for another 2 min at 16,000 × *g*.
15. Washing steps: alcohol containing Wash Buffer 1: once, spin 20 s at 6000 × *g* alcohol containing Wash Buffer 2: use twice, centrifuge 20 s at 6000 × *g*.
16. Additional centrifugation for 2 min at 16,000 × *g* to dry the filter. After that the filter is placed into a new 1.5 mL tube.
17. Elution step: Add 25–50 μL RNAse-free water or sterile TE buffer. Let sit for 1 min at +15 to +25 °C. Centrifuge for 1 min at 6000 × *g*.

    Store the eluted RNA: at −20 °C (short term) or at −80 °C (long term).
18. DNase treatment: add 1 U of RNase-free DNase in 50 mM Tris–HCl and 3 mM $MgCl_2$. Incubate for 15 min at room temperature. Stop the reaction by adding 1/10 volume of 50 mM EDTA and incubate at 75 °C for 1 min.

3.2 Assessment of RNA Quality

NGS applications require a minimum of 200 nucleotide length of each template, for oligonucleotide microarray-based gene expression profiling 4–500 nt templates are needed [7, 8]. In contrast, TaqMan qPCR assays can utilize template sequences as short as 65–90 nucleotides and RNA samples with very low RIN values (RIN < 1, 4) may be suitable for reverse transcription and PCR amplification. Therefore, PCR-based methods may be used for the assessment of RNA quality.

The fragment length of RNA extracted from formalin-fixed tissues may be as short as 100–200 nucleotides. Degradation is universal and not sequence specific, so target genes of interest and "housekeeping genes" should be affected to the same degree. Therefore, quantitation of a target sequence normalized to the levels of a constitutively expressed gene allows accurate determination of gene expression, a critical readout of retinoid activity [9].

In order to measure the ratio of intact to degraded RNA extracted from formalin-fixed paraffin-embedded tissues, an RT-qPCR-based degradation assay was developed. The method can be applied to any gene transcript of interest, as demonstrated here by way of a mouse β-actin assay. The method is based on the comparison of PCR-amplified products from cDNA templates generated from primers placed further toward the 3′ end. The length of a specific mRNA segment retrieved from a fixed tissue is proportional to the probability of generating sufficient intact cDNA templates for a downstream application, such as qPCR, microarrays, or NGS.

3.2.1 Assay Design

A key step is the design of appropriate primers and probe. The general guidelines to observe for new TaqMan assays to work at an acceptable efficiency are the following:

- Melting temperatures (T_M) of the primer pairs to be within 2 °C.
- Comparable GC contents at the 3′ ends of the primers, with preferably no less than 2 Gs or Cs within the last five nucleotides.
- A probe T_M 10 °C higher than that of the primers.
- Close proximity (no more than 6–8 nucleotides) of the forward primer 3′ end to the 5′ end of the TaqMan probe.
- A total amplicon length not exceeding 85–90 nucleotides.
- Avoid a G nucleotide in 5′ position of the probe.

In this example a standard Taqman qPCR assay is designed for the mouse β-actin gene with additional downstream reverse primers marked R2, R3. RNA from every sample to be assessed is reverse transcribed using all reverse primers (Table 1; *see* **Note 13**), while only the closest primer pair (F1035; Rl 1108) is used for subsequent PCR detection. The cDNAs generated with these sequential reverse primers (Rl–R3) will be compared to assess RNA intactness.

In this procedure each reverse primer "staggered" along the transcript is used in a separate RT reaction to generate cDNAs starting further toward the 3′ end from the location of the PCR primer pair (Fig. 1). The subsequent PCR reactions use the F-Rl primer pair to detect the 5′-most section of the transcript. Measurements of all cDNAs will then be referenced to the shortest cDNA generated by the Rl reverse primer. A delay in the amplification of cDNAs made with R3 and R2 compared to Rl will reflect the degree of RNA degradation of the mRNA template pool. Ultimately, each specimen from FFPE tissues should also be compared to a positive control of highly intact RNA, such as one extracted from fresh flash-frozen tissue of the same origin.

Table 1
Oligonucleotide components and products of a mouse b-actin RT-qPCR assay for the assessment of template RNA degradation

Primer name	Sequences	Product length
β-Actin F 1035 forward (+) primer	GCTCTGGCTCCTAGCACCAT	Compatible with each reverse primer
Mouse β-actin 1059 dual-labeled probe	ATCAAGATCATTGCTCCTCCTGAG CGC	Compatible with all primers
β-Actin R1 1108 reverse (−) primer	CCACCGATCCACACAGAGTAC	73 nt
β-Actin R2 1225 reverse (−) primer	GCTCAGTAACAGTCCGCCTAGAA	190 nt
β-Actin R3 1584 reverse (−) primer	CTTTTGGGAGGGTGAGGGAC	549 nt

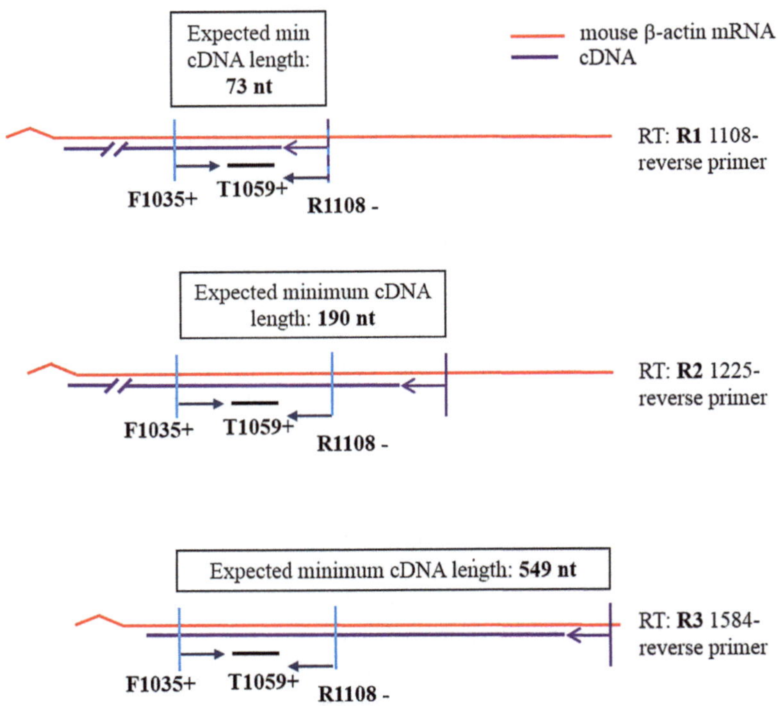

Fig. 1 cDNAs generated in parallel reverse transcription (RT) reactions by staggered gene-specific reverse primers (R1-R3) require increasing lengths of intact mRNA fragments to be amplified by a standard TaqMan real-time PCR assay (F1035+; R1108−)

3.2.2 Reverse Transcription (RT) Reaction Setup (Per Reaction)

• Nuclease-free water	1.93 μL
• 5× RT buffer	1.0 μL
• dNTP (10 mM)	1.0 μL
• β-actin R1/R2/R3 reverse primer (100 μM)	0.02 μL
• Reverse Transcriptase enzyme	0.05 μL
• RNA sample (*see* **Note 14**)	1 μL

Reaction conditions: 50 °C	*30 min*
72 °C	*5 min*
4 °C	*For storing cDNAs*

3.2.3 Quantitative Polymerase Chain Reaction Setup (Per Reaction; ***Note 15****)*

• Nuclease-free water	15.445 μL
• 10× PCR buffer	2.0 μL
• $MgCl_2$ (50 mM)	1.0 μL
• dNTP (10 mM)	1.2 μL
• β-actin forward primer (100 μM)	0.1 μL
• β-actin R1 reverse primer (100 μM)	0.08 μL
• β-actin dual-labeled probe (100 μM)	0.025 μL
• ROX reference dye	0.05 μL
• Taq polymerase (0.05–0.1 units/μL)	0.1 μL
Total	**20 μL/rxn added to each RT well**

Reaction conditions:	*94 °C*	*60 s*	*1 time*
	94 °C *60 °C*	*10 s* *30 s*	*40 cycles*

3.2.4 Evaluation of the PCR Reactions

The efficiency of any real-time PCR assay is not dependent on the length of the template used (for assay efficiency based on the slope of a standard curve *see* **Note 16**). Instead, a change in the amount of the template sequence is accurately reflected in the amplification curves crossing the sensitivity threshold (C_t, *see* Fig. 2).

Due to diminishing fractions of intact RNA recovered from highly degraded FFPE specimen, the amplification curves will show increasing cycle numbers (C_t), in proportion to the level of RNA degradation. As shown in the example in Table 2, when using fresh frozen tissue for RNA extraction there was an average of 8% increase in the first detectable cycle from an amplicon

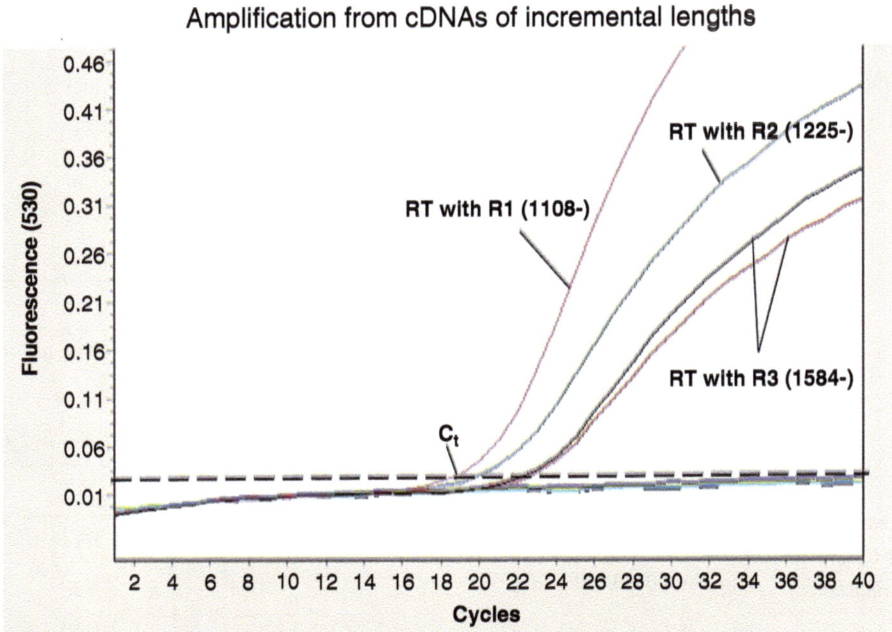

Fig. 2 Amplification from gene-specific cDNA fragments of various lengths reflect the level of degradation of RNA samples

Table 2
Test sample results measured in RNAs from frozen, non-fixed tissues compared to a recent (FFPE tissue 1) and an older (FFPE tissue 2) fixed, paraffin-embedded specimen

	R2 (190 nt) C_t/R1 (73 nt) C_t	R3 (549 nt) C_t/R1 (73 nt) C_t
RNAs from fresh frozen tissues	1.08 ± 0.009	1.14 ± 0.02
Example RNA from FFPE tissue 1	1.1	1.22
Example RNA from FFPE tissue 2	1.24	1.38

190 (generated with R2) nucleotides away from the F (forward) PCR primer, compared to the shortest one at only 73 nucleotides away (R1). Extending the minimum amplicon length to over 500 nt then increased the ratio of the first detected cycle numbers to 1.14. When RNA from a freshly prepared FFPE sample was used as template, the ratio of the 190–73 nt amplicons increased to 1.1, and the ratio of the 549–73 nt increased to 1.22. These ratios increased significantly in an older FFPE block to 1.24 and 1.38, respectively. These changes are due to the drop in the amount of transcribable mRNAs, rather than the decreased efficiency of the PCR reaction.

Based on these ratios we recommend to define arbitrary cutoff values based on the R2/R1 and R3/R1 ratios for either NGS or qPCR-based applications. Above this cutoff a sample should no longer be considered for analysis, because the amplification of a transcript less abundant than β-actin is unlikely (*see* **Note 17**). Samples with an R2/R1 ratio of <1.125 should be used for NGS studies and for microarray applications an R3/R1 ratio of <1.25 is favorable.

4 Notes

1. Some typical genes directly regulated through retinoic acid receptor activation include ACADM (MCAD), FACL2 (ACS), osteopontin (Spp1), Tgase K (TGM1), Survivin (BIRC5), CRBP I (Rbp1), IGF binding proteins 3, 6 [2, 10, 11].
2. Preparation of DEPC water: Add 1 mL of 0.1% Diethylpyrocarbonate to deionized water, for every 1 L of DEPC water to be prepared. Mix and incubate at room temperature for 1 h, before autoclaving. Cool down to room temperature before use. *N.B. Purine nucleotides in RNA are readily modified by carboxymethylation; thus, DEPC residues may be detrimental to any downstream applications,* i.e., *reverse transcription!*
3. Microtome sectioning makes the solid tissue material accessible to enzymatic action needed to free up nucleic acids from highly cross-linked protein structures; hence these sections should not be thicker than 5 μm each. Care should be taken to avoid carryover between specimens, which can be ensured by replacing the microtome blade after each sample. Take extra care if the blade is cleaned between samples with ethanol, as the blade is extremely sharp!
4. The processing of histological specimens traditionally involved Xylene for its hydrophobic solvent properties. Due to its toxicity, we recommend using a xylene substitute solution for deparaffinization.
5. The deparaffinization step and the first three steps of the RNA isolation are done in separate tubes to maximize yield. At the end, we pool the three preparations from the same block in **step 4**. Therefore, for each preparation only one-third of the required amount of reagents are used.
6. If using a vortex, avoid breaking up the tissue sections at this point, as small shreds will depart and get lost with the removal of the solvent.
7. The precipitation of the proteinaceous components in the specimen occurs via hydrophobic aggregation induced by trichloroacetic acid (TCA) causing proteins to denature while disrupting the solvation layers. As opposed to salts, TCA is more effective at lower concentrations than ammonium sulfate,

avoiding an unfavorable increase in sample volume. A further advantage is that TCA-induced precipitation is less selective, and independent of the native conformation of proteins [12]. In addition, follow-up desalting is not necessary. 260/280 ratios tend to be significantly lower, compared to RNA obtained using a column-based extraction method. However, the relevance of this observation to the success rate of the downstream PCR application is not clear.

8. The dissociation of nucleoprotein complexes by the TRIzol reagent is based on the phase separation principle occurring between a hydrophobic and hydrophilic phase and hence could be performed using a classic phenol-chloroform separation method [13]. However, the use of a TRIzol or TRI-reagent, originally made for the simultaneous separation of proteins, RNA, and DNA from whole tissues, is recommended. These reagents replace chloroform by 1-Bromo-3-chloropropane, which reduces the odds of DNA fragments mixing in with RNA, and is less toxic than chloroform. In either case, gloves should be worn at all times when working with these substances.
9. The precipitated RNA will form a pellet on the far side (away from the axis of the centrifuge) of the tube. It may be invisible to the naked eye, yet the sample should be carried forward as if it was visible.
10. The procedure may be put on hold at this point. RNA can be stored at −20 °C for weeks without compromising its quality.
11. Do not use a Speed-Vac, or overdry the pellet in any way. This will interfere with taking the RNA into solution again.
12. Formamide has been reported to protect RNA from degradation, and increases the specificity of PCR reactions [14, 15].
13. A number of applications may be suitable to design and check primers (i.e., NCBI Primer Blast, Oligo 7.0, Primer Express). It is advisable to optimize the F-R1 qPCR assay based on a dilution series of a cognate template, where the slope of the standard curve would be −3.33 at a 100% efficiency. However, in direct comparison of identical samples measured with alternate RT reactions, the increased cycle numbers will result from diminishing amounts of cDNA produced with R2 or R3 in the reverse transcription process.
14. In a typical reaction 50–100 ng of total RNA are the ideal starting amount. However, using an optical method (i.e., spectrophotometry) to normalize the amount of total RNA entered into the RT reaction, various organic chemicals make such reading unreliable, and thus obsolete. Therefore, a standard volume of the RNA sample is added to the reaction. Depending on the size of the tissue sample used for extraction the levels

of both housekeeping and regulated genes may vary significantly. Therefore, the use of the appropriate normalizer, or a combination of normalizers, is important [9].

15. Depending on the instrument used for qPCR measurements, the ROX reference dye may or may not be useful—please consult equipment manual. Instead of individual components, a ready-made PCR master mix may be used.

16. Real-time PCR assay efficiency mainly depends on the choice of primers and assay conditions. Assay efficiency can be tested by creating a serial dilution of a cDNA containing the target sequence of the PCR assay and determining the slope of the standard curve created this way. Assay efficiency should not vary due to the length of the cDNA, as long as no other inhibitor is introduced in the PCR reaction. This procedure allows the assessment of RNA quality irrespective of the quality of the qPCR assay used.

17. This cutoff is arbitrary, and the following estimations should be considered: Compared to a C_t value of 24 measured from the R1 cDNA, a 10% increase (or 2.4 cycles) is equivalent to the loss of over 75% of the gene transcripts. In comparison, a 50% increase in the C_t value or a 12 cycle shift means that only 0.025% of the transcripts are left intact, making any meaningful comparison of samples infeasible.

Acknowledgments

This study was supported by the European Union and the European Regional Development Fund (GINOP-2.3.2-15-2016-00020). The publication was also supported by the "Debrecen Venture Catapult Program" (EFOP-3.6.1-16-2016-00022). The authors thank Dr. Karen Uray for critical reading of the manuscript.

References

1. Mazumdar A, Medina D, Kittrell FS, Zhang Y, Hill JL, Edwards DE, Bissonnette RP, Brown PH (2012) The combination of tamoxifen and the rexinoid LG100268 prevents ER-positive and ER-negative mammary tumors in p53-null mammary gland mice. Cancer Prev Res (Phila) 5(10):1195–1202. https://doi.org/10.1158/1940-6207.CAPR-11-0524
2. Garattini E, Bolis M, Garattini SK, Fratelli M, Centritto F, Paroni G, Gianni M, Zanetti A, Pagani A, Fisher JN, Zambelli A, Terao M (2014) Retinoids and breast cancer: from basic studies to the clinic and back again. Cancer Treat Rev 40(6):739–749. https://doi.org/10.1016/j.ctrv.2014.01.001
3. Uray IP, Connelly JH, Thomazy V, Shipley GL, Vaughn WK, Frazier OH, Taegtmeyer H, Davies PJ (2002) Left ventricular unloading alters receptor tyrosine kinase expression in the failing human heart. J Heart Lung Transplant 21(7):771–782
4. Ribeiro-Silva A, Zhang H, Jeffrey SS (2007) RNA extraction from ten year old formalin-fixed paraffin-embedded breast cancer samples: a comparison of column purification and magnetic bead-based technologies. BMC Mol Biol 8:118. https://doi.org/10.1186/1471-2199-8-118
5. Shen Y, Li R, Tian F, Chen Z, Lu N, Bai Y, Ge Q, Lu Z (2018) Impact of RNA integrity

and blood sample storage conditions on the gene expression analysis. Onco Targets Ther 11:3573–3581. https://doi.org/10.2147/OTT.S158868

6. Haynes HR, Killick-Cole CL, Hares KM, Redondo J, Kemp KC, Moutasim KA, Faulkner C, Wilkins A, Kurian KM (2018) Evaluation of the quality of RNA extracted from archival FFPE glioblastoma and epilepsy surgical samples for gene expression assays. J Clin Pathol 71(8):695–701. https://doi.org/10.1136/jclinpath-2017-204969
7. Jackson AF, Williams A, Moffat I, Phillips SL, Recio L, Waters MD, Lambert IB, Yauk CL (2013) Preparation of archival formalin-fixed paraffin-embedded mouse liver samples for use with the Agilent gene expression microarray platform. J Pharmacol Toxicol Methods 68 (2):260–268. https://doi.org/10.1016/j.vascn.2013.02.008
8. Jaksik R, Iwanaszko M, Rzeszowska-Wolny J, Kimmel M (2015) Microarray experiments and factors which affect their reliability. Biol Direct 10:46. https://doi.org/10.1186/s13062-015-0077-2
9. de Kok JB, Roelofs RW, Giesendorf BA, Pennings JL, Waas ET, Feuth T, Swinkels DW, Span PN (2005) Normalization of gene expression measurements in tumor tissues: comparison of 13 endogenous control genes. Lab Investig 85(1):154–159. https://doi.org/10.1038/labinvest.3700208
10. Zhuang Y, Faria TN, Chambon P, Gudas LJ (2003) Identification and characterization of retinoic acid receptor beta2 target genes in F9 teratocarcinoma cells. Mol Cancer Res 1 (8):619–630
11. Uray IP, Dmitrovsky E, Brown PH (2016) Retinoids and rexinoids in cancer prevention: from laboratory to clinic. Semin Oncol 43 (1):49–64. https://doi.org/10.1053/j.seminoncol.2015.09.002
12. Rajalingam D, Loftis C, Xu JJ, Kumar TK (2009) Trichloroacetic acid-induced protein precipitation involves the reversible association of a stable partially structured intermediate. Protein Sci 18(5):980–993. https://doi.org/10.1002/pro.108
13. Maniatis T (1982) Isolation and quantification of DNA1. In: Protocol 13 molecular cloning a laboratory manual, vol 1. Cold Spring Harbor Laboratory Press, New York, p 74
14. Chomczynski P (1992) Solubilization in formamide protects RNA from degradation. Nucleic Acids Res 20(14):3791–3792
15. Sarkar G, Kapelner S, Sommer SS (1990) Formamide can dramatically improve the specificity of PCR. Nucleic Acids Res 18(24):7465

Chapter 17

Assessing Autophagy During Retinoid Treatment of Breast Cancer Cells

Sarah Parejo, Mario P. Tschan, Manuele G. Muraro, Enrico Garattini, Giulio C. Spagnoli, and Anna M. Schläfli

Abstract

Retinoids are derived from vitamin A through a multi-step process. Within a target cell, retinoids regulate gene expression by activating the retinoid acid receptors (RAR) and retinoid x receptors (RXR), which are ligand-dependent transcription factors. Besides its therapeutic use in dermatological disorders, all-*trans* retinoic acid (ATRA) is successfully utilized to treat acute promyelocytic leukemia (APL) patients. The use of ATRA in APL patients is the first example of clinically useful differentiation therapy. Therapeutic strategies aiming at cancer cell differentiation have great potential for solid tumors, including breast cancer. The few clinical studies conducted with ATRA in breast cancer are rather disappointing. However, these studies did not take into account the heterogeneity of the disease and were conducted on unselected cohorts of patients.

We recently showed that ATRA treatment of breast cancer cells induces autophagy, a highly conserved process aiming at degrading and recycling superfluous or harmful cellular components. In addition, autophagy inhibition significantly increases the therapeutic activity of ATRA. This finding is of fundamental importance, since autophagy has a dual role in cancer. Whereas autophagy may be a protective mechanism during the initial phases of cancer development, it may support cancer cell survival in already established tumors. Furthermore, autophagy can lower or enhance therapeutic efficiency, depending on the tumor type and the anticancer agent considered. Therefore, it is important to investigate the role of autophagy in the context of specific tumors and therapeutic approaches. Accurate autophagy studies are challenging given the dynamic nature of the process and the difficulty of measuring the rate of autophagosome degradation (autophagic flux). In this chapter, we provide protocols for a careful assessment of the autophagic flux in ATRA treated 2D and 3D breast cancer cultures.

Key words Autophagy, Retinoids, ATRA, Breast cancer, Bioreactor, mCherry-eGFP-LC3B, Autophagic flux, Immunofluorescence, Western blot

1 Introduction

Natural retinoids are a large family of molecules derived from and including vitamin A. Within a target cell, vitamin A can be metabolized into retinaldehyde, the precursor of retinoic acid (RA), 13-*cis* RA, 9-*cis*-RA, or all-*trans* RA (ATRA). Retinoids regulate gene expression by binding to ligand-dependent transcription

Swapan K. Ray (ed.), *Retinoid and Rexinoid Signaling: Methods and Protocols*, Methods in Molecular Biology, vol. 2019, https://doi.org/10.1007/978-1-4939-9585-1_17,

factors known as retinoid acid receptors (RARα, RARβ, and RARγ) and retinoid X receptors (RXRα, RXRβ, and RXRγ), which bind to the regulatory regions of RA-dependent genes as RAR/RXR heterodimers. While 9-*cis* RA can activate both RARs and RXRs, ATRA and 13-*cis*-RA are selective RAR agonists [1, 2]. Upon RA binding, the RAR/RXR heterodimer undergoes a conformational change, leading to the release of co-repressors and the recruitment of co-activators to initiate the transcription of target genes [2].

Retinoids are effective therapeutic agents in the treatment of a variety of visual and dermatological diseases [1]. In addition, retinoids are considered promising antitumor agents, given their effects on cell proliferation, differentiation, and apoptosis [3, 4]. The most successful implementation of retinoids in cancer therapy is represented by the use of ATRA in the differentiation therapy of acute promyelocytic leukemia (APL) [5]. Today, the concept of differentiation therapy is limited to the APL realm. However, differentiating agents have great potential, as they are associated with lower toxicity than standard cytotoxic chemotherapeutics [6].

Although significant progress in the treatment of breast cancer has been made in the last decades, the prognosis of metastatic disease remains poor [7]. Breast cancer stem cells (CSC) are known to evade conventional radiotherapy and chemotherapy and are thought to be responsible for the observed relapses. Therefore, development of therapies aiming at targeting CSCs is warranted [8]. One possibility is to induce CSC differentiation [6] into more mature and less chemotherapy resistant cells. In vitro studies have shown that ATRA indeed induces cell cycle arrest, differentiation, and apoptosis in certain breast cancer cell lines and stem cells [9, 10]. However, ATRA monotherapy is unlikely to represent a clinically useful option in breast cancer [11]. There is evidence that one of the major determinants of ATRA antiproliferative [12] and differentiating [9] responses is RARα expression. In breast cancer cell lines, the levels of RARα are variable and are correlated with sensitivity to the antiproliferative action of ATRA [12]. We recently demonstrated that ATRA treatment of responsive breast cancer cells induces autophagy and that inhibition of the process significantly increases the therapeutic activity of the retinoid [13].

Autophagy is a highly conserved and essential recycling process, through which superfluous or harmful cellular components are degraded. Three different subtypes of autophagy are known, i.e., macroautophagy, chaperone-mediated autophagy and microautophagy [14]. Macroautophagy (herein referred to as autophagy), which is the object of this chapter, is characterized by the formation of double-membrane vesicles, termed autophagosomes. The autophagosomes engulf cargo and fuse with lysosomes where the cargo itself is degraded and recycled (Fig. 1) [15]. Basal autophagy is crucial for the maintenance of cellular homeostasis in

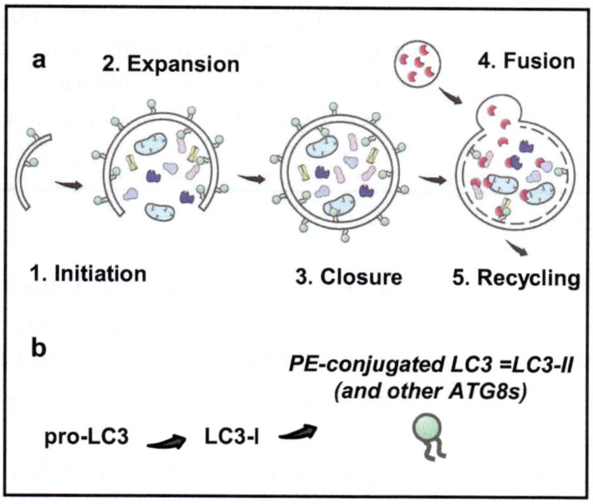

Fig. 1 A simplified scheme of the macroautophagy pathway. (**a**) Autophagy consists of several stages and is initiated by the formation of the isolation membrane. The isolation membrane expands and cargos are engulfed. The autophagosome closes and fuses with a lysosome to form an autolysosome. Finally, the content is degraded by the action of lysosomal hydrolases. Macromolecules are recycled to fuel the anabolic and metabolic pathways of the cell. (**b**) LC3B processing is shown. As all ATG8 family members, LC3B is processed from pro-LC3B by cleavage to give rise to LC3B-I. Subsequently, it undergoes phosphatidylethanolamine conjugation and incorporation into the inner and outer autophagosomal membrane

resting, proliferating, and differentiating cells [16, 17]. In cancer cells, the role of macroautophagy is complex and clearly depends on the stage, type, and the driving oncogene of the corresponding tumor [18, 19]. In healthy individuals, autophagy is deemed to be a tumor suppressive process, since it protects from genotoxic stress [20]. However, autophagy is also endowed with tumor-promoting and therapy-resistance potential [21–24] since in certain conditions it can favor cancer cell survival [25]. As for therapy-resistance, it must be emphasized that autophagy has also been shown to exert opposite effects, mediating the therapeutic activity of some anticancer agents. Regardless of its effect on cancer cell homeostasis, autophagy modulation is a promising therapeutic approach [26].

Careful assessment of autophagy is a challenging task. Since autophagy is a dynamic process, simple analysis of the steady-state expression of autophagy (ATG)-related genes is not sufficient to monitor autophagic activity [27]. Indeed, ATG protein levels may be low during a given autophagy phase, yet the process may be highly active [28]. Thus, an appropriate measurement of the autophagic flux, which is defined as the rate of degradation via autophagy [29], is of outmost importance. There are published guidelines for the accurate determination of autophagy and the concomitant use of multiple assays is advised [30]. This chapter provides detailed protocols on how to determine autophagy in retinoid treated breast cancer cultures using some of the available methods. Details for any given assay can be found in the respective sections.

2 Materials

2.1 Cell Culture

1. DMEM/F12 (Sigma-Aldrich, Buchs, Switzerland).
2. Fetal bovine serum (FBS) (Sigma-Aldrich, #F7524, Lot 044M3355 and BCBR6703V).
3. Trypsin-EDTA 0.25% (#T4049 Sigma-Aldrich).

2.2 Chemicals

1. Bafilomycin A1 (BML-CM110, Enzo-Life Sciences, Lausen, Switzerland).
2. All-*trans* retinoic acid (#R2625, Sigma-Aldrich).

2.3 Western Blotting

1. Lysis buffer: 8 M urea with 0.5% Triton-X100, supplemented with proteinase inhibitor cocktail (complete, Roche).
2. Sonicator (Sonopuls HD3200, Bandelin, Berlin, Germany).
3. 4× Laemmle Buffer (#1610747, BioRad).
4. 4–20% Mini-PROTEAN® TGX™ Precast Protein Gels (#4561096 BioRad, Cressier, Switzerland).
5. Trans-Blot® Turbo™ Mini PVDF Transfer Packs (#1704156).
6. Methanol (Merck, Switzerland).
7. Anti-LC3B antibody (Rabbit polyclonal #NB600-1384, LuBioScience, Luzern, Switzerland).
8. Anti-β-Catenin antibody (Rabbit polyclonal #9562, Cell Signaling).
9. Anti-Cleaved PARP antibody (Rabbit polyclonal #9541, Cell Signaling).
10. Anti-Cleaved Caspase-3 antibody (Asp175) (Rabbit polyclonal, #9661, Cell Signaling).
11. Anti-Rabbit-HRP.
12. Milk powder (Migros, Switzerland).
13. BSA powder (Sigma-Aldrich, Switzerland).
14. TBS and TBS-T (self-made according to standard protocols).
15. Trans-Blot® Turbo™ Transfer system (BioRad).
16. ChemiDoc™ MP system (BioRad).

2.4 Immunofluorescence

1. 18 × 18 mm cover slides (Menzel-Gläser, Thermo Scientific).
2. Microscopy slides (Menzel-Gläser, Thermo Scientific).

3. PBS (self-made according to standard protocol).
4. 4% Paraformaldehyde in PBS (#612103-VS64K, Thommen Furler AG).
5. Methanol (Merck, Switzerland).
6. Anti-LC3B antibody (Rabbit monoclonal, clone D11, #3686, Lausen, Switzerland).
7. FITC conjugated anti-rabbit (#115–166-003, Jackson Immunoresearch, Suffolk, UK).
8. Confocal microscope Olympus FluoView-1000, 63× objective.
9. SlowFade™ Gold Antifade Mountant with DAPI (#S36938, Fisher Scientific, Switzerland).
10. ImageJ software and plugin (modified and simplified version of http://imagejdocu.tudor.lu/doku.php?id=macro:gfp-lc3_macro to quantify dots.

2.5 mCherry-eGFP-LC3B Ratiometric FACS Analysis

1. pLVO-Puro-mCherry-eGFP-LC3B lentiviral construct (kind gift from Dr. María S. Soengas (CNIO, Spain).
2. FACS Buffer: PBS supplemented with 2% FBS.
3. BD™ LSR II flow cytometer (488 nm Laser and 530/30 filter for GFP and 561 nm laser and 610/20 filter for mCherry) with Diva Software.
4. FlowJo® analysis software (Version 10).

2.6 Perfused 3D Cultures

1. Collagen type 1Scaffold Sponge (Avitene™ UltraFoam™ Collagen Sponge (#1050050, Bard)).
2. 8 mm tissue punches (Xiomedica).
3. U-CUP disposable Kit with 8×4 adaptors (UCUP001 and URD08H04, Cellec Biotek AG).
4. Mesh (SEFAR FluorTex 09-590/47, available through Cellec Biotek AG).
5. Single-edged razor blades (#08410-, Brunschwig Chemie).
6. 4% Paraformaldehyde in PBS.
7. Thiazolyl Blue Tetrazolium Bromide (MTT, #M5655, Sigma-Aldrich).
8. Forceps IRES (B. Braun Medical AG, #OC022R).
9. U-CUP Rack (Cellec Biotek AG).
10. Programmable syringe pump PHD ULTRA™ 4400 (Harvard apparatus).
11. SZX16 Stereomicroscope (Olympus).
12. Axiophoto 2 microscope (Carl Zeiss).

3 Methods

3.1 Assessing Autophagic Flux by LC3B Western Blotting

LC3B is incorporated in autophagosomal membranes [31] (Fig. 1a, b) and is commonly used as a marker to monitor autophagy (*see* **Notes 1** and **2**). Importantly, we can discriminate between cytosolic (LC3B-I) and membrane-bound (LC3B-II, lipidated LC3B) form via Western blot (WB) analysis. To determine the autophagic flux, the autophagy pathway is usually blocked at the level of the autophagosome-lysosome fusion step using Bafilomycin A1 or other autophagosome-lysosome fusion/lysosome-activity inhibitors, such as hydroxy-chloroquine, NH_4Cl, or a combination of leupeptin, pepstatin A, and E64d. This allows determining whether a given increase in LC3B-II is due to a higher rate of vesicle formation or a block in vesicle degradation [30]. An example of WB analysis is illustrated in Fig. 2a. Treating ATRA-sensitive *SKBR3* breast cancer cells leads to an increase of the LC3B-II lipidated form, which is further enhanced by Bafilomycin A1 addition during the last 2 h of ATRA challenge (Fig. 2a). Significantly, calculation of the autophagic flux following vehicle- and ATRA treatment demonstrates that the retinoid causes an increase in LC3B flux (Fig. 2b).

1. Seed cells into wells of a 6-well plate (optimize cell count for your cell line and treatment duration, usually around 3×10^5 cells) containing 2 mL of media.
2. The next day treat cells with 2 mL of media containing DMSO (vehicle) or 1 μM ATRA (two wells/condition).
3. Treat cells for 24–48 h; 2 h before harvesting add 200 nM Bafilomycin A1 to one well for each condition and DMSO to the other wells.
4. 2 h post addition of Bafilomycin A1 harvest cells for WB analysis.
5. Remove media from cells.
6. Add 1 mL of cold PBS.
7. Scrape cells into 1 mL PBS.
8. Transfer cell suspension to 1.5 mL tubes.
9. Pellet cells by centrifugation at $300 \times g$ for 5 min at 4 °C.
10. Remove PBS.
11. Lyse cells in urea buffer.
12. Shortly vortex (avoid pipetting up and down).
13. Sonicate at 40% intensity for 2 s to shear the DNA.
14. Centrifuge 20 min at full speed at 4 °C.
15. Transfer supernatant to a fresh tube.

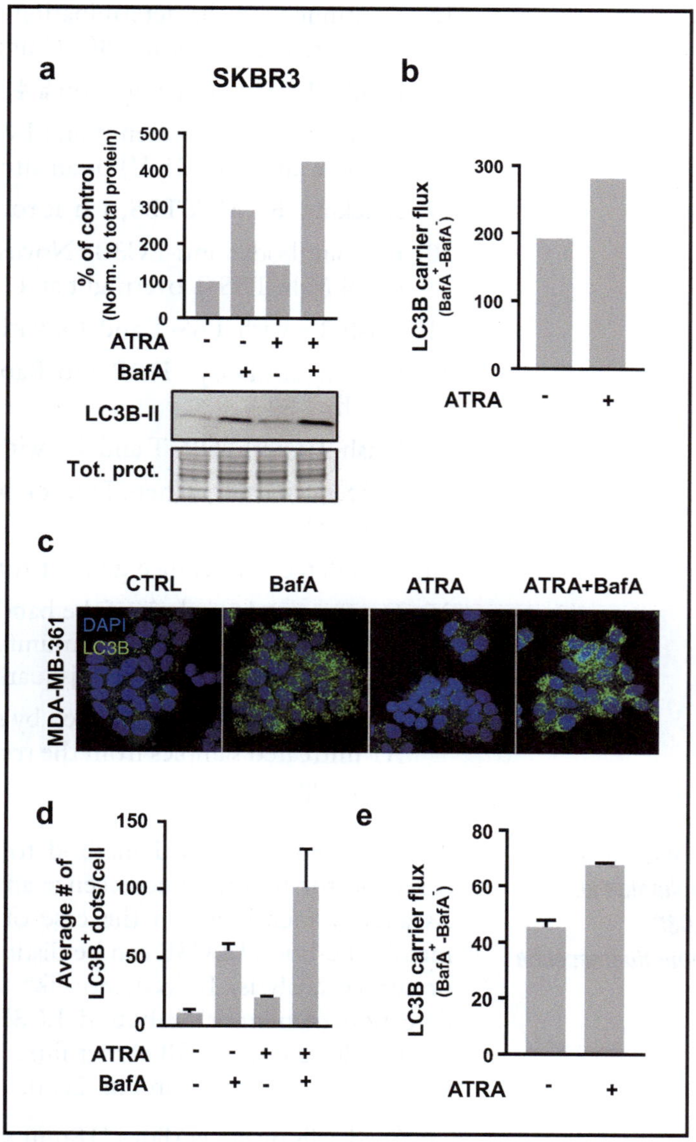

Fig. 2 Assessing LC3B carrier flux in ATRA treated breast cancer cells by Western blot and immunofluorescence analyses. *SKBR3* and *MDA-MB-361* breast cancer cells were treated with 1 μM ATRA for 24 h before Western blot or immunofluorescence analyses for LC3B was carried out. Bafilomycin A1 was added to determine autophagic flux. (**a**) The Western blot of LC3B in *SKBR3* cells is shown. LC3B-II was quantified and normalized for the total protein levels. (**b**) LC3B carrier flux was calculated based on the Western blot shown in (**a**). (**c**) Representative immunofluorescence pictures of LC3B in *MDA-MB-361* cells. (**d**) The LC3B-positive dots shown in (**c**) were quantified using the ImageJ software. (**e**) LC3B carrier flux was calculated based on the quantification shown in (**d**)

16. Continue with the determination of the protein concentration or freeze the lysates at −80 °C until use (*see* **Note 4**).
17. Load 12–25 μg of protein on a 4–20% or 12% gel.
18. Transfer to PVDF membrane by TurboBlot (10 min at 1.3 Ampère and max. 25 V) or an alternative methodology.
19. Blocking: 5% Milk TBS, 1 h at room temperature (RT).
20. First antibody: anti-LC3B (Novus NB600-1384, *see* **Note 5**) in 5% Milk TBS-T overnight at 4 °C or 2 h at RT
21. Wash 2× with TBS-T and 1× with TBS, 5 min each.
22. Second antibody: HRP-anti-Rabbit 1:10,000 in 5% Milk, 1–3 h at RT.
23. Wash 2× with TBS-T and 1× with TBS, 5 min each.
24. Detection using ChemiDoc or an alternative approach (*see* **Note 6**).
25. Also detect a reference gene or total protein.
26. Densitometric analysis of the bands using the ImageJ software (https://di.uq.edu.au/community-and-alumni/sparq-ed/sparq-ed-services/using-imagej-quantify-blots).
27. LC3B carrier flux is calculated by substracting the Bafilomycin A1 untreated samples from the respective Bafilomycin A1 treated samples.

3.2 Assessing Autophagic Flux by LC3B Immunofluorescence

Another frequently used method to measure autophagic flux is based on the immunofluorescence analysis of endogenous LC3B. As already mentioned in the case of WB analysis, cytosolic and membrane-bound LC3B can be distinguished also by immunofluorescence analysis. Indeed, dot-like staining is attributed to the detection of membrane-bound LC3B. Again, Bafilomycin A1 is used to determine LC3B carrier flux. Examples of LC3B immunofluorescence are shown in Fig. 2c, d.

1. Seed cells on cover slides (18mm × 18mm, *see* **Note 7**) inserted into the wells of a 6-well plate (optimize cell count for your cell line and treatment duration, usually around 3 × 10^5 cells).
2. The next day treat cells with DMSO (vehicle) or 1 μM ATRA (two wells/condition).
3. Treat cells for 24–48 h; 2 h before harvesting, add 200 nM BafilomycinA1 to one well for each condition and DMSO to the other wells.
4. 2 h post addition of Bafilomycin A1 start the immunofluorescence procedure.
5. Remove media from cells.
6. Wash 1× with RT PBS.
7. Cover the slide with 4% PFA (about 200 μL).

8. Incubate for 15 s (to avoid that cells detach when ice-cold MetOH is added).
9. Remove PFA and immediately add 750 μL −20 °C cold MetOH (avoid adding MetOH directly onto the cover slide).
10. Put at −20 °C for 10 min.
11. Remove MetOH.
12. Dry slides at RT.
13. Add 1 mL PBS to the wells.
14. Add 20–30 μL of the first Antibody solution LC3B (Cell Signaling #3868, Rabbit) 1:200 in PBS/1%BSA onto a clean surface (for example, on the lid of the 6-well plate).
15. Lift the cover slide with the aid of a syringe and forceps.
16. Get rid of excess PBS.
17. Add the cover slide onto the antibody-containing drop with the cells facing down.
18. Incubate 1–2 h at RT in a “humid chamber” to avoid evaporation.
19. Put the slides back to the wells for washing with the cells facing up.
20. Wash 2× with PBS/0.1% Tween.
21. Add 1 mL of PBS.
22. Prepare the second antibody solution (FITC conjugated anti-Rabbit 1:200 in PBS/1%BSA) and put 20–30 μL onto a clean surface (for example, lid of the 6-well plate).
23. Take the slides out of the well with the help of syringe and forceps.
24. Get rid of excess PBS.
25. Add the cover slide onto the antibody-containing drop with the cells facing down.
26. Incubate 1–2 h at RT in a “humid chamber.”
27. Put the slides back to the wells for washing with the cells facing up and wash 2× with PBS/0.1% Tween.
28. Add 1 mL of PBS.
29. Take slides out of the well.
30. Let the slides dry.
31. Add a small drop (5 μL) of mounting media containing DAPI onto the cells.
32. Take a microscopy slide and carefully cover the cells.
33. Let dry overnight (12–24 h).
34. Seal with nail polish.

35. Perform confocal microscopy analysis generating 5–10 pictures per condition.
36. Quantify the number of dots per picture using the ImageJ software. The Quantification is based on the plugin described in Chu et al., 2009 [32] and is performed as follow:
    (a) Split the channels.
    (b) Select the following under "Set measurements": Area, integrated density, mean gray value, limit to threshold, display label. Choose none for redirection and three decimal places.
    (c) Invert LUT.
    (d) Sharpen image.
    (e) Set the threshold.
    (f) Watershed.
    (g) Analyze particles with the following settings: size = 0-Infinity, circularity = 1, show = Ellipses display summarize.
37. Count the number of cells per picture.
38. Calculate the average number of dots/cell for each picture.
39. LC3B carrier flux is calculated by substracting the Bafilomycin A1 untreated samples from the respective Bafilomycin A1 treated samples.

3.3 mCherry-GFP-LC3B Ratiometric FACS Analysis to Determine Autophagic Flux

A method that does not require the addition of Bafilomycin A1 to measure autophagic flux is based on mCherry-eGFP-LC3B FACS analysis. Since eGFP is sensitive to the low pH observed in lysosomes, this method allows distinguishing autophagosome stages (red and green) from autolysosomes (red only) (Fig. 3a). It is well accepted that a high mCherry/GFP ratio reflects a high autophagic flux [33]. This ratio can be determined by ratiometric FACS analysis or confocal microscopy, where yellow and red dots are quantified. We use flow cytometry rather than microscopy because of (a) sensitivity and (b) simultaneous determination of flux in many samples at any given time. The original protocol describing ratiometric FACS analysis was published by Gump et al. in 2014 [33]. Our approach is similar and follows procedures and recommendations of the above paper. Whereas Gump et al. suggest performing experiments with flow-sorted single-cell clones or pooled clones with similar expression levels of the mCherry-eGFP-LC3B construct, our experience indicates that both approaches are equally effective. An example of ratiometric FACS analysis with single-cell clones expressing mCherry-eGFP-LC3B is shown in Fig. 3b. The three clones considered show similar increases of autophagic flux upon retinoid challenge (Fig. 3b),

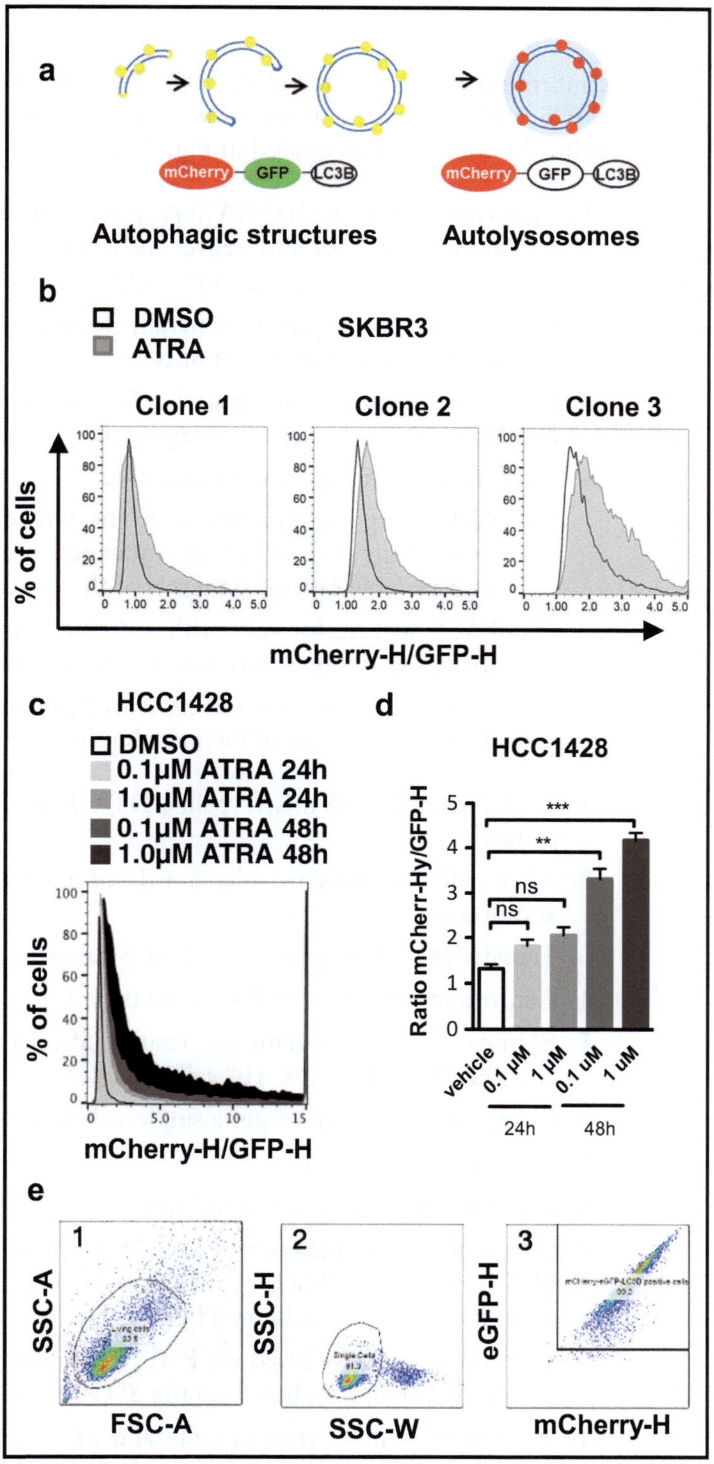

Fig. 3 Assessing autophagic flux using mCherry-GFP-LC3B tandem constructs and FACS analysis. (**a**) The principles of mCherry-GFP-LC3B FACS analysis are shown. (**b**) *SKBR3* cells stably expressing mCherry-GFP-LC3B were FACS sorted for single positive cells and expanded to a stably growing clonal population. Three clones were treated with 1 µM ATRA for 48 h before ratiometric FACS analysis for mCherry and GFP. (**c**) ATRA-sensitive *HCC1428* cells stably expressing mCherry-GFP-LC3B (population) were treated with ATRA for different amounts of time as indicated. Subsequently, ratiometric FACS analyses for mCherry and GFP were performed. (**d**) Quantification of the FACS data shown in (**c**). Kruskal-Wallis test followed by Dunn's correction for multiple comparisons was performed to compare all samples to untreated controls. $**p = 0.001$ to 0.01; $***p = 0.0001$ to 0.001. (**e**) Gating strategy of the experiment shown in (**d**). 1 = Living cells; 2 = Single cells; 3 = mCherry-eGFP-LC3B positive cells. Gate number 3 is set based on unstained control cells

confirming our published data [13]. An example of ratiometric FACS analysis using a population of cells expressing mCherry-eGFP-LC3B cells is illustrated in Fig. 3c, d.

3.4 Generation of Cell Lines Stably Expressing mCherry-eGFP-LC3B

1. Lentiviral vectors (pMD2G VSGV, pMDlg/pRRE, and pRSV-rev) are used to pack and deliver mCherry-eGFP-LC3B and a puromycin resistance gene to the cells of interest.
2. Selection in puromycin is generally conducted for 3–5 days starting 2 days post-infection of cells.
3. If transduction efficiency is low, FACS sorting for mCherry-eGFP-LC3B positive cells is suggested.
4. Thereafter several options are available:
   (a) Performing experiments with the whole population of cells which consists of an undefined number of clones with variable expression levels of the construct.
   (b) FACS sorting for cells with a defined range of the construct expression levels (low to intermediate).
   (c) Single-cell cloning to get clonal cell populations with very discrete fluorescence intensity levels.

3.5 Ratiometric FACS Analysis

1. Trypsinize cells (0.5 mL Trypsin-EDTA 0.25%/well of a 6-well plate).
2. After cell detachment, add 3 mL of media and pipet up and down.
3. Transfer the cell suspension to FACS tubes.
4. Centrifuge ($300 \times g$) at RT for 10 min.
5. Eliminate the supernatant and resuspend cells in FACS buffer (PBS/2% FBS, $0.5–1 \times 10^6$ cells/mL).
6. Pipette up and down to get a single-cell suspension.
7. Put FACS tubes on ice.
8. Perform FACS analysis immediately.
9. Set the following parameters to obtain FACS plots: SSC-A to FSC-A, SSC-H to SSC-W, mCherry-H to eGFP-H, a histogram for the ratio mCherry-H:eGFP-H.
10. Gate on living cells based on FSC-A and SSC-A, followed by gating on single cells based on the SSC-W and SSS-H pattern.
11. Take your control treated cells and set the voltage of the mCherry and eGFP lasers such that the peak ratio of the mCherry-H:eGFP-H is in the first third of the histogram (avoid having to many events on the right x-axis under any of the conditions. If so, consider adjusting the ratio).
12. After setting the conditions, record unstained cells. Compensation is not required (*see* **Note 8**).

13. Set a gate for mCherry-eGFP-LC3B positive cells.
14. Record the cells of interest.
15. Analyze the data using FlowJo (*see* **Note 9**).
16. Set the gates as previously indicated (*see* Fig. 3e for the gating strategy).
17. Compare the ratio of control and treated cells. Alternatively, you may wish to calculate the percentage of cells with high flux. Therefore, you arbitrary set the gate based on your control cells to have, e.g., 10% cells with high autophagic flux.

3.6 Autophagy Analysis of SKBR3 Breast Cancer Cells Cultured in 3D

The shortcomings of 2D cell cultures are well known. Thus, we validated our findings of increased autophagy upon ATRA treatment of *SKBR3* breast cancer cells [13] in a 3D cell culture system. We decided to use a 3D culture perfusion bioreactor (U-CUP system, Cellec Biotek) [34] to increase the relevance of our preclinical results. We established 3D breast cancer cell cultures on a collagen scaffold under perfusion and performed the experiments involving ATRA. Similar to conventional 2D cell culture experiments, ATRA treatment has an impact on the viability of *SKBR3* cells as determined by MTT assay, IHC staining for cleaved Caspase-3 (Fig. 4a) and WB analysis for cleaved Caspase-3 and cleaved PARP (Fig. 4b, c). Since ATRA triggers epithelial differentiation of *SKBR3* cells as shown previously [9], we further determined the levels and localization of β-Catenin by WB analysis and IHC respectively. In our experimental conditions, ATRA increases the amounts of β-Catenin at the level of tight junctions (Fig. 4a–c).

3.7 Set-Up Seeding and Culture Conditions

Before performing experiments with the U-CUP system, the cell number and the type of scaffold selected should be tested. The cells of interest should adhere to the scaffold with high efficiency and should distribute homogeneously inside the scaffold. In the following section, the general procedure necessary to establish the seeding efficiency is described. A detailed description on how to handle the U-CUP system is provided in the user manual.

1. Prepare a cell suspension of more than 2 mL containing different numbers of cells (e.g., 1, 2, 4 $10\times^{6}$).
2. Assemble the U-CUP with the scaffold inserted (if you have a collagen scaffold, insert a mesh on each side of the adaptor to avoid displacement of the scaffold).
3. Fill the U-CUP with 8 mL of media according to the manufacturer instructions.
4. Inject 2 mL of cells in the U-CUP according to the manufacturer instructions.
5. Let seeding program run for 24 or 48 h (the seeding program is usually faster than the culture program, flow rates may be cell

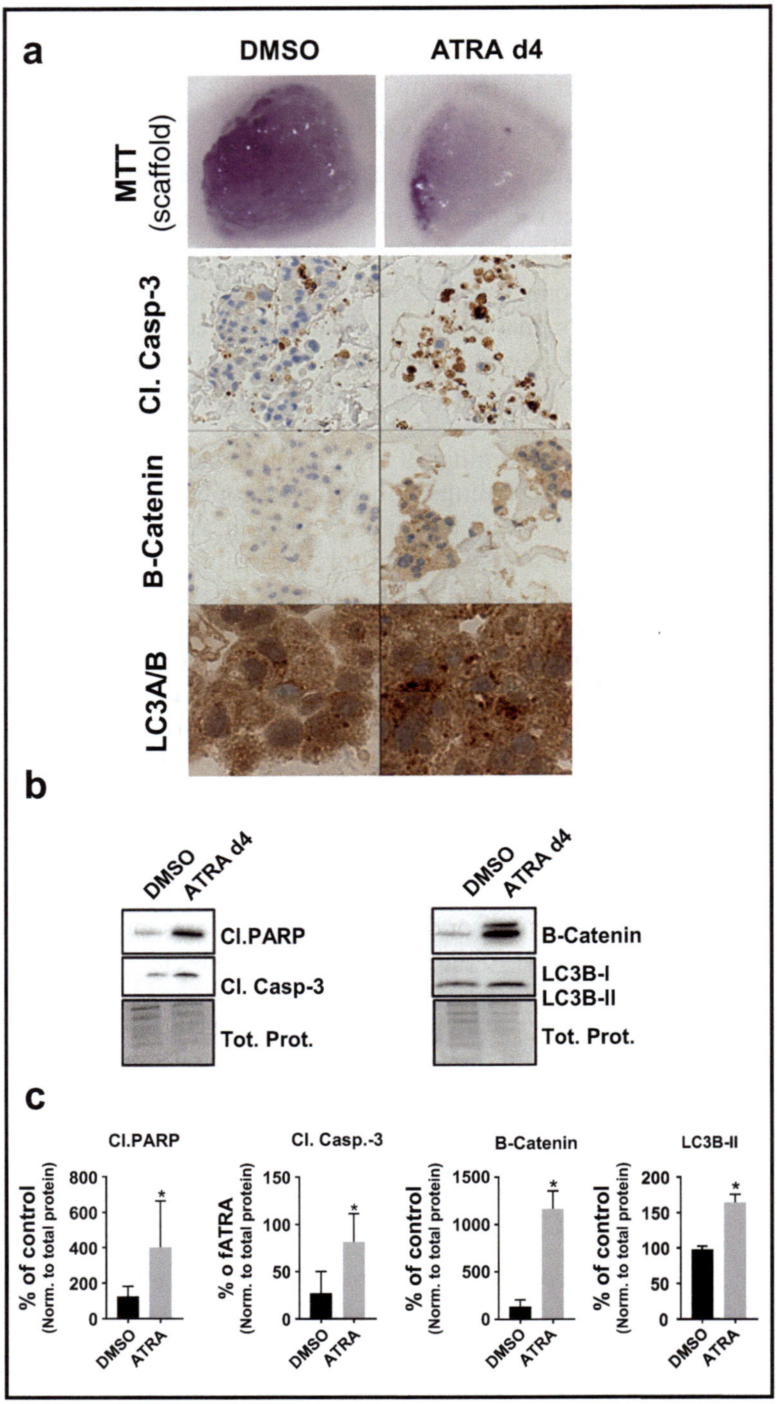

Fig. 4 ATRA response of 3D cultured *SKBR3* cells. ATRA-sensitive *SKBR3* cells were seeded onto a collagen scaffold in a U-CUP bioreactor and cultured for 10 days before addition of DMSO (Vehicle) or 1 μM ATRA to the 3D cultures. 4 days post ATRA treatment, scaffolds were harvested and cut in 3 pieces, which were processed for MTT staining, Immunohistochemistry (IH) and Western blotting (WB), respectively (**a**) The panel illustrates representative pictures of MTT stained scaffolds after 4 days of DMSO or ATRA treatment and IHC pictures thereof showing cleaved caspase-3 (Cl. Casp.-3), β-Catenin and LC3A/B. (**b**) Western blot analysis of cells harvested from the 3D cultures described above. Total protein is shown as a loading control. (**c**) Quantification of the Western blots shown in (**b**). Raw values were normalized to total protein and to the untreated control or ATRA of one Western blot. Mann-Whitney-U was used for statistical testing. $*p = 0.01–0.05$

line dependent, but are approximately 1.2–3.0 mL/min; in the case of SKBR3 cells: 1.2 mL/min).

Step 1: Infuse at 1.2–3.0 mL/min, target volume: 4 mL.

Step 2: Delay: 5′.

Step 3: Withdraw at 1.2–3.0 mL/min, target volume: 4.5 mL.

Step 4: Delay: 5′.

Step 5: Infuse at 1.2–3.0 mL/min, target: 4.5 mL.

Step 6: go to **step 2** (=repeat program from **step 2**).

6. Retrieve media from the bioreactor and count cells in the media (to assess the number of cells that did not adhere to the scaffold).
7. Cut the scaffold in two pieces (crescents) using a single-edge razor blade or scalpel.
8. Put each piece in a separate tube for different down-stream applications.

MTT staining

(a) Transfer ½ of the scaffold to medium containing 5 μg/mL MTT.

(b) Transfer the mixture to a 37 °C incubator.

(c) Incubate until the scaffold turns blue (note time point for further experiments).

(d) Fix in 4% PFA for 1 h at RT.

(e) Take pictures using a binocular.

(f) Estimate the distribution of cells across the scaffold.

Paraffin embedding for H&E staining

(a) Wash ½ of the scaffold in PBS.

(b) Transfer to 4% PFA.

(c) Dehydrate.

(d) Embed the collagen in paraffin for the microtome to cut cross-sections of the scaffold. Cut 3 μm slices at different levels.

(e) Stain for H&E.

(f) Take pictures using a microscope.

(g) Assess the quality of cells and the distribution across sections.

9. If the cell number and distribution are appropriate, set-up a long-term culture according to the needs and test the selected

parameters (e.g., viability by MTT, apoptosis/proliferation markers by WB and/or immunohistochemistry analyses; refer to the next section for details) on the day of interest.

3.8 ATRA Experiment with SKBR3 Breast Cancer Cells

1. Seed 3×10^6 cells in the U-CUP on a Collagen Sponge (Avitene™ UltraFoam™) using a flow rate of 1.2 mL/min.
2. 24 h after seeding set the flow rate to 0.3 mL/min.
3. Culture cells for 10 days and change the medium every 2–3 days according to the manufacturer instructions.
4. Retrieve 5 mL of medium from the bottom valve and inject 5 mL of medium containing 2 μM of ATRA (final concentration = 1 μM) or DMSO as a control.
5. Culture cells for 4 additional days.
6. Retrieve all medium.
7. Disassemble the U-CUP and harvest the scaffold.
8. Cut the scaffold in three pieces using a single-edge razor blade or scalpel.

MTT staining

See above for detailed procedure used.

Paraffin embedding for IHC and H&E staining

See above for detailed procedure used.

Of note: Paraffin embedding and IHC staining were performed by our translational research unit using the automated system BOND RX® (Leica Biosystems, Newcastle, UK). The protocol used for LC3B staining is published elsewhere [35].

Western blotting

(a) Wash the scaffold in PBS.

(b) Transfer the scaffold to 200 μL trypsin (*see* **Note 10**).

(c) Allow the cells to detach from the scaffold (Depending on the type of scaffold, it might be partially degraded).

(d) Remove the solid parts from the scaffold using forceps.

(e) Add 1 mL of FBS-containing medium.

(f) Pellet the cells by centrifugation.

(g) Wash cells in PBS.

(h) Pellet the cells by centrifugation.

(i) Lyse the cells with protein lysis buffer.

(j) Perform WB analysis using the protocol described above.

(k) The following antibodies and incubation conditions were used:

Anti-LC3B: 5% Milk, 3 h at RT or 4 °C overnight.

Anti-Cleaved Caspase-3: 5% Milk, 4 °C overnight.

Anti-Cleaved PARP: 5% BSA (more sensitive but side bands) or 5% Milk, 4 °C overnight.

Anti-β-Catenin: 5% BSA, 4 °C overnight.

4 Notes

1. It is important to note that LC3B-based methods do not allow specific measurement of autophagic activity. Therefore, it is highly recommended to complement LC3B monitoring using long-lived protein degradation (LLDPA) [36] or lactate dehydrogenase (LDH) [37] sequestration assays. These assays measure autophagic degradation of the total protein fraction or sequestration to autophagosomes of LDH, respectively. Alternatively, lysosome resistant probes, which emit different fluorescent light in basic and acidic environments, could be used to measure autophagic flux. One such probe is Keima. Coupled to a known autophagy cargo it allows monitoring the degradation process [38].
2. There is evidence for LC3B independent autophagy [39, 40]. Relying on LC3B as the sole marker may lead to underestimation of autophagic activity. To overcome this problem other members of the ATG8 family could be analyzed [39].
3. Another marker often used to monitor autophagy is p62 or sequestosome 1 (SQSTM1), one of the best-studied autophagy receptors. Autophagy receptors are involved in cargo recognition and mediate their selective degradation [41]. The p62 receptor itself can be degraded via autophagy [42]. Hence, under certain conditions [27], p62 levels are used to monitor autophagic flux. However, the results must be interpreted with caution, since p62 is upregulated transcriptionally by various stimuli [27]. For instance, ATRA treatment of APL-derived NB4 cells upregulates p62, in spite of the fact that autophagy is highly active [43]. Similar results are observed in the case of SKBR3 breast cancer cells treated with ATRA (unpublished).
4. It was reported that LC3B-II is sensitive to freeze-thaw cycles [30].
5. The antibody from Novus Biologicals (NB600-1384) [44] is reported to recognize LC3A and B, whereas the LC3B

antibody from Cell Signaling is specific for LC3B (Cell Signaling #3686) [45].

6. It is our experience that the LC3B antibody from Novus Biologicals shows significant variability in terms of detection sensitivity for LC3-I. Sometimes it is very difficult to detect LC3-I.
7. Some cell lines may not grow well on glass slides. In this case, slides may be coated with, e.g., fibronectin or poly-L-Lysin.
8. When we performed our first experiments using the tandem construct, we transduced our cells with eGFP-LC3B and mCherry-LC3B, respectively, to verify the requirement for compensation procedures. Based on our experience, there is no need for compensation procedures if the mCherry-eGFP-LC3B construct is used and measurement of the fluorescence intensity is performed on a BD™ LSR II flow cytometer with the 488 nm Laser and 530/30 filter for eGFP and 561 nm laser and 610/20 filter for mCherry.
9. If PC version 10 of FlowJo is used, the ratio calculated by the Diva Software cannot be processed and you have to calculate the ratio based on mCherry-H and eGFP-H (http://docs.flowjo.com/vx/experiment-based-platforms/plat-derived-overview/).
10. Depending on the type of scaffold used for 3D cultures, harvesting by using trypsin may not be suitable.

Acknowledgments

Deborah Shan-Krauer is gratefully acknowledged for excellent technical support. This research project was supported by the University of Bern Initiator Grant and grants from the Bernese Cancer League, the Werner and the Hedy Berger-Janser Foundation for Cancer Research (to A.M.S.) and Swiss Cancer Research (KFS-3409-02-2014 to M.P.T.).

Conflict of Interest: Manuele G. Muraro is a shareholder and scientific advisor of Cellec Biotek AG. Giulio C. Spagnoli is a Cellec Biotek AG shareholder.

References

1. Tang X-H, Gudas LJ (2011) Retinoids, retinoic acid receptors, and cancer. Annu Rev Pathol 6:345–364
2. Theodosiou M, Laudet V, Schubert M (2010) From carrot to clinic: an overview of the retinoic acid signaling pathway. Cell Mol Life Sci 67:1423–1445
3. Bushue N, Wan Y-JY (2010) Retinoid pathway and cancer therapeutics. Adv Drug Deliv Rev 62:1285–1298
4. Niles RM (2000) Recent advances in the use of vitamin a (retinoids) in the prevention and treatment of cancer. Nutrition 16:1084–1089

5. de Thé H (2018) Differentiation therapy revisited. Nat Rev Cancer 18:117–127
6. Pattabiraman DR, Weinberg RA (2016) Targeting the epithelial-to-mesenchymal transition: the case for differentiation-based therapy. Cold Spring Harb Symp Quant Biol 81:11–19
7. Di Leo A, Curigliano G, Diéras V et al (2015) New approaches for improving outcomes in breast cancer in Europe. Breast 24:321–330
8. Chen K, Huang Y, Chen J (2013) Understanding and targeting cancer stem cells: therapeutic implications and challenges. Acta Pharmacol Sin 34:732–740
9. Zanetti A et al (2015) All-trans-retinoic acid modulates the plasticity and inhibits the motility of breast cancer cells: role of Notch1 and transforming growth factor (TGFβ). J Biol Chem 290(29):17690–17709
10. Ginestier C, Wicinski J, Cervera N et al (2009) Retinoid signaling regulates breast cancer stem cell differentiation. Cell Cycle 8:3297–3302
11. Garattini E, Paroni G, Terao M (2012) Retinoids and breast cancer: new clues to increase their activity and selectivity. Breast Cancer Res 14:111
12. Centritto F, Paroni G, Bolis M et al (2015) (2015) cellular and molecular determinants of all-trans retinoic acid sensitivity in breast cancer: luminal phenotype and RARα expression. EMBO Mol Med 7(7):950–972
13. Brigger D, Schläfli AM, Garattini E et al (2015) Activation of RARα induces autophagy in SKBR3 breast cancer cells and depletion of key autophagy genes enhances ATRA toxicity. Cell Death Dis 6:e1861
14. Dikic I, Elazar Z (2018) Mechanism and medical implications of mammalian autophagy. Nat Rev Mol Cell Biol 19:349–364
15. Tanida I (2011) Autophagy basics. Microbiol Immunol 55:1–11
16. Cecconi F, Levine B (2008) The role of autophagy in mammalian development: cell makeover rather than cell death. Dev Cell 15:344–357
17. Mizushima N, Levine B, Cuervo AM et al (2008) Autophagy fights disease through cellular self-digestion. Nature 451:1069–1075
18. Guo JY, White E (2016) Autophagy, metabolism, and cancer. Cold Spring Harb Symp Quant Biol 81:73–78
19. Amaravadi R, Kimmelman AC, White E (2016) Recent insights into the function of autophagy in cancer. Genes Dev 30:1913–1930
20. Chen H-Y, White E (2011) Role of autophagy in cancer prevention. Cancer Prev Res (Phila) 4:973–983
21. Strohecker AM, White E (2014) Autophagy promotes BrafV600E-driven lung tumorigenesis by preserving mitochondrial metabolism. Autophagy 10:384–385
22. Guo JY, White E (2013) Autophagy is required for mitochondrial function, lipid metabolism, growth, and fate of KRASG12D-driven lung tumors. Autophagy 9:1636–1638
23. Degenhardt K, Mathew R, Beaudoin B et al (2006) Autophagy promotes tumor cell survival and restricts necrosis, inflammation, and tumorigenesis. Cancer Cell 10:51–64
24. Guo JY, Chen H-Y, Mathew R et al (2011) Activated Ras requires autophagy to maintain oxidative metabolism and tumorigenesis. Genes Dev 25:460–470
25. Thorburn A, Thamm DH, Gustafson DL (2014) Autophagy and cancer therapy. Mol Pharmacol 85:830–838
26. Levy JMM, Towers CG, Thorburn A (2017) Targeting autophagy in cancer. Nat Rev Cancer 17:528–542
27. Klionsky DJ, Abdalla FC, Abeliovich H et al (2012) Guidelines for the use and interpretation of assays for monitoring autophagy. Autophagy 8:445–544
28. Barth S, Glick D, Macleod KF (2010) Autophagy: assays and artifacts. J Pathol 221:117–124
29. Loos B, du TA, Hofmeyr J-HS (2014) Defining and measuring autophagosome flux—concept and reality. Autophagy 10:2087–2096
30. Klionsky DJ, Abdelmohsen K, Abe A et al (2016) Guidelines for the use and interpretation of assays for monitoring autophagy (3rd edition). Autophagy 12:1–222
31. Kabeya Y (2000) LC3, a mammalian homologue of yeast Apg8p, is localized in autophagosome membranes after processing. EMBO J 19:5720–5728
32. Chu CT, Plowey ED, Dagda RK et al (2009) Autophagy in neurite injury and neurodegeneration: in vitro and in vivo models. Methods Enzymol 453:217–249
33. Gump JM, Thorburn A (2014) Sorting cells for basal and induced autophagic flux by quantitative ratiometric flow cytometry. Autophagy 10:1327–1334
34. Hirt C, Papadimitropoulos A, Muraro MG et al (2015) Bioreactor-engineered cancer tissue-like structures mimic phenotypes, gene expression profiles and drug resistance patterns observed "in vivo". Biomaterials 62:138–146
35. Berezowska S, Galván JA (2017) Immunohistochemical detection of the autophagy markers LC3 and p62/SQSTM1 in formalin-fixed and paraffin-embedded tissue. In: Pellicciari C,

Biggiogera M (eds) Histochemistry of single molecules: methods and protocols. Springer, New York, NY, pp 189–194

36. Dupont N, Leroy C, Hamaï A et al (2017) Chapter 3: long-lived protein degradation during autophagy. In: Galluzzi L, Bravo-San Pedro JM, Kroemer G (eds) Methods in enzymology. Academic Press, Cambridge, MA, pp 31–40
37. Luhr M, Szalai P, Engedal N (2018) The lactate dehydrogenase sequestration assay—a simple and reliable method to determine bulk autophagic sequestration activity in mammalian cells. J Vis Exp 137:e57971
38. Katayama H, Kogure T, Mizushima N et al (2011) A sensitive and quantitative technique for detecting autophagic events based on lysosomal delivery. Chem Biol 18:1042–1052
39. Engedal N, Seglen PO (2016) Autophagy of cytoplasmic bulk cargo does not require LC3. Autophagy 12:439–441
40. Ra E, Lee S, Park B (2015) Possible roles of LC3-independent autophagy in elimination of Shigella in human colon cells (INC4P.343). J Immunol 194(1 Supplement):125.22
41. Birgisdottir ÅB, Lamark T, Johansen T (2013) The LIR motif—crucial for selective autophagy. J Cell Sci 126:3237–3247
42. Ichimura Y, Komatsu M (2010) Selective degradation of p62 by autophagy. Semin Immunopathol 32:431–436
43. Trocoli A, Bensadoun P, Richard E et al (2014) p62/SQSTM1 upregulation constitutes a survival mechanism that occurs during granulocytic differentiation of acute myeloid leukemia cells. Cell Death Differ 21:1852–1861
44. Zois CE, Giatromanolaki A, Sivridis E et al (2011) "Autophagic flux" in normal mouse tissues: focus on endogenous LC3A processing. Autophagy 7:1371–1378
45. Martinet W, Schrijvers DM, Timmermans J-P et al (2013) Immunohistochemical analysis of macroautophagy: recommendations and limitations. Autophagy 9:386–402

Chapter 18

Nonradioactive and Radioactive Telomerase Assays for Detecting Diminished Telomerase Activity in Cancer Cells after Treatment with Retinoid

Swapan K. Ray

Abstract

Detection of any decrease in telomerase activity in cancer cells and tumor tissues is an important part in assessing overall therapeutic outcomes of a treatment agent in the laboratory and clinical settings. Almost 85% of cancers have activation of telomerase activity that promotes cell proliferation and discourages differentiation to sustain growth of the cancers. Retinoids are highly regarded as the anti-proliferation and pro-differentiation agents that cause down regulation of telomerase activity in the cancer cells. Two (nonradioactive and radioactive) telomeric repeat amplification protocol (TRAP) assays are optimized and fully described for detection of the diminished or abolished telomerase activity in a very low amount of protein extracts from cancer cells after treatment with a natural retinoid or a synthetic retinoid. These highly optimized and improved nonradioactive and radioactive TRAP assays can also be used for determining the presence or absence of telomerase activity in a small amount of any tumor tissue. The results from these TRAP assays can also help decide appropriate therapeutic options for the cancers with or without telomerase activity.

Key words Differentiation, Proliferation, Retinoid treatment, Telomerase activity in cancers, Telomerase assays

1 Introduction

As many as 85% cancers show telomerase activation and varying degrees of telomerase activity that supports unlimited cell proliferation without cell differentiation and thereby sustains growth of the cancers [1]. So, inhibition of telomerase activity in cancers continues to be an important therapeutic goal for arresting cell cycle, promotion of differentiation, and induction of apoptosis [2]. Telomere is a noncoding repetitive hexanucleotide DNA sequence (TTAGGG) that provides protection to the ends of all eukaryotic chromosomes for their stability and contribution to cell division. Telomerase is an enzyme complex, which is composed of RNA and proteins. The structure of telomerase in humans has two important

Swapan K. Ray (ed.), *Retinoid and Rexinoid Signaling: Methods and Protocols*, Methods in Molecular Biology, vol. 2019, https://doi.org/10.1007/978-1-4939-9585-1_18, © Springer Science+Business Media, LLC, part of Springer Nature 2019

components: human telomerase reverse transcriptase (hTERT) and telomerase RNA template. Activation of telomerase adds telomere to chromosomal ends and maintains their telomeric length to continue cell division in most of the cancers. The original telomeric repeat amplification protocol (TRAP) assay, a clever combination of conventional activity of telomerase with amplifying power of polymerase chain reaction (PCR), was developed for detecting telomerase activity in different cancer cells and tumor tissues [3, 4]. In the design of original TRAP assay, telomerase activity adds the hexanucleotide TTAGGG repeats to a telomerase substrate (TS) primer followed by PCR amplification using a reverse primer (named CX primer) for production of 6 base-pair (bp) DNA ladder, which can be resolved by polyacrylamide gel electrophoresis (PAGE).

Several commercial vendors have developed TRAPeze assay kit to make the detection of telomerase activity faster than before, but TRAPeze assay kit is more about convenience of readymade reagents and less about reliability in showing many bands of high molecular sizes [5]. Telomerase activity is proportional to the total amount of DNA that telomerase synthesizes and should be detected in a reliable telomerase assay. The processivity of the enzyme is proportional to the length of the DNA ladder that a robust telomerase assay should bring to light. The intensity of the DNA bands and the length of the DNA ladder exhibit the amount of active telomerase molecules and their processivity, respectively, in a sample. Commercially available TRAPeze assay kit does not fully unveil these properties of telomerase in the samples. Although TRAPeze kit is more cost effective and less time consuming than the original TRAP assay for detecting telomerase activity, its 36 bp internal standard for determining telomerase activity in tumor tissues may potentially lead to false negative results [6].

Our research team has surmised that original TRAP assay still remains superior and we have tried to get used to it several times for detection of any changes in telomerase activity in cancer cells after treatment with retinoid and other chemotherapeutic agents [7–9]. However, it is a familiar feeling in our research group that the original TRAP assay needs substantial optimization with custom-made components for increasing its reliability and versatility. Also, a clear step-by-step description is needed for executing the complete procedure comfortably for detecting telomerase activity in a very low amount of protein extract from the cancer cells or tumor tissues. So, a nonradioactive TRAP assay is optimized and described in detail for the investigators who want to avoid radiation hazards and use it in their experiments with the cancer cells or tumor tissues with relatively abundant amount of telomerase activity. Also, an improved radioactive TRAP assay is developed and described fully for using it by the investigators who have proper laboratory setup, feel no problem in cautiously dealing with radioactive materials, and also have no other choice but to employ radioactive TRAP assay for detecting very low telomerase activity in specific cancer cells or tumor tissues.

Retinoids, which are vitamin A derivatives, are widely known to act using the retinoic acid receptors and retinoid X receptors in the cells for modulation of expression of various genes that ultimately provide anticancer effects. We have been widely using the natural retinoid all-*trans* retinoic acid (ATRA) [10, 11] and the synthetic retinoid *N*-(4-hydroxyphenyl) retinamide (4HPR) [12, 13] for their capability of inhibition of cell proliferation and induction of cell differentiation in preclinical cancer therapeutic studies. Presence of telomerase activity promotes cell proliferation and prohibits cell differentiation in cancers. So, the decreased or abolished telomerase activity due to treatment with a retinoid in cancers is a clear indication of inhibition of cell proliferation and induction of differentiation. Our studies showed that ATRA [8] and 4HPR [14] could decrease the expression hTERT, the catalytic subunit of telomerase, in different human cancer cells. Transcription of the *hTERT* gene may be regulated by multiple mechanisms. An important mechanism of action of ATRA could be a rapid decrease in histone acetylation at the *hTERT* gene promoter leading to suppression of *hTERT* gene transcription and thereby decrease in telomerase activity, which could be considered as one manifestation of the many anticancer effects of ATRA [15].

Human leukemia-60 (HL-60) cell line is a popular cell culture model of cancer for evaluating efficacy of different chemotherapeutic agents [16–18] as well as for studying cancer cell differentiation [19]. Therefore, HL-60 cell line has been employed here as a cell culture model of cancer for developing the improved nonradioactive and radioactive TRAP assays for detecting the diminished telomerase activity indicating inhibition of cell proliferation and induction of cell differentiation due to treatment with a retinoid (ATRA or 4HPR). A telomerase assay needs to be highly reliable, robust, and versatile for confirming the presence or absence of telomerase activity in a very low amount of protein extract from very a low number of cancer cells. The nonradioactive TRAP assay described here requires only 6 μg of protein extract for detecting telomerase activity (Figs. 1 and 2). On the other hand, radioactive TRAP assay that is described subsequently should be used for additional sensitivity of detection of telomerase activity in as low as 1000 cancer cells (Fig. 3) or in as low as 1 μg of protein extract from cancer cells (Figs. 4 and 5). The natural retinoid ATRA at a relatively high dose clearly diminishes telomerase activity in cancer cells (Fig. 4) but the synthetic retinoid 4HPR at a relatively low dose is highly potent in diminishing or almost abolishing

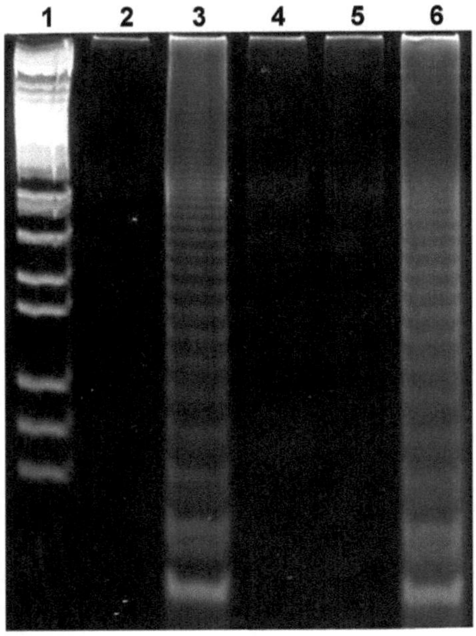

Fig. 1 The optimized nonradioactive TRAP assay products when resolved on the 10% polyacrylamide non-denaturing gel by electrophoresis. Protein extracts from the cells were used in the nonradioactive TRAP assay and the TRAP assay products were loaded on *lane 1* (*Hae* III digested pBR322 DNA as molecular size marker, it is not a TRAP assay product), *lane 2* (lysis buffer), *lane 3* (protein extract from 293 HEK cells as telomerase positive control), *lane 4* (protein extract from human fibroblast cells as telomerase negative control), *lane 5* (RNase A treated protein extract from HL-60 cells), and *lane 6* (protein extract from HL-60 cells)

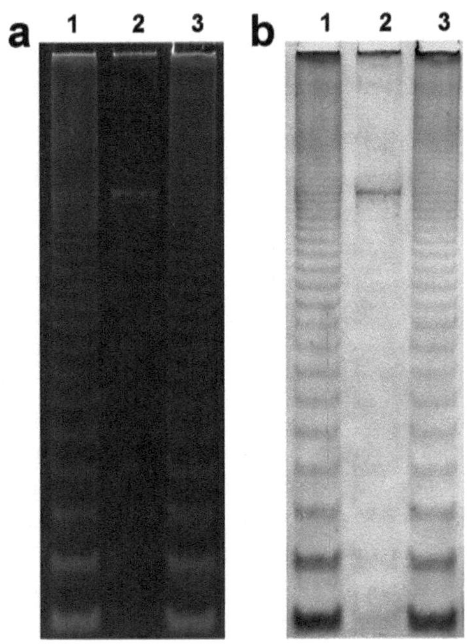

Fig. 2 Nonradioactive TRAP assay with protein extracts from HL-60 cells after treatments with ATRA and Ara-C. Protein extracts from the drug untreated and drug treated cells were subjected to nonradioactive TRAP assay. The TRAP assay products were loaded onto the 10% polyacrylamide non-denaturing gel and resolved by electrophoresis. (**a**) Polaroid film Positive image of the gel. (**b**) Polaroid film Negative image of the gel. *Lane 1* shows telomerase activity in control (drug untreated) HL-60 cells. *Lane 2* shows the diminished telomerase activity in HL-60 cells due to treatment with 1 μM ATRA for 7 days. *Lane 3* shows almost unaltered telomerase activity in HL-60 cells after treatment with 10 nM Ara-C for 7 days. When the nonradioactive TRAP assay is used, the Negative image of the gel is better than its Positive image for detection and analysis of telomerase activity

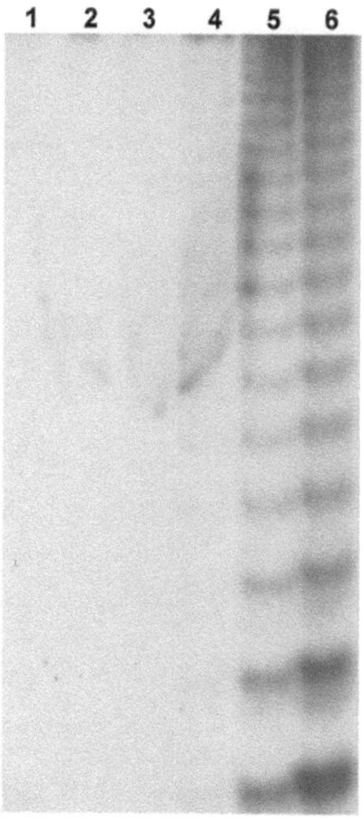

Fig. 3 Radioactive TRAP assay with protein extracts from different number of HL-60 cells. Protein extract from the cells was used for determination of protein concentration and radioactive TRAP assay. Different amounts of this protein extract reflecting different numbers of HL-60 cells were employed in the radioactive TRAP assay and the TRAP assay products were loaded onto the 10% polyacrylamide non-denaturing gel and resolved by electrophoresis: *lane 1*, lysis buffer; *lane 2*, protein extract from 1 cell; *lane 3*, protein extract from 10 cells; *lane 4*, protein extract from 100 cells; *lane 5*, protein extract from 1000 cells; and *lane 6*, protein extract from 10,000 cells. Distinctly resolved many DNA bands on the autoradiogram indicated that this assay worked to reveal the presence of telomerase activity and its processivity in the samples. This radioactive TRAP assay has been optimized to detect telomerase activity in the protein extract from about 1000 HL-60 cells

telomerase activity in cancer cells (Fig. 5). Taxol is a non-retinoid drug that has very potent cancer chemotherapeutic effects but it does not have any ability in affecting telomerase activity in cancer cells. Because the levels of telomerase activity significantly vary depending on cancer cells and tumor tissues, these improved non-radioactive and radioactive TRAP assays can be readily modified using more or less amount of protein extract to adjust the level of sensitivity of detection of telomerase activity in cancers before and after treatment with a retinoid. Most importantly, these TRAP assays are highly reliable and robust for assessing both the activity and the processivity of the telomerase molecules in the samples.

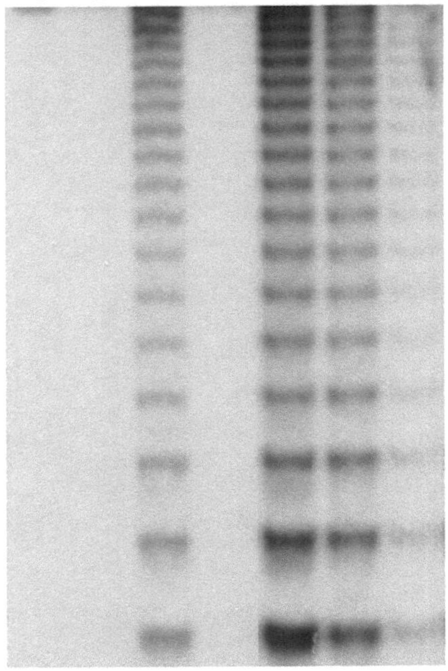

Fig. 4 Radioactive TRAP assay for detecting the diminished telomerase activity in HL-60 cells after treatment with different doses of ATRA. Exponentially growing HL-60 cells were treated with 1 μM ATRA or 10 μM Ara-C for 7 days. Protein extracts from cells were subjected to radioactive TRAP assay and the TRAP assay products were loaded onto the 10% polyacrylamide non-denaturing gel and resolved by electrophoresis. The gel was dried and autoradiographed: *lane 1*, lysis buffer; *lane 2*, protein extract from human skin fibroblasts (used as negative telomerase activity); *lane 3*, protein extract from 293 HEK cells (used as positive telomerase activity); *lane 4*, RNase A treated protein extract from HL-60 cells; *lane 5*, protein extract from HL-60 cells (untreated control); *lane 6*, protein extract from HL-60 cells treated with 1 μM ATRA for 7 days; and *lane 7*, protein extract from HL-60 cells treated with 10 μM ATRA for 7 days. Detection of distinctly resolved many DNA bands on the autoradiogram indicated that this TRAP assay could clearly reveal activity and processivity of the telomerase molecules in the control sample. This radioactive TRAP assay using 1 μg/μL protein extract is highly sensitive to detect only slightly diminished telomerase activity in HL-60 cells treated with 1 μM ATRA for 7 days. On the contrary, nonradioactive TRAP assay using 6 μg/μL protein extract was not sensitive enough and showed the highly diminished telomerase activity in HL-60 cells treated with 1 μM ATRA for 7 days (*lane 2*, Fig. 2). However, this radioactive TRAP assay showed the highly diminished telomerase activity in HL-60 cells treated with 10 μM ATRA for 7 days

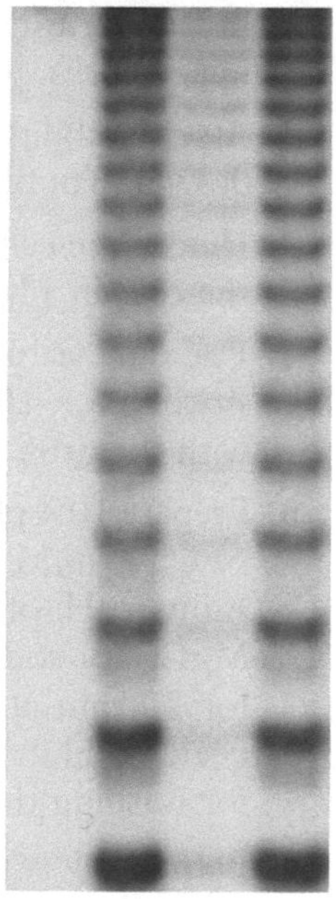

Fig. 5 Radioactive TRAP assay for detecting changes in telomerase activity in HL-60 cells after treatments with a synthetic retinoid drug and a non-retinoid drug. Exponentially growing HL-60 cells were untreated, treated with 3 μM 4HPR for 24 h, and treated with 0.5 μM Taxol for 24 h. Protein extracts from cells were subjected to radioactive TRAP assay and the TRAP assay products were loaded onto the 10% polyacrylamide non-denaturing gel and resolved by electrophoresis. The gel was dried and autoradiographed: *lane 1*, RNase A treated protein extract from HL-60 cells; *lane 2*, protein extract from HL-60 cells (untreated control); *lane 3*, protein extract from HL-60 cells treated with 3 μM 4HPR for 24 h; and *lane 4*, protein extract from HL-60 cells treated with 0.5 μM Taxol for 24 h. Distinctly resolved many DNA bands, as recognizable on the autoradiogram, very clearly demonstrated the activity and processivity of the telomerase molecules in the samples. The synthetic retinoid 4HPR is capable of greatly diminishing the telomerase activity, while the non-retinoid drug Taxol is not able to change telomerase activity in HL-60 cells

2 Materials

Use analytical grade chemicals and molecular biology grade biochemicals in the nonradioactive and radioactive TRAP assays. Prepare all the reagents using the sterile distilled water (dH_2O). After preparations, store the reagents at room temperature or as indicated otherwise. Follow the manufacturer guidelines for proper handling and disposing of the hazardous chemicals and radioactive waste materials.

2.1 Culturing and Harvesting of Cells

1. Use 1× RPMI 1640 medium supplemented with 10% fetal bovine serum (FBS) and 1% penicillin/streptomycin for growing the cells.
2. Assess viability of the cells by testing their ability to exclude 0.5% trypan blue dye.
3. Determine cell density in the cell culture medium by counting the cells in a hemocytometer or a Coulter counter.
4. Use only viable and exponentially growing cells at a final density of 0.5×10^6 cells/mL in the experiments and treat the cells with the drug at specific dose for a desired time period.
5. Prepare 10× phosphate-buffered saline (PBS, pH 7.4): 1.37 M NaCl, 27 mM KCl, 100 mM Na_2HPO_4, and 18 mM KH_2PO_4. Adjust pH of the solution with HCl. Do not add $CaCl_2$, $MgCl_2$, or sodium azide to make PBS. Sterilize by autoclaving for 20 min at 15 psi and store at room temperature. When needed, dilute 10× PBS to 1× PBS with sterile dH_2O and use for washing the cells (*see* **Note 1**).
6. Prepare the wash buffer: 10 mM 4-(2-hydroxyethyl)-1-piperazineethanesulfonic acid (HEPES)-KOH (pH 7.5), 1.5 mM $MgCl_2$, 10 mM KCl, and 1 mM dithiothreitol (DTT). After sterilization, store the wash buffer at 4 °C (*see* **Note 2**).

2.2 Cell Lysis and Protein Estimation

1. Use cell pellets or tissues (*see* **Note 3**).
2. Prepare the lysis buffer: 10 mM Tris–HCl (pH 7.5), 1 mM $MgCl_2$, 1 mM ethylene glycol-bis(β-aminoethyl ether)-*N,N,N′,N′*-tetraacetic acid (EGTA), 0.1 mM phenylmethylsulfonyl fluoride (PMSF), 5 mM 2-mercaptoethanol (2-ME), 0.5% 3-[(3-*ch*olamidopropyl)dimethyl*a*mmonio]-1-propane*s*ulfonate (CHAPS), and 10% glycerol (*see* **Note 4**).
3. Use Coomassie-Plus protein reagent (Pierce Biotechnology) for estimation of the protein extracted using the lysis buffer.
4. After protein estimation, make small (10 μL) aliquots of the protein, quick-freeze on dry ice, and store at −80 °C (*see* **Note 5**).

2.3 TRAP Reaction

1. CX primer (5′-CCC TTA CCC TTA CCC TTA CCC TAA-3′).
2. Molten wax or AmpliWax PCR Gem (Perkin Elmer).
3. Mixture of dNTPs.
4. TS primer (5′-AAT CCG TCG AGC AGA GTT-3′).
5. T4 gene 32 protein (T4g32 protein), which is used to stabilize ssDNA, increase DNA yield, and promote processivity of a reverse transcriptase.
6. *Taq* DNA polymerase (Boehringer Mannheim).
7. PCR Machine, also known as DNA thermal cycler.

2.4 PAGE and SYBR Green Staining

1. 10× Tris-borate-EDTA (TBE).
2. 40% Acrylamide/bisacrylamide (at 19:1 ratio).
3. 10% Ammonium persulfate (prepare freshly or use the stock aliquot from −20 °C).
4. *N,N,N,N′*-Tetramethyl-ethylenediamine (TEMED).
5. Bio-Rad 2-D gel apparatus.
6. SpeedVac machine.
7. 6× DNA loading dye mixture: 0.25% (w/v) bromophenol blue, 0.25% (w/v) xylene cyanol FF, and 15% (w/v) Ficoll 400. Prepare the solution using nuclease-free sterile dH_2O and glasswares.
8. DNA size marker: *Hae* III digested pBR322 DNA (Boehringer Mannheim).
9. SYBR (an acronym for Synergy Brands, Inc.) Green (Molecular Probes).

2.5 Processing of PAGE Gel with Radioactive PCR Products

1. Blotting papers (a little bigger size than the size of the gel).
2. Saran wrap.
3. Gel dryer.

2.6 Autoradiography or PhosphoImaging of the Dried PAGE Gel Containing Radioactive PCR Products

1. X-Ray cassettes.
2. X-Ray films.
3. Intensifying screens.
4. −80 °C Freezer.
5. Kodak film processor.
6. PhosphoImager cassette.
7. PhosphoImager.

3 Methods

Perform all the steps of nonradioactive and radioactive TRAP assays at room temperature unless mentioned otherwise.

3.1 Nonradioactive TRAP Assay

1. Wash 1×10^6 cells once in PBS and pellet cells at 10,000 × *g* for 1 min at 4 °C.
2. Resuspend the cells in ice-cold wash buffer and pellet cells as described above.
3. Resuspend the cells in 20 μL of ice-cold lysis buffer.
4. Incubate on ice for 30 min.

5. Centrifuge in a microcentrifuge (13,000 × *g* at 4 °C) for 20 min.
6. Transfer the supernatant (protein extract) to a fresh tube. Measure protein concentration (which is typically around 10 μg/μL). Make small (10 μL) aliquots of protein extract, quick-freeze on dry ice, and store at −80 °C.
7. Lyophilize 0.1 μg of CX primer (5′-CCC TTA CCC TTA CCC TTA CCC TAA-3′) onto the bottom of the PCR tube, seal it with 7–10 μL of molten wax or add one AmpliWax PCR Gem (Perkin Elmer) to the tube, and heat in a water bath at 65 °C for 5 min to melt the wax. Allow 5 min to solidify the wax at room temperature and store the tubes at 4 °C.
8. Perform TRAP reaction (above the wax barrier of PCR tube) in 50 μL of 20 mM Tris–HCl (pH 8.3), 1.5 mM $MgCl_2$, 63 mM KCl, 0.005% (v/v) Tween-20, 1 mM EGTA, 50 μM dNTPs, 0.1 μg of TS primer (5′-AAT CCG TCG AGC AGA GTT-3′), 0.1 mg/mL bovine serum albumin (BSA), 0.5 μg of T4g32 protein, 2 Units of *Taq* DNA polymerase (Boehringer Mannheim), and 1 μL of cell extract (6 μg protein/μL). When desired, perform pretreatment of the cell extract before adding it to the assay tube (*see* **Note 6**). To achieve the final concentrations of these reagents, simply make following additions and run the nonradioactive TRAP reaction (total 50 μL) in the assay tube:

   39 μL of sterile dH_2O.

   5 μL of 10× TRAP buffer (200 mM Tris–HCl, pH 8.3, 15 mM $MgCl_2$, 630 mM KCl, 0.05% Tween-20, 10 mM EGTA).

   1 μL of 4 dNTPs (a mixture of 10 mM each).

   1 μL of TS primer (100 ng/μL).

   1 μL of BSA (5 μg/μL, GIBCO-BRL).

   1 μL of T4g32 protein (0.5 μg/μL, Boehringer Mannheim).

   1 μL of *Taq* DNA polymerase (2 Units/μL, Boehringer Mannheim).

   1 μL of 1× lysis buffer or cell extract (6 μg protein/μL).

   A master mix without *Taq* DNA polymerase and cell extract can be made and saved in small (300 μL) aliquots at −20 °C (*see* **Note 7**).
9. Mix all the reagents gently with a pipette tip, do not vortex, and immediately place the tube in the PCR machine. For close and uniform contact of the tube with PCR machine, add three drops of mineral oil to each tube hole before placing the tube in it.

10. Program the PCR machine for the following operations (programming of PCR machine for 40 cycles will give only qualitative results):

    1 cycle at 20 °C for 60 min for extension of TS primer by telomerase.

    1 cycle at 90 °C for 3 min to inactivate telomerase.

    40 cycles each at 94 °C for 45 s, 50 °C for 45 s, and 72 °C for 1 min 30 s.

    1 cycle at 72 °C for 2 min.

    Soak at 4 °C for at least 20 min.

11. Make a 10% polyacrylamide non-denaturing gel by adding the ingredients as follows:

    21 mL of sterile dH_2O.

    1.5 mL of 10× TBE.

    7.5 mL of 40% acrylamide/bisacrylamide (at 19:1 ratio).

    240 μL of 10% ammonium persulfate (freshly prepare or use the stock at −20 °C).

    32 μL of TEMED.

12. Immediately pour the gel (20 cm × 16 cm × 0.75 mm that fits in Bio-Rad 2-D gel apparatus), use a 20-teeth comb, and allow overnight to polymerize and age the gel at room temperature. Mark the wells with a blue marker pen and remove the comb carefully without damaging the individuality of the lanes. Traces of very small gel fragments may be present in the wells and these, if not removed, may interfere with resolution of DNA bands resulting in deformed DNA bands after the PAGE. To avoid this problem, clean the wells by flushing water in and out. Then, place the gel into the apparatus and add 1.5 L of 0.5× TBE.
13. Take the PCR tubes out of the DNA thermal cycler, make a hole in the wax barrier with the yellow tip (*see* **Note 8**).
14. Add 1 μL of 6× DNA loading dye mix to each 5 μL of sample. Load 20 μL of sample into each well. Use 750 ng of *Hae* III digested pBR322 DNA (Boehringer Mannheim) as the DNA size marker.
15. Perform PAGE at 175 volts for 45 min and then at 280 volts for 1 h 45 min or until the blue dye moves to the end of the gel.
16. Cut out excess gel parts with a sharp razor. For detection of nonradioactive PCR products [20], use SYBR Green (Molecular Probes) to stain (freshly prepare 1× SYBR Green solution in 150 mL of 0.5× TBE) the polyacrylamide gel for 20 min, and rinse once with water. Do not destain the gel anymore to avoid diffusion and fading of bands. Immediately photograph the gel

on a UV (preferably 254 nm epi-illuminator or 305 nm trans-illuminator) box using yellow filter and Polaroid film (Positive/Negative) 667.

3.2 Radioactive TRAP Assay

In radioactive TRAP assay, some initial steps are similar to nonradioactive TRAP assay (as described above). However, to avoid the inconvenience of going back and forth, all the steps of radioactive TRAP assay are fully described below.

1. Wash 1×10^6 cells once in PBS, and pellet at 10,000 × *g* for 1 min at 4 °C.
2. Resuspend the cells in ice-cold wash buffer [10 mM HEPES-KOH (pH 7.5), 1.5 mM $MgCl_2$, 10 mM KCl, 1 mM DTT] and pellet as above.
3. Resuspend the cells in 20 μL of ice-cold lysis buffer [10 mM Tris–HCl (pH 7.5), 1 mM $MgCl_2$, 1 mM EGTA, 0.1 mM PMSF, 5 mM 2-ME, 0.5% CHAPS, 10% glycerol].
4. Incubate on ice for 30 min.
5. Centrifuge at 13,000 × *g* in a microcentrifuge at 4 °C for 20 min.
6. Transfer the supernatant to a fresh tube. Measure protein concentration (typically around 10 μg/μL). Make small (10 μL) aliquots, quick-freeze on dry ice, and store at −80 °C.
7. Lyophilize 0.1 μg of CX primer (5′-CCC TTA CCC TTA CCC TTA CCC TAA-3′) onto the bottom of the PCR tube, and seal it with 7–10 μL of molten wax or add one AmpliWax PCR Gem (Perkin Elmer) to the tube and heat in a water bath at 65 °C for 5 min to melt the wax. Allow 5 min to solidify the wax at room temperature and store the tubes at 4 °C.
8. Perform TRAP reaction (above wax barrier of PCR tube) in 50 μL of 20 mM Tris–HCl (pH 8.3), 1.5 mM $MgCl_2$, 63 mM KCl, 0.005% (v/v) Tween-20, 1 mM EGTA, 50 μM dNTPs, 0.1 μg of TS primer (5′-AAT CCG TCG AGC AGA GTT-3′), 0.1 mg/mL BSA, 1 μg of T4g32 protein, 2 Units of *Taq* DNA polymerase (Boehringer Mannheim), 1 μL of cell extract (1 μg protein/μL), and 1 μL of α-^{32}PdCTP (2 μCi/μL). When desired, perform pretreatment of the cell extract before adding it to the assay tube (*see* **Note 6**). To achieve the final concentrations of these reagents, simply make following additions and run the radioactive TRAP reaction (total 50 μL) in the assay tube:

   38 μL of dH_2O.

   5 μL of 10× TRAP buffer (200 mM Tris–HCl, pH 8.3, 15 mM $MgCl_2$, 630 mM KCl, 0.05% Tween-20, 10 mM EGTA).

1 μL of 4 dNTPs (a mixture of 10 mM each).

1 μL of TS primer (100 ng/μL).

1 μL of BSA (5 μg/μL, GIBCO-BRL).

1 μL of T4g32 protein (1 μg/μL, Boehringer Mannheim).

1 μL of *Taq* DNA polymerase (2 Units/μL, Boehringer Mannheim).

1 μL of 1× lysis buffer or cell extract (1–6 μg protein/μL).

1 μL of α-^{32}PdCTP (2 μCi/μL).

A master mix without *Taq* DNA polymerase, cell extract, and α-^{32}PdCTP can be made and saved in small (300 μL) aliquots at −20 °C (*see* **Note 9**).

9. Mix gently with a pipette tip, do not vortex, and immediately place the tube in the PCR machine. For close and uniform contact, add three drops of mineral oil to each tube hole before placing the tube in it.
10. Program the PCR machine for the following operations (for reliable quantitative results):

    1 cycle at 20 °C for 30 min for extension of TS primer by telomerase.

    1 cycle at 94 °C for 45 s to inactivate telomerase.

    30 cycles each at 94 °C for 30 s, 50 °C for 30 s, and 72 °C for 45 s.

    1 cycle at 94 °C for 30 s, 50 °C for 30 s, and 72 °C for 1 min.

    Soak at 4 °C for at least 20 min.

11. Make a 10% polyacrylamide on-denaturing gel by adding the ingredients as follows:

    21 mL of dH_2O.

    1.5 mL of 10× TBE.

    7.5 mL of 40% acrylamide/bis-acrylamide (at 19:1 ratio, it can be saved at room temperature).

    240 μL of 10% ammonium persulfate (freshly prepare or use stock from −20 °C).

    32 μL of TEMED.

12. Immediately pour the gel (20 cm × 16 cm × 0.75 mm that fits in Bio-Rad 2-D gel apparatus), use a 20-teeth comb, and allow overnight to polymerize and age the gel at room temperature. Mark the wells with a blue marker pen and remove the comb carefully without damaging the individuality of the lanes. Traces of very small gel fragments may be present in the wells and these, if not removed, may interfere with resolution of

DNA bands resulting in deformed DNA bands after the PAGE. To avoid this problem, clean the wells by flushing water in and out. Then, place the gel into the apparatus and add 1.5 L of 0.5× TBE.

13. Take the PCR tubes out of the DNA thermal cycler and make a hole in the wax barrier with the yellow tip.
14. Add 1 μL of 6× dye mix to each 5 μL of sample (PCR products). Mix the sample by pipetting up and down gently, spin briefly, and load 20 μL of the sample into each well.
15. Perform PAGE at 175 volts for 45 min, and at 280 volts for 1 h 45 min or until the blue dye moves to the end of the gel.
16. Remove the gel sandwich from the electrophoresis apparatus and lay on a diaper with the short plate side up.
17. Remove one side spacer and the very gently insert a long metal spatula between the two glass plates to pry them apart with a gentle rocking motion. The gel should stick to the bottom plate. Remove the second side spacer and any bits of polyacrylamide around the gel.
18. Hold a piece of dry blotting paper (which is a little bigger than the size of the gel) and lay it on the top of the gel beginning at one end and working slowly toward the other. Take care to prevent air bubbles from forming between the paper and the gel. Smear the hands all over the paper.
19. Allow the gel little time (about 2 min) to absorb and stick to the blotting paper. Peel the blotting paper up off the plate.
20. Turn over the blotting paper with the gel. Cover it with a Saran wrap.
21. Preheat the gel dryer to 75 °C.
22. Place the blotting paper with the gel and Saran wrap onto the platform of preheated gel dryer. Cover with the plastic flap. Set following buttons: temperature (75 °C), pressure, and time (1.5 h). Turn on vacuum pump.
23. Place the dried gel in the X-ray cassette in a dark room (*see* **Note 10**).
24. Autoradiograph overnight or more at −80 °C under pressure (use a weight of about 10 pounds). Develop the film in Kodak film processor.
25. Alternately, expose the film to PhosphoImager cassette overnight, image, and perform quantitative analysis of the results using an appropriate software.

4 Notes

1. 1× PBS has the ionic strength, osmolarity, and pH that mimic our body fluids. Because PBS is isotonic to human body fluids, it does not cause undesirable damage and toxicity to the human cells during their pre-washing with it.
2. Prepare a sterile 2× wash buffer. Prior to use, add DTT and then sterile dH_2O to make 1× wash buffer.
3. If telomerase activity needs to be detected in tumor tissue, wash approximately 100 mg of frozen (−80 °C) tissue in ice-cold lysis buffer briefly (20 s) to remove the outer layer of lysed cells, contaminants, and any blood from the surface of the tissue. Remove lysed cells by aspiration and freeze the tissue again in liquid nitrogen. Then, partially thaw the tissue, make thin slices of the tissue on sterile disposable petri dishes with disposable surgical blades, and immediately transfer to a 1.5-mL Eppendorf tube containing 200 μL of ice-cold lysis buffer. Use a matching disposable pestle to rotate at about 450 rpm to homogenize the tissue using a Kontes homogenizer (battery-operated). Centrifuge the lysate at 16,000 × *g* for 20 min at 4 °C in a refrigerated microcentrifuge.
4. Prepare a sterile 2× lysis buffer. Prior to use, add PMSF and 2-ME and then sterile dH_2O to make 1× lysis buffer.
5. Telomerase activity remains stable even after multiple freeze-thaws.
6. Pretreatment of the cell extract when desired: (a) heat (65 °C for 10 min), (b) RNase A (10 μL of cell extract with 0.5 μg of RNase A at 23 °C for 10 min), (c) phenol (1:1 extraction), (d) protease (50 μL of cell extract with 5 μg of Bromelain or 0.2 mg/mL of proteinase K at 37 °C for 10 min followed by protease removal).
7. When needed, thaw the aliquot of master mix and use 48 μL/assay tube. Place all the tubes on ice. Do not freeze the left over. Discard if there is any extra. Always freshly add 1 μL of *Taq* DNA polymerase and 1 μL of cell extract to the assay tube. Use only the highest calibrated pipette for measuring very small volume of enzyme or cell extract.
8. When necessary, transfer PCR products to a fresh tube and reduce the volume to about 20 μL in a SpeedVac at low heat.
9. When needed, thaw the aliquot of master mix and use 47 μL/assay tube. Place all the tubes on ice. Do not freeze the left over. Discard if there is any extra. Always freshly add 1 μL of *Taq* DNA polymerase, 1 μL of cell extract, and 1 μL of α-^{32}PdCTP to the assay tube. Now, the assay tube contains

radioactive materials. Therefore, use all safety precautions for handing the tube.

10. Go to a dark room. Turn off even the dim lights before exposing gel to the X-ray film for direct contact. Put an intensifying screen on the top of the film, and close the cassette tight.

Acknowledgments

The work was supported in part by the R01 grants (CA-091460 and NS-057811) from the National Institutes of Health (Bethesda, MD, USA).

References

1. Akincilar SC, Unal B, Tergaonkar V (2016) Reactivation of telomerase in cancer. Cell Mol Life Sci 73:1659–1670
2. Zhang L, Chen Z, Zuo W, Zhu X, Li Y, Liu X, Wei X (2016) Omacetaxine mepesuccinate induces apoptosis and cell cycle arrest, promotes cell differentiation, and reduces telomerase activity in diffuse large B-cell lymphoma cells. Mol Med Rep 13:3092–3100
3. Kim NW, Piatyszek MA, Prowse KR, Harley CB, West MD, Ho PL, Coviello GM, Wright WE, Weinrich SL, Shay JW (1994) Specific association of human telomerase activity with immortal cells and cancer. Science 266:2011–2015
4. Piatyszek MA, Kim NW, Weinrich SL, Hiyama K, Hiyama E, Wright WE, Shay JW (1995) Detection of telomerase activity in human cells and tumors by a telomeric repeat amplification protocol (TRAP). Methods Cell Sci 17:1–15
5. Wu YY, Hruszkewycz AM, Delgado RM, Yang A, Vortmeyer AO, Moon YW, Weil RJ, Zhuang Z, Remaley AT (2000) Limitations on the quantitative determination of telomerase activity by the electrophoretic and ELISA based TRAP assays. Clin Chim Acta 293:199–212
6. Holt SE, Norton JC, Wright WE, Shay JW (1996) Comparison of the telomeric repeat amplification protocol (TRAP) to the new TRAP-eze telomerase detection kit. Methods Cell Sci 18:237–248
7. Das A, Banik NL, Ray SK (2007) Differentiation decreased telomerase activity in rat glioblastoma C6 cells and increased sensitivity to IFN-gamma and taxol for apoptosis. Neurochem Res 32:2167–2183
8. Das A, Banik NL, Ray SK (2008) Retinoids induced astrocytic differentiation with down regulation of telomerase activity and enhanced sensitivity to taxol for apoptosis in human glioblastoma T98G and U87MG cells. J Neuro-Oncol 87:9–22
9. Das A, Banik NL, Ray SK (2009) Retinoids induce differentiation and downregulate telomerase activity and N-Myc to increase sensitivity to flavonoids for apoptosis in human malignant neuroblastoma SH-SY5Y cells. Int J Oncol 34:757–765
10. Karmakar S, Banik NL, Patel SJ, Ray SK (2007) Combination of all-*trans* retinoic acid and taxol regressed glioblastoma T98G xenografts in nude mice. Apoptosis 12:2077–2087
11. Karmakar S, Banik NL, Ray SK (2008) Combination of all-*trans* retinoic acid and paclitaxel-induced differentiation and apoptosis in human glioblastoma U87MG xenografts in nude mice. Cancer 112:596–607
12. Janardhanan R, Banik NL, Ray SK (2008) N-(4-Hydroxyphenyl)retinamide induced differentiation with repression of telomerase and cell cycle to increase interferon-gamma sensitivity for apoptosis in human glioblastoma cells. Cancer Lett 261:26–36
13. Janardhanan R, Butler JT, Banik NL, Ray SK (2009) N-(4-Hydroxyphenyl) retinamide potentiated paclitaxel for cell cycle arrest and apoptosis in glioblastoma C6 and RG2 cells. Brain Res 1268:142–153
14. Das A, Banik NL, Ray SK (2008) N-(4-Hydroxyphenyl) retinamide induced both differentiation and apoptosis in human glioblastoma T98G and U87MG cells. Brain Res 1227:207–215

15. Phipps SM, Love WK, White T, Andrews LG, Tollefsbol TO (2009) Retinoid-induced histone deacetylation inhibits telomerase activity in estrogen receptor-negative breast cancer cells. Anticancer Res 29:4959–4964
16. Ray S, Ponnathpur V, Huang Y, Tang C, Mahoney ME, Ibrado AM, Bullock G, Bhalla K (1994) 1-β-D-Arabinofuranosylcytosine-, mitoxantrone-, and paclitaxel-induced apoptosis in HL-60 cells: improved method for detection of internucleosomal DNA fragmentation. Cancer Chemother Pharmacol 34:365–371
17. Bullock G, Ray S, Reed J, Miyashita T, Maria Ibrado A, Huang Y, Bhalla K (1995) Evidence against a direct role for the induction of c-jun expression in the mediation of drug-induced apoptosis in human acute leukemia cells. Clin Cancer Res 1:559–564
18. Bullock G, Ray S, Reed JC, Krajewski S, Ibrado AM, Huang Y, Bhalla K (1996) Intracellular metabolism of Ara-C and resulting DNA fragmentation and apoptosis of human AML HL-60 cells possessing disparate levels of Bcl-2 protein. Leukemia 10:1731–1740
19. Birnie GD (1988) The HL60 cell line: a model system for studying human myeloid cell differentiation. Br J Cancer Suppl 9:41–45
20. Morin PA, Smith DG (1995) Nonradioactive detection of hypervariable simple sequence repeats in short polyacrylamide gels. BioTechniques 19:223–228

Index

Swapan K. Ray (ed.), *Retinoid and Rexinoid Signaling: Methods and Protocols*, Methods in Molecular Biology, vol. 2019, https://doi.org/10.1007/978-1-4939-9585-1, © Springer Science+Business Media, LLC, part of Springer Nature 2019

MIX
Papier aus verantwortungsvollen Quellen
Paper from responsible sources
FSC® C105338

If you have any concerns about our products,
you can contact us on
ProductSafety@springernature.com

In case Publisher is established outside the EU,
the EU authorized representative is:
Springer Nature Customer Service Center GmbH
Europaplatz 3, 69115 Heidelberg, Germany

Printed by Libri Plureos GmbH
in Hamburg, Germany